RADON MEASUREMENTS by ETCHED TRACK DETECTORS

Applications in Radiation Protection, Earth Sciences and the Environment

RADON MEASUREMENTS
by
ETCHED TRACK DETECTORS

Applications in Radiation Protection, Earth Sciences and the Environment

Editors

Saeed A. Durrani

School of Physics and Space Research
University of Birmingham, England

Radomir Ilić

J. Stefan Institute, University of Ljubljana &
Faculty of Civil Engineering
University of Maribor, Slovenia

World Scientific
Singapore • New Jersey • London • Hong Kong

Published by

World Scientific Publishing Co. Pte. Ltd.

P O Box 128, Farrer Road, Singapore 912805

USA office: Suite 1B, 1060 Main Street, River Edge, NJ 07661

UK office: 57 Shelton Street, Covent Garden, London WC2H 9HE

Library of Congress Cataloging-in-Publication Data

Radon measurements by etched track detectors : applications in radiation protection, earth sciences, and the environment / editors, Saeed A. Durrani, Radomir Ilić.

p. cm.

Includes bibliographical references and index.

ISBN 9810226667

1. Radon -- Measurement -- Technique. 2. Nuclear track detectors. I. Durrani, S. A. II. Ilić, Radomir.

QC796.R6R33 1997

539.7'7--dc21 97-10524

CIP

British Library Cataloguing-in-Publication Data

A catalogue record for this book is available from the British Library.

Printed in Singapore by Uto-Print

To all those who have contributed to a better understanding of radon, its effects, and its applications.

Contributors

Gustav ÅKERBLOM, Swedish Radiation Protection Institute, S-17116 Stockholm, Sweden.

Miguel BALCÁZAR, Instituto Nacional de Investigaciones Nucleares, Apartado Postal 18-1027, Mexico D.F. 11801, Mexico.

T. Keith BALL, British Geological Survey, Keyworth, Nottingham NG12 5GG, U.K.

Saeed A. DURRANI, School of Physics and Space Research, University of Birmingham, Edgbaston, Birmingham B15 2TT, U.K.

Robert L. FLEISCHER, Department of Earth and Environmental Sciences, Rensselaer Polytechnic Institute, RM G-17, West Hall PIR, Troy, New York 12180-3590, U.S.A.

József HAKL, Institute of Nuclear Research, P.O.Box 51, H-4001 Debrecen, Hungary.

Ilona HUNYADI, Institute of Nuclear Research, P.O.Box 51, H-4001 Debrecen, Hungary.

Radomir ILIĆ, J. Stefan Institute, University of Ljubljana, POB 100, 1000 Ljubljana, Slovenia.

Gilbert JÖNSSON, Department of Physics, University of Lund, S-22100 Lund, Sweden.

Hameed A. KHAN, Pakistan Institute of Nuclear Science and Technology, P.O. Nilore, Islamabad, Pakistan.

Hans MELLANDER, Swedish Radiation Protection Institute, Box 60204, S-17116 Stockholm, Sweden.

Jon C.H. MILES, National Radiological Protection Board, Chilton, Didcot, Oxon OX11 0RQ, U.K.

Michel M. MONNIN, Centre National de la Recherche Scientifique - UMR 5569, Université Montpellier II, Place Eugene Bataillon, 34095 Montpellier Cedex 5, France.

Colin R. MUIRHEAD, National Radiological Protection Board, Chilton, Didcot, Oxon OX11 0RQ, U.K.

Jean F. PINEAU, R & D Department, ALGADE, Route Nationale 20, 87250 Bessines-sur-Gartempe, France.

Aziz A. QURESHI, Pakistan Institute of Nuclear Science and Technology, P.O. Nilore, Islamabad, Pakistan.

Imtinan E. QURESHI, Pakistan Institute of Nuclear Science and Technology, P.O. Nilore, Islamabad, Pakistan.

Jean L. SEIDEL, Centre National de la Recherche Scientifique - UMR 5569, Université Montpellier II, Place Eugene Bataillon, 34095 Montpellier Cedex 5, France.

Mehdi SOHRABI, National Radiation Protection Department, Atomic Energy Organization of Iran, P.O.Box 14155-4494, Tehran, Islamic Republic of Iran.

Tomaž ŠUTEJ, Health Inspectorate, Ministry of Health, Parmova 33, 1000 Ljubljana, Slovenia.

Luigi TOMMASINO, ANPA - Agenzia Nazionale per la Protezione dell'Ambiente, Via Vitaliano Brancati 48, 00144 Rome, Italy.

András VÁRHEGYI, Mecsekurán Ore Mining Ltd., POB 65, H-7614 Pécs, Hungary.

Preface

The idea of writing this book first came to one of us (RI) during a Nuclear Tracks Conference at Odessa (Ukraine) in September 1991, attended by both co-editors. The realization that rose to our consciousness during that conference was that while Solid State Nuclear Track Detection (SSNTD) technique was the most widely used method for radon studies nowadays, yet there was no book available globally, devoted exclusively or largely to the methodology of, and dealing with results obtained in, radon investigations by this technique. So, after a period of cogitation and consultation, we decided to fill that gap. That the decision, conceived in a spirit of idealism and service to science, would lead us to months and years of 'blood, toil, tears and sweat' is a situation not unknown to authors/editors of multidisciplinary books — but more of that later.

To make the proposed book comprehensive and representative of all the many different types of radon studies in progress globally, we decided to enlist the support of, and elicit contributions from, a number of the most prominent and active research workers in the SSNTD discipline as well as in various fields of radon investigations. We are pleased to record here that all the co-authors approached by us — and all of them fabulously busy persons in their respective fields of research — volunteered unreserved and enthusiastic support in response to our appeal for help. Some, conscious of their own busyness and other preoccupations, proposed names of colleagues to form a small 'task force' in their subject areas; this, of course, was readily accepted by the co-editors. The book, as it stands, has 18 sections comprising four chapters, and covers, we hope, all the important subdisciplines that constitute the field of radon measurements and their applications as well as the necessary foundations for such work, viz. the basic properties of radon gas itself and — a *sine qua non* for the book — the basic concepts and procedures of the SSNTD technique. In the choice of authors and the subjects to be covered by this book, we have borne vividly in mind the chosen subtitle: *Applications in Radiation Protection, Earth Sciences and the Environment.* Each subdiscipline has been treated by an expert (or a small team of experts) actively engaged in the given subject area.

The intended/envisaged readership of the book includes all those involved in radon surveys and measurements as well as those concerned with analysing and assessing the effects of radon on human health. It also encompasses all those who are applying, and those who wish to learn about the applications of, radon measurements to the amelioration of human condition or society, whether by trying to predict earthquakes or by seeking to locate uranium or oil/gas reserves. Thus we have sought to make available up-to-date reviews that should be of use not only to physical scientists, geologists, environmentalists and epidemiologists, but also to technical personnel in the field and in the laboratory as well as managers of private firms (or public utilities) carrying out or overseeing radon measurement services. In addition, many postgraduate students and PhD research workers in radon studies and related fields should, we believe, find parts of this book to be particularly helpful and instructive.

We hope that the present volume will form an essential reference text for university students pursuing courses or research studies, as well as other workers, in the areas of radiation physics, environmental studies, geology, health and safety, dosimetry and instrumentation, medical physics and biophysics, epidemiology, etc.

We, the co-editors, are most grateful to all our co-authors who have, throughout the execution of this project, given us magnificent help and constant cooperation. One of us (SAD) took it upon himself to check (and sometimes interfere with!) every comma, semicolon, colon, and the spelling of every word — not to mention double–checking every calculation and numerical value/physical derivation of every equation or formula written or put forward by all the authors, and to draw their attention to any modifications proposed by him. What constituted perhaps going a bit OTT (Over The Top) was, however, his whim to attempt to make the style and linguistic expression of all contributions — made by authors of a motley group of nationalities speaking diverse mother tongues — conform to the dictates and ideals of Mr HW Fowler and his amusing if somewhat quirky masterpiece *Modern English Usage* (Clarendon Press, Oxford). He (SAD) was touched and gratified to witness the grace and good humour with which all his co-authors appeared to tolerate his fads and his whims (often modelled on those of the said HW Fowler).

The laborious task of checking the minutiae of the thousands of references in the book was undertaken by RI. The Herculean labour, the challenges, and the pitfalls encountered in this field are best known to journal editors and others of that ilk. While he (RI) has done his human best, he wishes to put on record here that, in the final analysis, the references — in common with the general content of each section — are the responsibility of the respective authors themselves. But both co-editors wish to emphasize that the credit for the depth, breadth and masterly coverage of their allotted fields belongs fully to the eminent author or authors of the respective section(s). It should also perhaps be put on record here that the detailed organization of the entire project — which involved keeping in almost continuous touch with all the authors, using all the present-day rapid communication methods over many months, and much more besides — was also carried out almost single-handedly by RI. The whole project was rounded off in a final, almost round-the clock, bout of work over a period of ten days (and nights!) spent by both co-editors at the J. Stefan Institute, Ljubljana.

If we had to single out one individual, out of our co-authors, for special thanks, it would undoubtedly be Bob Fleischer. At all stages of the proposal, gestation and compilation of this book his sagacious advice and practical help were always willingly, promptly and constructively proffered to us, whenever requested. Also, in the case of the *Glossary of Terms*, we should like to assign credits as follows: Mehdi Sohrabi for his extensive contributions in the field of radiation protection; Gustav Åkerblom and Keith Ball for defining all the geological terms; and, once again, Bob Fleischer for his overall advice and critical appraisal of the form and contents of most of the entries.

We take this opportunity to express our gratitude to our respective institutions, viz. the University of Birmingham and the J. Stefan Institute, for their help and support. In particular, without the logistic support of the J. Stefan Institute, the compilation and production of this volume would have been quite impossible. We also acknowledge the financial support, in the final stages of compilation, given by the University of Maribor to enable one of us (SAD) to take up residence for ten days in the lovely, and at the time (early December 1996; in SAD's eyes: alluringly) snow-laden town of Ljubljana. Thanks are also due to the two editors of World Scientific, Dr You-Huan Yeo and Ms Kim Tan, for their help in the publication of this book.

Finally, it is a matter of great pleasure for us to thank two individuals at the J. Stefan Institute who gave outstanding and unstinted support to the co-editors, with great competence, cheerfulness and loyalty. We mean Uršula Turšič, who prepared the entire camera-ready copy and diligently incorporated the thousands of amendments and alterations, proposed by the authors and by ourselves; and Bojan Žefran, who demonstrated great technical expertise in the design and assembly of the whole book and in converting it to the LaTeX format from the various types of computer discs submitted by the authors, and, with utmost ingenuity, retouching, and often redrawing, the figures and tables emanating from many sources. We give both of them a big thank you.

And a big THANK YOU also to our two respective families who put up with nearly two years of our varying degrees of neglect while we chased that visionary bird, longingly (and perhaps naively) called by us "scientific creativity".

(at the time of writing, both at)
J. Stefan Institute, Ljubljana, Slovenia — Saeed A. Durrani and Radomir Ilić
6 December 1996

Contents

1.

INTRODUCTION

R. L. Fleischer[1]

1.1 Radon: Overview of Properties, Origin, and Transport

1.1.1 Introduction

Two isotopes of radon were discovered in 1900 - thoron (later recognized as ^{220}Rn) by R.B. Owens and E. Rutherford, and radon (later identified as ^{222}Rn) by F. E. Dorn (see Cook, 1961). Shortly thereafter, using liquid air, radon was liquefied (Rutherford and Soddy, 1902) - presumably in minute quantities. Although properties have been quoted for condensed radon (Cook, 1961; van Loef, 1981), discussion of such a phase is largely academic. The internal energy generation in bulk ^{222}Rn from alpha decay would heat it from absolute zero to its boiling temperature at 211 K in an estimated 4 ms. Instead, throughout this book we will be concerned with trace concentrations of radon - typical indoor radon activities of 37 Bq m^{-3} (1 pCi l^{-1}) being lower by a factor of more than 10^{21}! Hence, we must think of radon, not as a substance, but as isolated atoms whose transport is brought about by other gases, in which they may diffuse - being within a fluid that may itself be moving. Radon is a unique natural element in being (1) a gas, (2) noble, and (3) radioactive in all of its isotopes. Gases are of special interest, first, because they are mobile and can carry messages over significant distances within the earth and in the atmosphere and, second, because inhalation can be a problem to health. The fact that radon is noble insures that it is not immobilized by chemically reacting with the medium that it permeates. As a result, free radon normally diminishes only by its radioactive decay as it moves from its source. Finally, its radioactivity allows radon to be measured with remarkable sensitivity. For humans the greatest importance of the radioactivity of radon is that in high concentrations it can be a health hazard, a cause of lung cancer.

A caveat should be added on radon's chemical inertness, which, like that of lighter noble gases, is not complete. Radon can form a fluoride (Fields *et al.*, 1962), which by analogy with the behavior of xenon may be RnF_4. Radon can also be trapped by physical absorption on finely divided carbon, a behavior that is utilized in the charcoal canister type of radon measurements that will be described later. In the earth, radon isotopes are born by the decay of radium, and radium production in turn comes from uranium or thorium decay - usually (but not always) from trace impurities in minerals. As we will see later, by far the greatest importance among radon isotopes is attributed to ^{222}Rn because it is the longest lived of the three naturally produced isotopes. Radon is both a hazard and a help. The possibilities of transport of ^{222}Rn within the earth, its waters, and

[1]Rensselaer Polytechnic Institute, Troy, New York, USA

atmosphere make it a useful tracer for a remarkable variety of geophysical, geochemical, hydrological, and atmospheric purposes. These applications range from exploration for resources such as uranium and hydrocarbon deposits, to studying gas flow and mixing in the atmosphere, to recognizing fluid transport within the earth, to attempting to predict seismic and volcanic events through premonitory changes in radon concentrations in the earth.

These desirable functions of radon are countered by destructive health effects: sustained exposure of humans to substantial concentrations of radon decay products can produce lung cancer. Radon measurements play a critical role in monitoring human health and safety, both in homes and in mines. Radon is a ubiquitous hazard - and radon problems are aggravated where it is naturally concentrated in certain geological materials or is sociologically intensified by undesirable home construction practices.

1.1.1.1 Role of etched track detectors

This volume describes the diverse uses of solid nuclear track detectors in measuring radon and its daughters. It is now well known that in virtually all insulating solids the passage of heavily ionizing particles produces linear tracks of atomically disturbed material (Fleischer *et al.*, 1975). These latent tracks can be transformed into characteristically shaped holes (etched tracks) by the simple process of exposure to suitable chemical reagents. What is important here is that some of the solid track detectors, such as the polymeric materials allyl diglycol carbonate (CR-39), bisphenol-A polycarbonate (Lexan, Makrofol), and cellulose nitrate (LR 115, CN 85), record tracks of the low-energy alpha particles that radon and certain radon daughters emit.

Such solids are therefore useful as integrating detectors of radon; they are especially suitable for determining long-term average radon levels simply, ruggedly, and reliably. Long-term measurements over a year, or at least many months, are of clear relevance for quantifying human exposure in homes - where years of cumulative exposure are needed to bring about lung cancer and where the usual abrupt short-term variations in concentration need to be averaged. Most geological, geophysical, and geochemical processes vary more slowly with time; and integrating methods have the advantage of averaging the shorter-term variations that are caused by extraneous factors such as soil moisture, as well as pressure fluctuations in the ground that are caused by moving atmospheric pressure systems. Some recent examples of radon dosimetry are given by Tommasino *et al.* (1990) and Furlan and Tommasino (1993).

Relevant characteristics of etched track detectors are documented in a vast general literature, in the journal variously called *Nuclear Track Detection* (1977-78), *Nuclear Tracks* (1979-83), *Nuclear Tracks and Radiation Measurements* (1984-1993), and *Radiation Measurements* (1994-), in every-two-year conference proceedings of *Nuclear Tracks in Solids*, and more compactly in several monographs (Durrani and Bull, 1987; Fleischer *et al.*, 1975; Spohr, 1990; Wagner and van den Haute, 1992). Chapter 2 of this volume documents the characteristics and properties of detectors that are important for radon detection.

1.1.2 Physical and Chemical Properties of Radon

Radon is a mobile, chemically inert radioactive element. Its atomic number of 86 makes it a noble element - and therefore both non-reactive chemically and atomically mobile at normal temperatures. The three naturally produced isotopes ^{222}Rn (called "radon"), ^{220}Rn (called "thoron" because it is derived from thorium), and ^{219}Rn (from the ^{235}U decay series) all decay by emitting alpha particles. These particles are simple to measure with solid state track detectors, and they allow the abundances of the important radon atoms to be measured.

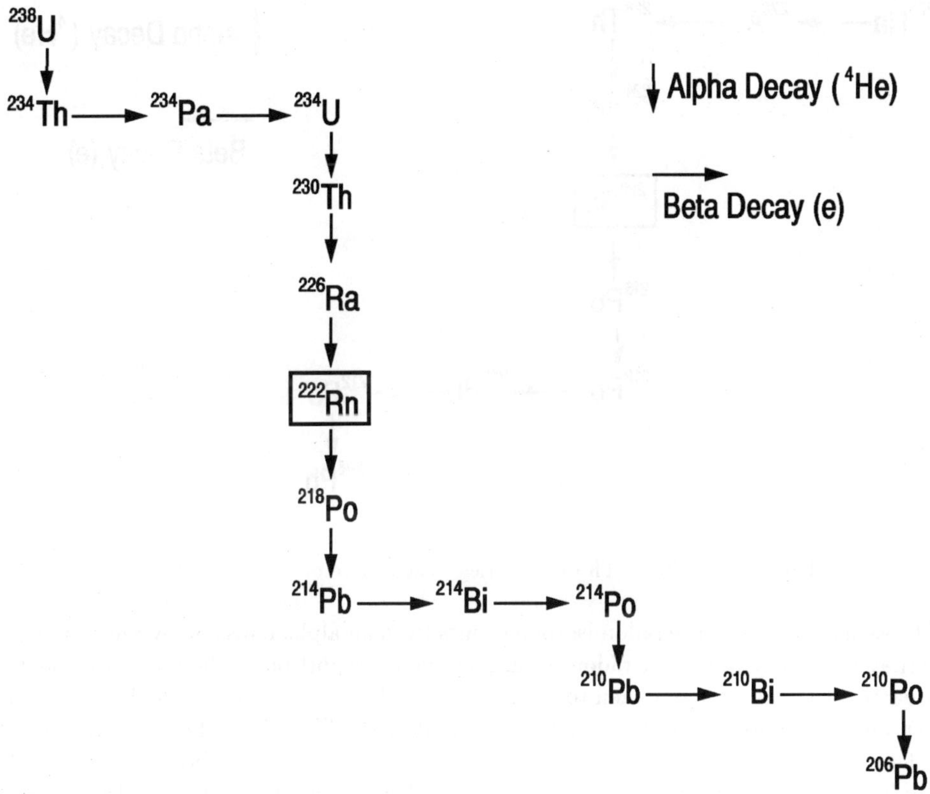

Figure 1.1.1: ^{238}U decay series - the source of ^{222}Rn (radon).

1.1.2.1 Radioactive decay chains

The three natural isotopes are produced from radium decay as steps in lengthy sequences of decays that originate from uranium or thorium isotopes - ^{222}Rn from ^{238}U; ^{220}Rn from ^{232}Th; and ^{219}Rn from ^{235}U. Their respective half lives ($\tau_{1/2}$) are 3.82 d, 55.6 s, and 3.96 s (and average or mean lives τ of 5.51 d, 80.2 s, and 5.71 s). The decay series of radon and thoron (with only the most common branches included) are shown in Figs. 1.1.1 and 1.1.2. Two dozen other radon isotopes can be produced (!), but none of them come from present-day natural processes. In these figures horizontal shifts represent electron emission (β-decay) which increases the atomic number by l, but does not alter the atomic mass; vertical processes are ^{4}He emission (alpha decay), which lowers the atomic number by 2 and the atomic mass by 4.

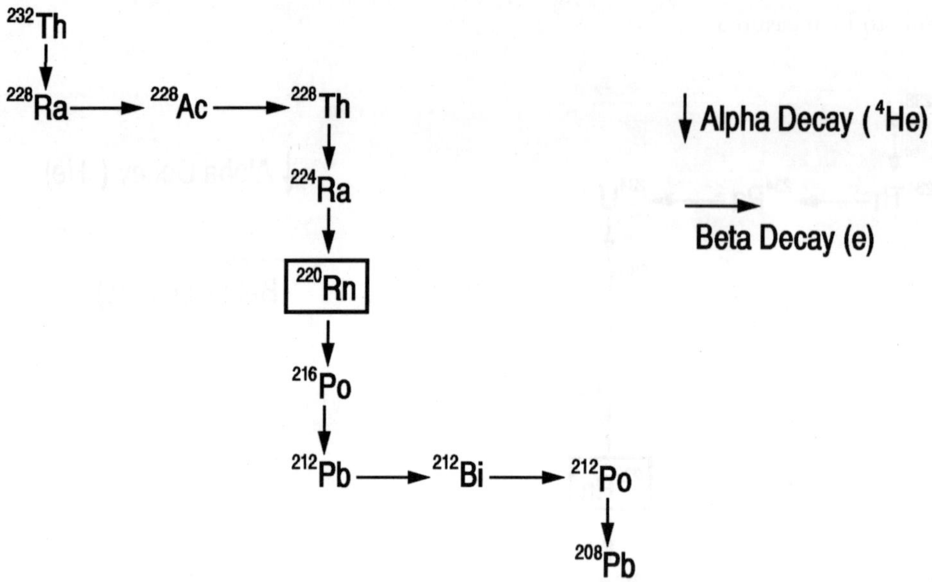

Figure 1.1.2: ^{232}Th decay series - the source of ^{220}Rn (thoron).

In each decay chain the radon isotope results from an alpha emission by radium; hence one may rigorously infer that radon production is in proportion to the radium present in the earth. Usually (since progenitor nuclides to radium are found in condensed solids) ^{222}Rn and ^{219}Rn are produced where U is located, and ^{220}Rn where Th is. However, one must be alert to exceptions. For example Levinson and Coetzee (1978) have described a site where Ra has been dissolved, transported, and redeposited elsewhere, with the result that radon is generated far from the progenitor uranium deposit. If radon was being sampled in the ground as a means of exploration for uranium, an unknowing prospector would be doomed to disappointment at Levinson and Coetzee's site.

The relative importance of the three radon isotopes increases with their mean lives and relative abundance. ^{219}Rn is the shortest lived, and is virtually always produced in much smaller amounts than is ^{222}Rn, since the natural ^{235}U/^{238}U ratio of these ultimate progenitors is 0.00719. Hence ^{219}Rn is largely ignored. ^{220}Rn too is short lived relative to ^{222}Rn and consequently moves a much smaller distance from its source than does ^{222}Rn. In air, for a diffusion constant, D, of 0.1 cm^2 s^{-1}, the mean distances of diffusive motion

$\sqrt{D\tau}$ (i.e. the distance diffused over a mean life) are 2.2 m for ^{222}Rn and 0.029 m for ^{220}Rn. Hence in circumstances where signals from relatively distant resources or processes in the earth are sought, ^{222}Rn is by far the dominant nuclide, and ^{220}Rn provides only an uninteresting local background that one wants to exclude during detection.

Exceptionally high Th/U ratios, however, can lead to greatly enhanced ^{220}Rn relative to ^{222}Rn. One renowned example is in Kerala State in India, where monazite ($ThPO_4$) sands can produce copious ^{220}Rn release that is of potential importance to health (Eisenbud, 1987).

1.1.2.2 The importance of prompt daughters

Health hazards of radon and thoron are not primarily due to those isotopes directly, but to the short-lived daughters that can be inhaled. Because radon is a noble gas with a lifetime that is long relative to breathing times, most of it that is inhaled is exhaled again rather than decaying, or becoming lodged in the lungs and later decaying. In contrast, the immediate, promptly - decaying daughters of ^{222}Rn (^{218}Po, ^{214}Pb, ^{214}Bi, and ^{214}Po), which wish to be part of a condensed phase, attach to surfaces, typically of aerosols, which can be inhaled; they then deposit on epithelial surfaces within the lung, and shortly decay. Since the longest lived of the four, ^{214}Pb, has a half life of less than 27 minutes, the whole sequence of decays can be completed before the normal clearance processes of the lung can sweep them away. The result is that the sensitive surfaces of the bronchi are irradiated by these decays - the most energetic and destructive of which are the heavily ionizing, short-range alpha particles from the polonium isotopes ^{214}Po and ^{218}Po. A third alpha-active polonium isotope ^{210}Po contributes relatively little to the radon measurements and effects discussed in this volume because its decay requires the prior decay of 22-year half-life ^{210}Pb.

Thoron (^{220}Rn) is a somewhat different case. Its first daughter ^{216}Po is so prompt (mean life 0.22 s) that it decays wherever ^{220}Rn decays. There then follows 15.2 h mean life ^{212}Bi, which lives on the average long enough that most of it will be transported from the bronchial portions of the lungs[2], except at points of stagnation that exist at bifurcations in the bronchial tree. Its primary health effect is likely to be at such points, where ultimately the daughter ^{212}Po will alpha decay.

1.1.2.3 Ranges of alpha particles and recoil nuclei

For understanding the physical behavior of radon and its decay products, it is helpful to know the penetrating power of alpha particles that are emitted in decay and also those of the residual nuclei that recoil from the alpha decays. Table 1.1.1 lists ranges in air and water of alpha particles emitted by ^{220}Rn and ^{222}Rn and their prompt alpha-active daughters. Their alpha-decay energies are also given. Ranges in typical plastics are similar to those in water. For example, for the range of the 5.49 MeV ^{222}Rn alpha

[2]The mucous lining of the bronchi is propelled upward and outward at a rate of about 1.4 cm min^{-1}, thus skimming particles off on a time scale that is comparable with the half lives of the prompt ^{222}Rn daughters (Hatch and Gross, 1964).

particle approximate values are 39 μm in water, 32 μm in polyethylene terephthalate (of density 1.2 g cm^{-3}), 31 μm in polyethylene, 37 μm in Lexan polycarbonate, but 23 μm in soda lime glass (density $\approx$ 2.5 g cm^{-3}).

Table 1.1.1: Alpha particle energies and ranges.

Nuclide	Alpha Energy (MeV)	Alpha Range (μm of water)	Alpha Range (cm of air)
^{222}Rn	5.49	39	4.08
^{218}Po	6.00	44	4.67
^{214}Po	7.69	66	6.91
^{220}Rn	6.29	48	5.01
^{216}Po	6.78	41	5.67
^{212}Po (64%)	8.78	83	8.53
^{212}Bi (36%)	6.05	45	4.73

Calculations are for 6°C at sea level.

Table 1.1.2: Ranges of alpha recoil nuclei in silica.

Decaying Nuclide	Recoil Nuclide	Range in SiO_2 (nm)
^{226}Ra	^{222}Rn	26
^{222}Rn	^{218}Po	30
^{218}Po	^{214}Pb	34
^{214}Po	^{210}Pb	45
^{224}Ra	^{220}Rn	31
^{220}Rn	^{216}Po	35
^{216}Po	^{212}Pb	38
^{212}Po (64%)	^{208}Pb	52
^{212}Bi (36%)	^{208}Tl	35

Recoiling nuclei resulting from alpha decay have far lower energies ($\sim$ 100 keV) and consequently very short ranges, $\sim$ hundreds of atomic distances. Values computed by the method of Lindhard *et al.* (1963) are given in Table 1.1.2 for SiO_2. Other common minerals will have similar values. For example, if the range in SiO_2 is 30 nm, the expected value is 27 nm for muscovite mica and 21 nm for zircon, but 68 nm for polyethylene.

1.1.2.4 Distribution of radon between phases

When radon is present in a gas such as air, nitrogen, or natural gas, it will to some extent penetrate water until it reaches an equilibrium concentration that decreases with increasing temperature (Faul, 1954). Some values for the ratio of radon in water to radon in gas (atoms per unit volume) are 0.53 (0°C), 0.43 (6°C), 0.28 (15°C), 0.22 (26°C), 0.16 (40°C), and 0.11 (100°C).

Where water is in direct contact with mineral particles that contain radium, radon can be implanted in water by alpha recoil (Kigoshi, 1971) to levels that far exceed the equilibrium with nearby gas. For example, Hess *et al.* (1982) found as much as $1.9 \cdot 10^6$ Bq m^{-3} (52,000 pCi l^{-1}) in water, while the corresponding air concentration was only 150 Bq m^{-3} (4 pCi l^{-1}). Water with such supersaturation is prone to rapid outgassing when given sufficient agitation, as for example in washing machines or showers within homes. Such emission can be a significant health hazard where ground water comes from a radium-rich aquifier, as has been noted in parts of the state of Maine in the U.S.A. by Hess *et al.* (1982).

Radon partition between air and various fatty acids and triglycerides has also been measured. Nussbaum and Hursh (1958) found preferential solubility by volume of mostly 3x to 8x, on the average (at human body temperature) of 5.8x (± 1.5) for 23 materials. The value for human fat was 6.3x.

1.1.2.5 Diffusion

Radon diffusion rates in air and water are vastly different. As Table 1.1.3 shows, the diffusion constant D in water is lower by a factor of 10,000. There is a mean distance of motion of a radon isotope before it disappears by decay $\sqrt{D\tau}$, where τ is the mean life of the isotope. Thus the average ^{222}Rn atom moves about 1.6 m in dry porous soil, but only 1.6 cm in water-saturated soil.

Table 1.1.3: Mean Diffusion distances of radon isotopes in different media.

Medium	Mean Distance		Diffusion Constant Assumed*
	^{222}Rn (cm)	^{220}Rn (Tn) (cm)	(cm^2 s^{-1})
Air	220	2.85	10^{-1}
Porous Soil	155	2.0	$5 \cdot 10^{-2}$
Water	2.2	0.0285	10^{-5}
Saturated Porous Soil	1.55	0.020	$5 \cdot 10^{-6}$

* From a compilation by Tanner (1964)

Data on the transition from $D = 0.05$ cm^2 s^{-1} for porous soil to $5{\cdot}10^{-6}$ cm^2 s^{-1} for saturated porous soil with percent water content in soil are reviewed by Nazaroff and Nero (1988). Although data scatter fairly widely, typical values of D in cm^2 s^{-1} are $5{\cdot}10^{-2}$ (0%), $3{\cdot}10^{-2}$ (20%), $1.5{\cdot}10^{-2}$ (40%), $7{\cdot}10^{-3}$ (60%), 10^{-3} (80%), $3{\cdot}10^{-4}$ (90%), and $2{\cdot}10^{-5}$ (100%).

The mean diffusion distances given in Table 1.1.3 are important in setting limits on the effective motion of radon (^{222}Rn) and thoron (^{220}Rn) in the absence of special transport mechanisms. Thus the radon inputs to the foundations of homes come primarily from soil within 1 or 2 meters of the house - a very local source.

1.1.2.6 Permeation

The permeation characteristics for radon in various substances are of great interest for two primary reasons. One factor is the obvious need for barriers to seal homes against invasion by radon. Are walls and foundations tight? What plastic sheet, caulks or sealing compounds are effective barriers? The second issue is less obvious - the use of membrane barriers to keep all other alpha emitting isotopes except ^{222}Rn out of a detector air space, so that ^{222}Rn can be measured by itself (Ward *et al.*, 1977). This application is part of the established Terradex methodology that is standard in the United States for accurate long-term indoor radon measurements (Fleischer *et al.*, 1980a).

Table 1.1.4, most of which comes from Fleischer (1988a), gives permeabilities, k, for plastic sheets and caulking compounds. A delay time $\tau_m(=V\,\delta/kA)$ is given for ^{222}Rn entering an air space of volume V through a membrane of area A and thickness δ for one set of assumed δ, A, and V values that is used in ^{222}Rn detectors. For ^{222}Rn detection systems a delay time that is long relative to the 1.34 min mean life of ^{220}Rn is desired (so that no thoron is counted) and which is not much larger than that of ^{222}Rn (5.5 days). Polyethylene is one commonly used choice. For barriers against radon entering a home, thick tightly adhering low-permeability materials are desired. Amongst the caulks, acrylic latex is by far the most effective of the four that have been reported (Fleischer, 1992).

1.1.3 Sources and Transport of Radon

Where is radon generated and how is it released? How does it move? These are the subjects of this section. Table 1.1.5 gives rough representative numbers for the concentrations of radon at various locations. ^{222}Rn is most abundant in mineral grains, is 25x lower in soil gas in the ground, a further 1000x less in homes, still lower in the air over the ground (by 5x), and another 100x lower over oceans. The trends give a coherent picture: radon originates from radium decay in soil grains; a small fraction of it is freed to move through the pore spaces into the atmosphere over land; and much of that has decayed before being transported by the winds to over the oceans. Homes, because they lie at the interface between the soil and the continental atmosphere, have radon concentrations that are intermediate between those of the soil gas and the free atmosphere.

Table 1.1.4: Radon permeabilities and transit times for various materials.

Substance	Permeability (ref.) $[10^{-8}\mathrm{cm}^2\,\mathrm{s}^{-1}]$	Delay Time* τ_m
Membrane Materials		
Cellulose acetate (Cellit-T)	0.75 (h)	19 days
Cellulose nitrate	12.5 (g)	1.1 days
Celgard (porous filter, impregnated with silicone oil)	150 (fm)	2.3 h
Dimethylsiloxane-bisphenol polycarbonate copolymer (MEM-213)	125-840 (ft)	0.4-2.7 h
Fiberglass	450 (r)	46 min
Hydrate cellulos (cellophane)	0.97 (h)	15 days
Polycarbonate	0.38 (g)	38 days
(Makrofol G)	2.4 (h)	6 days
(Makrofol KG)	0.55 (h)	6 days
Polyester	0.20 (g)	26 days
Polyethylene terephthalate (Mylar)	0.084 (g)	170 days
(Melinex-0)	0.30 (h)	48 days
Polyethylene	0.34 (r)	42 days
	7.8 (h)	1.8 days
2°C	2.8 (ft)	5.1 days
23°C	5.5 (fm)	2.6 days
43°C	16.5 (f)	21 h
Polyvinylchloride	5.0 (r)	2.9 days
(Dom pak)	0.58 (h)	25 days
(Folpak)	0.61 (h)	23 days
Polyvinylidene chloride (Saran)	<0.6 (ft)	>24 days
Rubber, natural	635 (g)	32 min
Caulking Compounds		
Acrylic latex (GE)	0.4 (f)	36 days
Polyurethane (Bondex)	80 (f)	4.3 h
Silicone (GE Silpruf)	140 (f)	2.4 h
Unknown type (Geocel)	30 (f)	11 h

$^*V\delta/A$ = 0.0123 cm^2 assumed (the value that happens to characterize a standard Terradex cup)

f = Fleischer (1992)

fm = Fleischer and Mogro Campero (1978)

ft = Fleischer and Turner (1979, 1981) unpublished

g = Giridhar *et al.* (1982)

h = Hafez and Somogyi (1986)

r = Ramachandran *et al.* (1987).

1.1.3.1 Sources

Radon isotopes are produced in minerals by the decay of Ra that itself is the product of a natural chain of decays that begins with ^{238}U, ^{235}U, or ^{232}Th. Although Table 1.1.5 made use of an average radon production rate, minerals vary widely in their contents of these progenitors of radon, ranging from quartz (SiO_2) with $\leq 3{\cdot}10^{-12}$ wt fraction of uranium (Fleischer, 1988b) to UO_2 with 0.88. Typical continental rocks have $\sim 10^{-6}$; sub-oceanic basalts $\sim 10^{-7}$. Granitic rocks and alum shales tend to be higher than the usual continental rocks. They are responsible for high radon readings in portions of Maine and Sweden respectively.

Table 1.1.5: Typical ^{222}Rn concentration.

Location	Concentration (atoms cm^{-3})
Air over oceans	0.04
Air near the earth's surface	4
Typical home in the US	20
Soil Gas	20,000
Interior of typical mineral	500,000

Within a given rock type the different minerals contain very different concentrations of uranium, which often is preferentially located in certain accessory minerals such as zircon ($ZrSiO_4$), apatite ($Ca_5(F,Cl)(PO_4)_3$), sphene ($CaTiSiO_5$), and allanite [H (Ca,Fe)(Al, $Fe)_3Si_3O_{13}$]. As an example of a "rock" where the uranium concentrations have been measured, in the Vaca Muerta meteorite values (in wt ppm) extend from $6{\cdot}10^{-4}$ in anorthite ($Ca\ Al_2Si_2O_8$), to $2{\cdot}10^{-3}$ in hypersthene [$(Mg, Fe)SiO_3$)], $3{\cdot}10^{-2}$ in tridymite (SiO_2) and rutile (TiO_2), 1 in chromite ($FeCr_2O_4$), $\sim$90 in whitlockite (or merrilite) [$Ca_3(PO_4)_2$], and up to 4,000 ppm in zircon (Fleischer *et al.*, 1965). Variability within different zircons and within individual zircons was extensive: 1-10 ppm in one and 800 to 4,000 in a second!

The heterogeneous distribution of uranium has implications for the processes responsible for the release of radon. These are aided by uranium and radium being close to mineral surfaces, as will be made evident in the next section.

1.1.3.2 Release mechanisms

Table 1.1.5 implied that most radon that is produced by the decay of radium never escapes from the mineral in which it is born; rather usually it is wedged tightly in place within the crystal lattice for the few days until it in turn decays. The small fraction of radon that escapes is either released promptly as it is born or within the few days before it decays. One process is established in each of those alternatives.

The first possibility for escape is direct ejection of the radon atom by recoil from alpha emission, proposed and documented by Kigoshi (1971). Conservation of momentum reveals that emission of an alpha particle with 4.78 MeV by ^{226}Ra gives the residual ^{222}Rn nucleus a recoil energy of 86 keV, which Table 1.1.2 shows is sufficient to move the recoiling radon through 26 nm of SiO_2. Alternatively, if the radium was within 26 nm of the surface of the mineral and the recoil properly directed, as in Figure 1.1.3, the recoil can actually leave the grain and enter interstitial space. Two results can be imagined. If the pore space contains water, as Kigoshi imagined, the ejected recoil most often will be brought to rest in the liquid as sketched in the upper part of Fig. 1.1.3. The radon atom is then free to diffuse from the water or be transported by it.

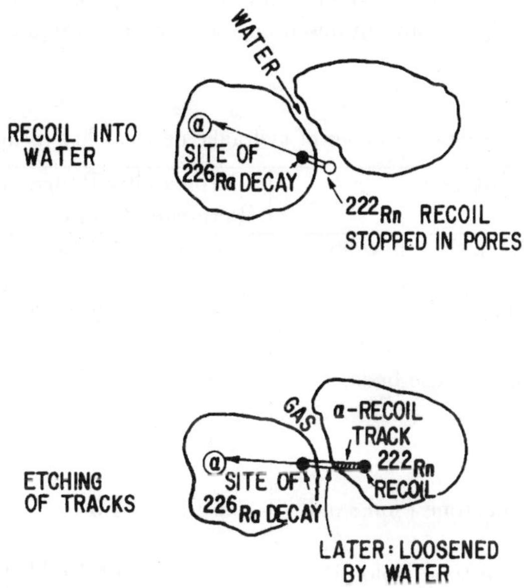

Figure 1.1.3: Models of two radon-release mechanisms: (top) the recoiling ^{222}Rn nucleus is stopped by water in the intergranular material (after Kigoshi, 1971); (bottom) the nucleus recoils into an adjacent grain and the damage track is later removed chemically by the intergranular liquid, releasing the recoil nucleus (Fleischer, 1980, 1988b).

The second possibility, shown in the lower part of Fig. 1.1.3, is that the interstitial space is dry (i.e. filled only with soil gas) and not wide enough to stop the recoiling radon. In this case, it enters a neighboring grain - apparently immobilizing itself. It is, however, less isolated than if it failed to exit its grain of origin, the reason being that radiation damage extends to it from where it entered its new grain. This damage turns out to be etchable by exposure to water (Fleischer, 1980). Hence, if the originally dry grains become wet before the radon has decayed, it can be released into the interstitial space. It is not obvious that this effect should exist; but since the original finding in quartz (Fleischer, 1980), it has been demonstrated also for muscovite mica [$KAl_3Si_3O_{10}(OH_2)$], diopside ($CaMgSi_2O_6$), orthoclase ($KAlSi_3O_8$), apatite, sphene, and obsidian (Fleischer, 1982).

1.1.3.3 Radon transport

Radon moves by two basic means, diffusion and forced flow. Diffusion inevitably occurs, even though its extent may be limited. Hence diffusive migration sets a lower limit on the transport of radon. Forced flow depends on pressure gradients, which may or may not be present in a given situation. Numerous hypothetical origins of such gradients have been proposed, but only occasionally are they dependably known to be active. In any such case diffusive effects are superimposed on those of pressure gradients.

Table 1.1.6: Representative values of hydraulic permeability (Scheidegger, 1960).

Substance	Permeability Range (Permeability in cm^2)
Black slate powder	$4.9 \cdot 10^{-10}$ - $1.2 \cdot 10^{-9}$
Silica powder	$1.3 \cdot 10^{-10}$ - $5.1 \cdot 10^{-10}$
Sand (loose beds)	$2.0 \cdot 10^{-7}$ - $1.8 \cdot 10^{-6}$
Soils	$2.9 \cdot 10^{-9}$ - $1.4 \cdot 10^{-7}$
Sandstone ("oil sand")	$5.0 \cdot 10^{-12}$ - $3.0 \cdot 10^{-8}$
Limestone, dolomite	$2.0 \cdot 10^{-11}$ - $4.5 \cdot 10^{-10}$
Brick	$4.8 \cdot 10^{-11}$ - $2.2 \cdot 10^{-9}$
Bituminous concrete	$1.0 \cdot 10^{-9}$ - $2.3 \cdot 10^{-7}$
Cork board	$3.3 \cdot 10^{-6}$ - $1.5 \cdot 10^{-5}$
Fiber glass	$2.4 \cdot 10^{-7}$ - $5.1 \cdot 10^{-7}$

1.1.3.3.1 Diffusive transport in soil and rock. Diffusion in gas and water was considered in subsection 1.1.2.5. Diffusive motion in the ground is in those same phases, but is controlled by the extent of open, connected porosity in soil and rock. In general, higher porosity enables more extensive diffusive transport; but it is critical to consider only that fraction of the pore space that allows through motion. Most porosity in soil is interconnected; but in rocks that is not necessarily true. For a group of 32 soils that extend from clays to silty sands, values of D ranged from 0.005 to 0.062 $cm^2\ s^{-1}$, with an average of 0.03 (Silker and Kalkwarf, 1983). The considerable variability implies that where a precise value is wanted, the actual soil of interest should be measured using multiple samples. The values quoted may be compared to the free-air value of 0.1 $cm^2\ s^{-1}$.

Rocks can vary from almost crack- and pore-free to highly fractured. Values of hydraulic permeability will give indications as to relative levels.

1.1.3.3.2 Forced transport of radon in soil and rock. Pressure gradients can cause flow of the gas or liquid in interstial space in the earth, the ease of such flow being given by the hydraulic permeability $K[=\eta(dV/dt)/A(dP/dx)]$ where η is the viscosity, $(dV/dt)/A$ the volume moved per unit time and per unit area, and dP/dx the pressure gradient. K is in units of $(\text{length})^2$, usually either cm^2 or darcys, where 1 darcy = 10^{-8} cm^2. Table 1.1.6 indicates the wide spectrum of values that are found for hydraulic permeabilities (Scheidiger, 1960). Note that K is distinct from k as used in subsection 1.1.2.6.

For a non-uniform material the K values depend on the scale of the measurement, since flow on a large enough scale between two measuring points may be facilitated by occasional easy-flow paths. As an example, Fleischer *et al.* (1980b) examined permeabilities along drill holes, including core samples from the holes. K values for 15 cores were $<10^{-10}$ to $2{\cdot}10^{-9}$ cm^2, while different 15 to 20 m vertical regions along holes gave 10^{-9} to $2{\cdot}10^{-6}$ cm^2 - ranges that just barely overlap.

1.1.3.3.3 Detection of vertical transport of radon in the earth. How does one learn whether transport is occurring? The procedure is to note what gradients exist in the ground in the absence of flow and then look for changes. Since in a uniform earth the primary reproducible gradients near the surface will be vertical, those are the gradients that are usually sought. Near the surface, ^{222}Rn escapes into the atmosphere, lowering the natural concentration to near zero at the surface and producing a near-surface gradient whose profile is set by diffusion and the finite mean life τ of ^{222}Rn. The result is shown in Fig. 1.1.4 by the middle curve. The near-surface zero radon concentration grows and by exponential decay approaches the concentration C_∞ at depth, with the characteristic distance $\sqrt{D\tau}$, which is 1.2 m for $D = 0.03$ $cm^2\ s^{-1}$.

The radon lost to the atmosphere in each mean life is equivalent to an upward flux at the surface at a velocity $\sqrt{D/\tau}$, which in this example is $2.5{\cdot}10^{-4}$ $cm\ s^{-1}$ (or 1.2 m/5.5 days). Equivalently, for every 5.5 days mean life, radon equal to that in 1.2 m of the ground is drained into the air. This is the velocity that any systematic upflow (or downflow) must exceed decisively in order to be easily recognized.

The effects of both diffusion and superimposed flow, as were calculated by Grammakov (1936), are shown by the other two curves in Fig. 1.1.4. Upflow by an amount $7\sqrt{D/\tau}$

lifts the radon gradient closer to the surface, and downward flow at the same speed lowers the gradient, with the result that it extends to greater depths than with no flow. If radon is measured (as it commonly is) at 0.5 to 1.0 m depths, the values are strongly affected by flows of this magnitude - increased by upflow and decreased by downflow. In contrast, at >10 m depth only minor charges would occur, ones that probably could not be recognized experimentally.

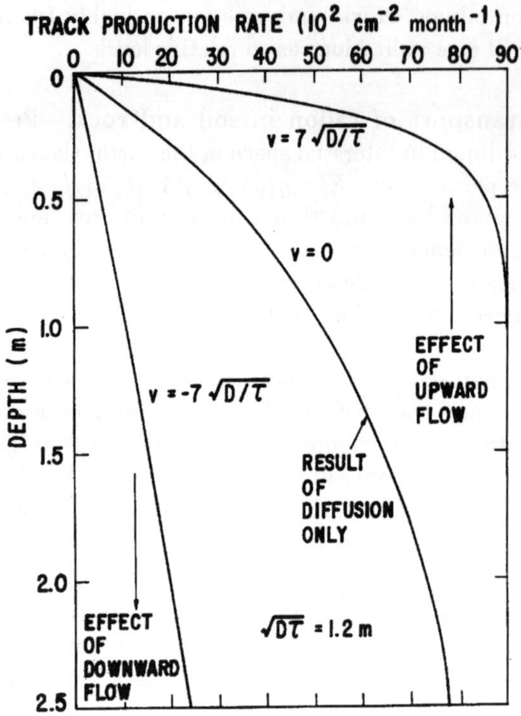

Figure 1.1.4: Radon concentration versus depth calculated for zero vertical flow and flows (i.e. velocities v) per unit porosity of $\pm$ 7 $\sqrt{(D/\tau)}$, where the mean diffusion distance $\sqrt{(D\tau)}$ is 1.2 m in the ground and the mean diffusion time τ is 5.5 d. The horizontal scale and the coefficient 7 are calculated to correspond to typical field values at Thoreau, New Mexico. In the absence of transport, at typical hole depths used for monitoring of radon, the concentration is about one-third the value C_∞ at great depth, taken in this example as 28,500 Bq m^{-3} (770 pCi l^{-1}).

If flow were constant (so that the method of looking for changes would not work), it could be recognized nevertheless if the emanation rate was measured in the laboratory and the reference no-flow radon profiles inferred. This procedure has been followed at one petroleum-rich site in Oklahoma (Fleischer and Turner, 1984).

1.1.3.3.4 Mechanisms of forced transport in the earth. What physical processes in the earth might produce pressure gradients and thereby effect radon flow? A number of observations of flow have been made and mechanistic models proposed. Although most of these processes no doubt occur at some times and places, it is difficult to prove for any single observation that in fact a particular mechanism was clearly responsible.

The mechanisms differ depending on the geophysical and geochemical phenomena - earthquakes, volcanic eruptions, oil prospecting, and uranium prospecting. Although these will be elaborated in the later sections about these phenomena, they merit mention here in order to show that multiple effects are plausible.

Three main categories are gas emission, stresses, and fluid convection. Gases can be released by oil deposits (Fleischer and Turner, 1984), and they can be generated and pressurized by hot regions such as volcanos. Stress can be generated by the build-up of strains that precede earthquakes (Dobrovolsky *et al.*, 1979; Fleischer, 1981) and volcanic eruptions, and the stress gradients in turn can encourage fluid flow in the earth. Fluid convection could produce flows if there were sufficiently steep geothermal gradients and high enough permeabilities (Mogro-Campero and Fleischer, 1977). Such flows could deliver signals from uranium deposits from greater depths than would allow detection of ore solely by diffusion of radon. Here again volcanic effects (in this case thermal) could reveal impending activity.

A fourth contributor to pressure-induced flow is atmospheric pumping in the near-surface soil by the pressure fluctuations that are associated with weather systems (Clements and Wilkening, 1974). Track-based radon measurements are normally taken over weeks in order to average and thus minimize such effects.

Uranium and petroleum exploration, volcanic effects, earthquakes effects and geothermal exploration will be discussed more fully in this volume by Khan *et al.*, by Monnin and Seidel, by Fleischer, and by Balcázar (Chapter 4) respectively.

1.1.4 Radon Daughters in Air

The prompt-decay products of ^{222}Rn are metal atoms (Pb, Bi, and Po) that will condense on any solid or liquid surface that they contact. In typical air they will attach to small particles that are normally present. The abundance of these particles is $\sim 10^{-3}$ cm^{-3} in "clean" air and rises to $\sim 10^5$ in urban areas (Junge, 1963). Indoors, there are many surfaces which compete with aerosols as depositories for daughter radioactivity, and aerosols too may precipitate on these surfaces. The result is that indoor air rarely will approach the state of retaining its full complement of short-lived daughters and is usually significantly below equilibrium (where the activity of each daughter equals that of the parent Rn). The fractions of equlibrium indoors have been measured in nine countries, as reviewed by Ramachandran *et al.* (1995); values average 0.47 ($\pm$ 0.14) with the extremes being 0.27 and 0.80; Scandinavia, Canada, and the United States average 0.49 ($\pm$0.03).

Although lung doses are due to the daughters rather than ^{222}Rn directly, ^{222}Rn is most often measured in homes. This is appropriate since (1) radon measurements are far easier and therefore less expensive, and (2) because the equilibrium factors are relatively constant.

Radon daughters and their health effects in underground mines are quite different matters, however. In a mine tunnel, at a wall that is being worked, fresh radon is released that initiallly has none of its daughters accompanying it. As a result, different parts of a mine will have widely differing fractions of radon daughters present. Daughter measurements at multiple mine locations are therefore obligatory.

1.1.5 Concluding Note

The following sections of this volume will describe methods of measurements and their various areas of application, from health to geophysics and geochemistry.

References

Clements W. E. and Wilkening M.H. (1974) Atmospheric pressure effects on ^{222}Rn transport across the earth-air interface. *J. Geophys. Res.* **79**, 5025-5029.

Cook G. A. (1961) *Argon, Helium, and the Rare Gases.* Interscience, New York.

Dobrovolsky I. P., Zubkov S. I. and Miachkin V. I. (1979) Estimation of the size of earthquake preparation zones. *Pure Appl. Geophys.* **117**, 1025-1044.

Durrani S. A. and Bull R. K. (1987) *Solid State Nuclear Track Detection: Principles, Methods and Application.* Pergamon Press, Oxford.

Eisenbud M. (1987) *Environmental Radioactivity.* Third edition, Academic Press, San Diego.

Faul H. (1954) *Nuclear Geology.* John Wiley & Sons, New York.

Fields P. R., Stein L. and Zirin M. H. (1962) Radon fluoride. *J. Amer. Chem. Soc.* **84**, 4164-4165.

Fleischer R. L. (1980) Isotopic disequilibrium of uranium: Alpha-recoil damage and preferential solution effects. *Science* **207**, 979-981.

Fleischer R. L. (1981) Dislocation model for radon response to distant earthquakes. *Geophys. Res. Lett.* **8**, 477-480.

Fleischer R. L. (1982) Alpha recoil damage and solution effects in minerals: Isotopic disequilibrium and radon release. *Geochim. Cosmochim. Acta* **46**, 2191-2201.

Fleischer R. L. (1988a) Radon in the environment - opportunities and hazards. *Nucl. Tracks Radiat. Meas.* **14**, 421-435.

Fleischer R. L. (1988b) Alpha recoil damage: relation to isotopic disequilibrium and leaching of radionuclides. *Geochim. Cosmochim. Acta* **52**, 1459-1466.

Fleischer R. L. (1992) Permeability of caulking compounds to ^{222}Rn. *Health Phys.* **62**, 91-95.

Fleischer R. L., Naeser C. W., Price P. B., Walker R. M. and Marvin U. B (1965) Fossil particle tracks and uranium distributions in minerals of the Vaca Muerta meteorite. *Science* **148**, 629-632.

Fleischer R. L., Price P. B. and Walker R. M. (1975) *Nuclear Tracks in Solids: Principles and Applications.* Univ. of California Press, Berkeley.

Fleischer R. L. Giard W. R., Mogro-Campero A., Turner L. G., Alter H. W. and Gingrich J. E. (1980a) Dosimetry of environmental radon: Methods and theory for low-dose integrated measurements. *Health Phys.* **39**, 957-967.

Fleischer R. L., Hart H. R. Jr. and Mogro-Campero A. (1980b) Radon emanation over an ore body: Search for long-distance transport of radon. *Nucl. Instr. Meth.* **173**, 169-181.

Fleischer R. L. and Mogro-Campero A. (1978) Mapping of integrated radon emanation for detection of long-distance migration of gases within the earth: Techniques and principles. *J. Geophys. Res.* **83**, 3539-3549.

Fleischer R. L. and Turner L. G. (1984) Correlations of radon and carbon isotopic measurements with petroleum and natural gas at Cement, Oklahoma. *Geophysics* **49**, 810-817.

Furlan G. and Tommasino L., eds. (1993) *Proc. 2nd Workshop on Radon Monitoring in Radioprotection, Environmental and/or Earth Sciences.* Trieste, 1991, World Scientific, Singapore.

Girdhar J., Ragharayya M. and Padmanabhan N. (1982) Radon permeability of some membranes. *Health Phys.* **42**, 723-725.

Grammakov A. G. (1936) On the influence of some factors in the spreading of radioactive emanations under natural conditions. *Zh. Geofiz.* **6**, 123-148.

Hafez A-F. and Somogyi G. (1986) Determination of radon and thoron permeability through some plastics by track technique. *Nucl. Tracks* **12**, 1-6, 697-700.

Hatch T. F. and Gross P. (1964) *Pulmonary Deposition and Retention of Inhaled Aerosols.* Academic Press, New York.

Hess C. T., Weiffenbach C. V. and Norton S. A. (1982) Variations of airborne and water borne Rn-222 in houses in Maine. *Environ. Int.* **8**, 59-66.

Junge C. E. (1963) *Air Chemistry and Radioactivity.* Academic Press, New York.

Kigoshi K. (1971) Alpha-recoil ^{234}Th: Dissolution into water and the ^{234}U/^{238}U disequilibrium in nature. *Science* **173**, 47-48.

Levinson A. A. and Coetzee G. L. (1978) Implications of disequilibrium in exploration for uranium ores in the surficial environment using radiometric techniques-a review. *Minerals Sci. Eng.* **10**, 19-27.

Lindhard J., Scharff M., and Schiott H. E. (1963) Range concepts and heavy ion ranges. *Mat. Fys. Medd. Dan. Vid. Selsk.* **33**, 14, 1-42.

Loef J. J. van (1981) On the thermophysical properties of liquid radon-222. *Physica* **103 B**, 362-364.

Mogro-Campero A. and Fleischer R. L. (1977) Subterrestrial fluid convection: A hypothesis for long-distance migration of radon within the earth. *Earth Planet. Sci. Lett.* **34**, 321-325.

Nazaroff W. W. and Nero A. V. Jr. (1988) *Radon and its Decay Products in Indoor Air.* Wiley - Interscience, New York.

Nussbaum E. and Hursh J. B. (1958) Radon solubility in fatty acids and triglycerides.

J. Phys. Chem. **62**, 81-84.

Ramachandran T. V., Lalit B. Y. and Mishra U. C. (1987) Measurement of radon permeability through some membranes. *Nucl. Tracks Radiat. Meas.* **13**, 81-84.

Ramachandran T. V., Subba Ramu M. C. and Nambi K. S. V. (1995) Simultaneous measurements of radon and its progeny using SSNTD's and evaluation of internal doses due to inhalation. *Bull. Rad. Protect.* **18**, 109-113.

Rutherford E., and Soddy F. (1902) Condensation of radioactive emanations. *Phil. Mag.* **5**, 561-576.

Scheidegger A. E. (1960) *The Physics of Flow Through Porous Media.* Univ. of Toronto Press, Toronto.

Spohr R. (1990) *Ion Tracks and Microtechnology.* Friedr. Vieweg u. Sohn, Braunschweig.

Silker W. B. and Kalkwarf D. R. (1983) Radon diffusion in candidate soils for covering uranium mill tailings. US Nucl. Reg. Commission report NUREG/CR-2924, PNL-4434, 49 pages.

Tanner A. B. (1964) Radon migration in the ground: A review. In *The Natural Radiation Environment* (eds. J.A.S. Adams and W. M. Lowder), Univ. of Chicago Press, Chicago, pp. 161-190.

Tommasino L., Furlan G., Khan H.A. and Monnin M., eds. (1990) *Proc. Int. Workshop on Radon Monitoring in Radioprotection, Environmental Radioactivity and Earth Sciences.* Trieste, 1989, World Scientific, Singapore.

Wagner G. A. and van den Haute P. (1992) *Fission Track Dating.* Kluwer, Dordrecht, The Netherlands.

Ward W. J., Fleischer R. L. and Mogro-Campero A. (1977) Barrier technique for separate measurement of radon isotopes. *Rev. Sci. Instr.* **48**, 1440-1441.

G. Åkerblom[1] and H. Mellander[1]

1.2 Geology and Radon

1.2.1 Introduction

The noble gas isotope ^{222}Rn is formed by the decay of ^{226}Ra. As ^{226}Ra is one of the nuclides formed in the disintegration series from ^{238}U, the amount of radon formed in rocks and soils depends on their uranium content. But that alone is not decisive in determining the radon concentration of air or water in the pores of the soil. It is also determined by the extent to which the radon atoms formed actually emanate from the mineral grains and whether radon can leave the pore space either by diffusion or together with a flow of air or water. In addition, radon concentration in the soil air is significantly affected by the occurrence of water in the pores.

Apart from ^{222}Rn, the isotopes ^{220}Rn, also called thoron, and ^{219}Rn, occur in nature. ^{220}Rn is formed in the decay of ^{224}Ra in the disintegration series starting with ^{232}Th, and ^{219}Rn in the series from ^{235}U. The concentration of ^{219}Rn in nature is so low that it does not constitute a health hazard and neither, to any appreciable extent, do levels of ^{220}Rn. On the other hand, ^{220}Rn *can* constitute a health hazard under certain conditions, and sometimes occurs in such concentrations that it must be taken into account in measuring ^{222}Rn.

This section deals with radon in soil air and ground water and the factors determining its concentration and transport. It is largely based on a chapter on radon theory in the handbook for surveys of ground radon carried out in Sweden (Åkerblom *et al.*, 1990). This in turn, is based on results from various research projects, both international (Bigu *et al.*, 1984; Karner, 1964; Reimer and Tanner, 1991; Rose *et al.*, 1990; Tanner, 1964, 1980; Wilkening, 1980, etc.) and Swedish (Åkerblom *et al.*, 1983; Hesselbom, 1985; Kristiansson and Malmqvist, 1984). A lot of know-how comes from the extensive material on ground radon from uranium exploration, in which radon has been used as a geophysical means of locating deposits. A further significant source is the authors' own experience. Studies of ground radon prior to planning land utilisation and building work, began in Sweden in 1979 (Åkerblom, 1986; Åkerblom and Wilson, 1982a,b). Now, in 1995, radon-risk maps on a scale of 1:50,000 or greater have been produced for some 80% of Sweden's surface area, and detailed investigations of the radon situation are carried out regularly before building (Åkerblom, 1994). In the United States, Radon Potential Maps have been drawn up for most counties (Gundersen *et al.*, 1991). In many countries all round the world mappings of radon risk areas or radon-prone areas are in progress. The International Committee on Radiological Protection has recommended to the member states to produce maps that

[1]Swedish Radiation Protection Institute, Stockholm, Sweden

delineate radon-prone areas (ICRP, 1993).

1.2.2 Uranium - Radium - Radon in Rocks and Soils

Uranium and thorium with their disintegration products, and the radioactive isotope of potassium, ^{40}K, account for by far the greatest part of naturally occurring ionizing radiation in the human environment. Uranium, thorium and potassium are lithophile elements. This means that they occur primarily together with magmata rich in silica (Fig. 1.2.1), i.e. the ones forming granites and syenites, and the hypabysal and volcanic rocks associated with granites and syenites, e.g. the hypabysal rocks pegmatite and aplite and the volcanic rocks rhyolite and porphyry. Uranium and thorium can be mobilised and transported by geological processes such as metamorphosis and with hot hydrothermal solutions, to be precipitated in fissures and other permeable structures. Precipitation often takes place in a reducing environment. Examples of such environments are rocks rich in calcium and fracture filling, as well as magnesium, rocks rich in iron, and products of weathering. For instance, uranium often occurs in contact with basic rocks, in skarn in iron ores and in clay gouge. Uranium is easily leached even at a normal pH of around 5-7, and many of the large uranium ore deposits are formed by ground water that has transported uranium which has been precipitated in a reducing environment. When the bedrock weathers and is exposed to water erosion, the heavy minerals bearing uranium and thorium are often separated. They can accumulate together with iron, titanium and zircon and form placer deposits. Areas with large recent deposits of heavy minerals include the coasts of Florida, Brazil, India and Australia, but they also occur as layers and deposits in sandstones and quartzites.

Thorium does not readily dissolve in water. The small quantity in streams and river water is precipitated almost immediately on reaching the sea. Uranium, on the other hand, has higher concentrations in sea-water than river-water. It is removed from sea-water by adsorption of clay particles and precipitation, e.g. with phosphate. Thus clays and shales nearly always have higher uranium and radium concentrations than other sedimentary rocks. Examples of sedimentary rocks with relatively high uranium concentrations are certain black shales (e.g. alum shale in Sweden and Norway, and Chattanooga shale in the USA) and sedimentary deposits of phosphate in Morocco, Jordan and Florida.

Soil consists of products broken down from the bedrock. The topsoil, or A-horizon, can be more or less mixed up with organic material. There are several different ways in which soil is formed:

- The bedrock is subjected to chemical and mechanical weathering (conditioned by temperature, running water and wind). Soil formed by chemical- and temperature-conditioned weathering remains directly on the surface of the bedrock. Examples are lateritic soils.

- The bedrock is subjected to downward crushing and wear from an immense covering of glacial ice. Till is formed in that way, as a direct product of the underlying bedrock.

- Eruption products from volcanoes form layers of ash and stone.
- Precipitation products from sea- and fresh-water. The precipitation may be chemical or biological. Examples of such soils are lime sludge, peat and diatomaceous earth.
- Loose soil is transported, sorted and redeposited by moving water, waves and wind, often with different means of transport combining. Examples of soils formed by water transport and sorting are river deposits of gravel, sand and clay. Examples of redeposited soils are sand along sea coasts, while examples of deposits formed by wind transport are loess and sand dunes.

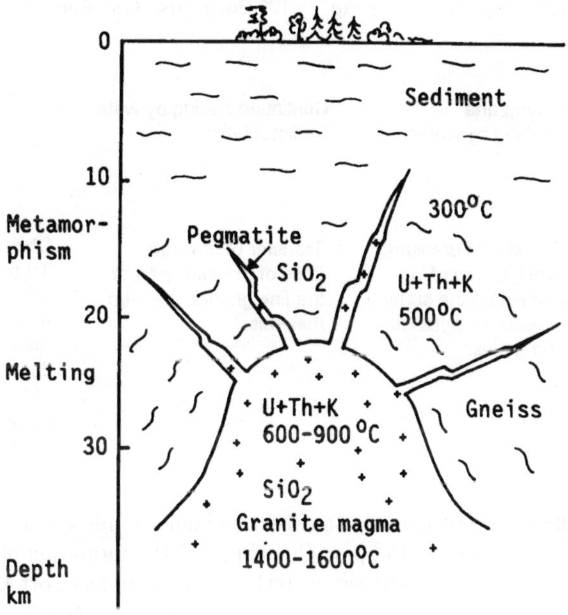

Figure 1.2.1: An intrusion of a granitic magma moves upward through the Earth's crust. The elements K, U and Th are enriched in the upper part of the dome and often in pegmatites and aplites. Metamorphosis partly melts the sediments and older intrusive rocks which are transformed into gneiss. In this, water-rich solutions form pegmatitic veins, sills and dykes, which often have a higher concentration of U, Th and K than the surrounding rock.

Throughout the process of soil formation, changes take place in the original uranium and radium concentrations in the bedrock. Moreover, uranium and radium have different chemical properties and are affected to a different extent by chemical leaching. There is often a separation of uranium and radium. Fig. 1.2.2 illustrates the glacial process by which a granite rich in uranium is mechanically and chemically broken down (Åkerblom *et al.*, 1990).

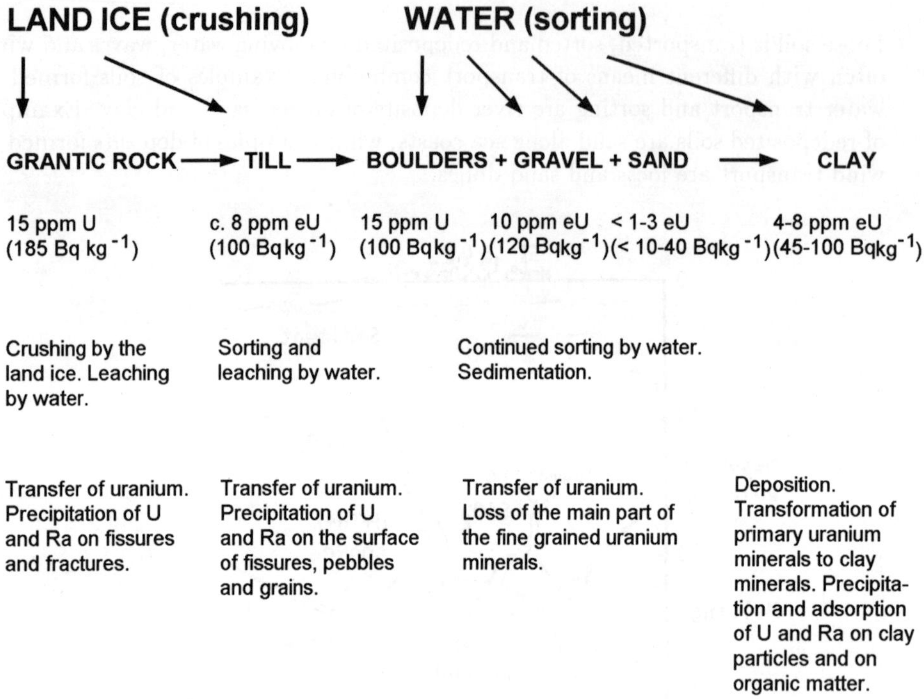

Figure 1.2.2: The effects of crushing by the sheet of land ice, on a granite bedrock with an assumed uranium concentration of 15 ppm. Crushing and the formation of till allow water (melted ice, running water and lake- and sea water) to transport and sort the broken rock, forming soils such as gravel, silt and clay. Uranium mineral is removed in the process. Uranium and radium are continuously leached into this water, percolating water and ground water. At the same time, uranium and radium are precipitated on fracture surfaces in the bedrock, stones and minerals, and on the surfaces of the grains. Some of the uranium and radium in the water is reabsorbed on the sedimentary clay particles. In the figure eU stands for equivalent uranium.

Like radium, thorium is not readily dissolved and the disintegration period from ^{228}Ra (the first decay product of ^{232}Th) to ^{208}Pb is about 20 years (three half-lives of ^{228}Ra). There is thus no time for any significant disintegration disequilibrium between ^{232}Th and its daughter products. The relationship between the concentrations of ^{232}Th and its daughters is not altered to any extent in chemical decomposition of the bedrock or soil horizon.

In chemical weathering, uranium can be released and removed by percolating or flowing water, to be precipitated once again in a reducing environment, e.g. soil horizons fairly rich in calcium, or at the water table. In podsol soils, uranium is depleted in the topsoil and deposited in the rust-coloured B-horizon at a depth of one, or a few, decimetres (Minell, 1983, 1990). The B-horizon gets its pigmentation from precipitated iron- and manganese hydroxide and organic matter, which act as a reducing environment for uranium and radium leached from the topsoil (Fig. 1.2.3).

During mechanical transport by water, uranium, and to a certain extent radium, are leached from the material transported. In addition, uranium and radium are transported in the form of uranium-rich clay particles that were originally grains of uranium material. Uranium often occurs in the bedrock as very small grains of uranium mineral, e.g. coffenite $[U(SiO_4)_{1-x}(OH)_{4x}]$ and uraninite (U_3O_8), or absorbed on the grains or included

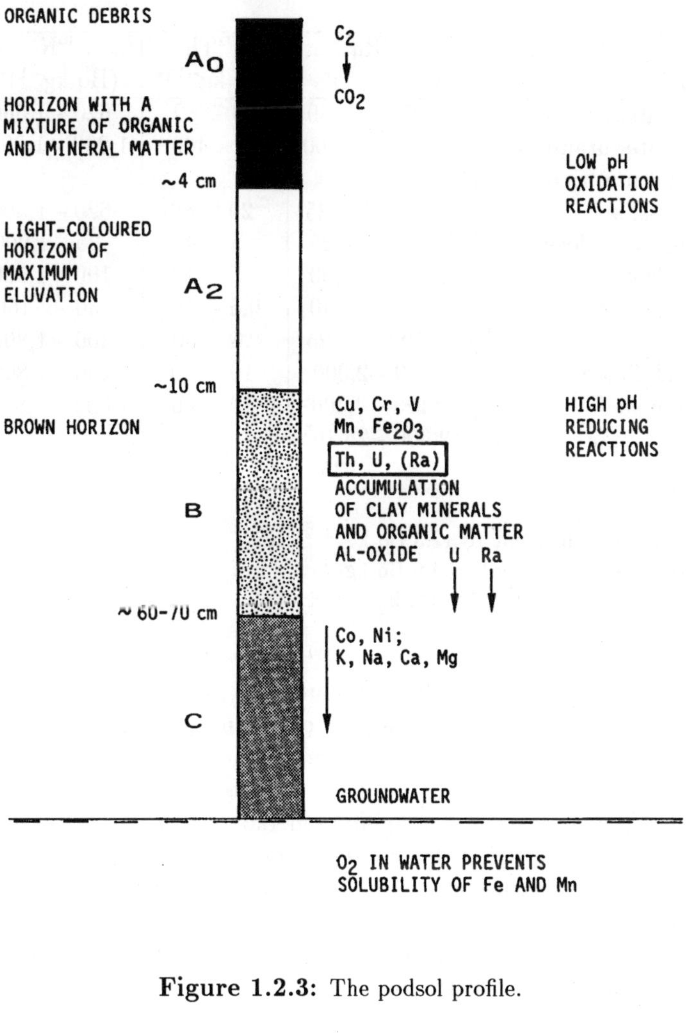

Figure 1.2.3: The podsol profile.

in the lattice of zircon, apatite, titanite and other accessory minerals. Common to these is the fact that they are much smaller than the other mineral grains in the rock, in a granite mainly quartz, feldspar and glimmer. The uranium-bearing mineral grains are released during erosion and transport. The soft mineral grains are converted into clay particles and accompany the current of water, while hard grains such as zircon, monazite and titanite accompany the sand fraction. As these hard grains are heavier than most other minerals, they can be enriched by mechanical sorting and are often included in layers of heavy minerals, consisting otherwise of iron and titanium mineral.

Table 1.2.1: Range of concentrations of ^{226}Ra, ^{232}Th and ^{40}K in some common rocks and uranium ores. Higher and lower concentrations occur occasionally. Averages from Kogan *et al.*, (1971).

Rock	^{226}Ra (Bq kg^{-1})*	^{232}Th (Bq kg^{-1})*	^{40}K (Bq kg^{-1})*
Granite, normal	25 - 80	20 - 80	620 - 1,500
Granite, uranium-thorium rich	100 - 500	40 - 400	1,200 - 1,900
Gneiss, sedimentary	25 - 125	20 - 80	620 - 1,500
Gabbro, dolerite	1 - 25	5 - 40	25 - 1,000
Sandstone	1 - 60	5 - 40	100 - 1,500
Limestone	5 - 40	0.5 - 10	30 - 160
Shale	10 - 125	5 - 60	300 - 1,800
Black Shale	10 - 2,000	10 - 50	300 - 1,800
Alum shale	125 - 4,300	10 - 40	1,000 - 1,800
Uranium ore	12,000 - 25·10^{5}		
Average	35	38	775

* 1 ppm U is equivalent to 12.3 Bq kg^{-1} ^{226}Ra
1 ppm Th is equivalent to 4.0 Bq kg^{-1} ^{232}Th
1 % K is equivalent to 310 Bq kg^{-1} ^{40}K

An effect of sorting by water is that uranium and radium concentrations gradually decrease, and more so the greater the degree of sorting; so that they are very low in well-sorted, fine- grain soils such as quartz sand. Concentrations also increase with an increase in the clay mineral content.

As a result of the separation of uranium and radium that takes place when soil is formed, transported and sorted, radon concentrations in ground water and soil air may differ substantially in different soil types. Thus they vary in different horizons in the same soil profile.

Table 1.2.1 gives the uranium concentrations in some typical rocks and uranium deposits and Table 1.2.2 the concentrations of radium and radon in the soil air of some soils.

Table 1.2.2: Normal concentrations of ^{226}Ra and ^{222}Rn in soils, measured at a depth of 1 metre.

Soil type	^{226}Ra (Bq kg^{-1})*	^{222}Rn (Bq m^{-3})
Till or residual soils with normal radium content	15 - 65	5,000 - 30,000
Till or residual soils with fragments of granite	130 - 125	10,000 - 60,000
Till with fragments of uranium - rich granite	125 - 360	10,000 - 200,000
Gravel	30 - 75	10,000 - 150,000
Sand	5 - 35	2,000 - 20,000
Silt	10 - 50	5,000 - 60,000
Clay	10 - 100	10,000 - 100,000
Soils containing fragments of alum shale	175 - 2,500	50,000 - >10^6

* 12.3 Bq kg^{-1} ^{226}Ra is equivalent to 1 ppm uranium

1.2.3 Formation and Emanation of Radon

There are three differing theories about how radon and thoron emanation takes place from the mineral lattice or molecule in which the radon or thoron atom is formed. As emanation occurs in a similar manner for both radon and thoron, only the principles of radon disintegration and emanation are dealt with here (see also section 1.1).

When ^{226}Ra disintegrates, a radon atom and an alpha particle are formed, the latter being ejected from the decaying radium atom. When the alpha particle is ejected, a recoil effect arises. This dislodges the radon atom from the place in the mineral lattice or molecule where the radium atom was. The distance the radon atom can then move in a mineral grain of normal density has been, from kinematics put at (0.02-0.07 μm). It is this very movement of the radon atom that enables radon to emanate from a mineral grain (Fleischer, 1980; Kigoshi, 1971; Tanner, 1980).

The radon atom could be ejected from the grain as a result of the movement, provided it was close to the surface, and was kicked in an outward direction. In the same way, the radon atom could be ejected to a micro- fissure in the mineral grain. Further transport in the micro-fissure is by diffusion. The process is illustrated in Fig. 1.2.4.

Other researchers have assumed that radon could emanate from the mineral grain by diffusion through the mineral lattice or by leaching by water through radiation damage in the mineral grain (Fleischer 1980, 1982).

Radon and thoron concentrations in soil air or ground water are determined by the radium concentration in surrounding soil and bedrock, how many of the radon atoms emanate to the pore space from the mineral grain in which they are formed, and the porosity and water content of the soil. In addition, the radon concentration is affected by how much of the radon is removed or supplied by diffusion and transported by the movement of the soil air or water or of ascending gases such as CO_2 and CH_4. The permeability of the soil and rock are of decisive importance for transporting radon.

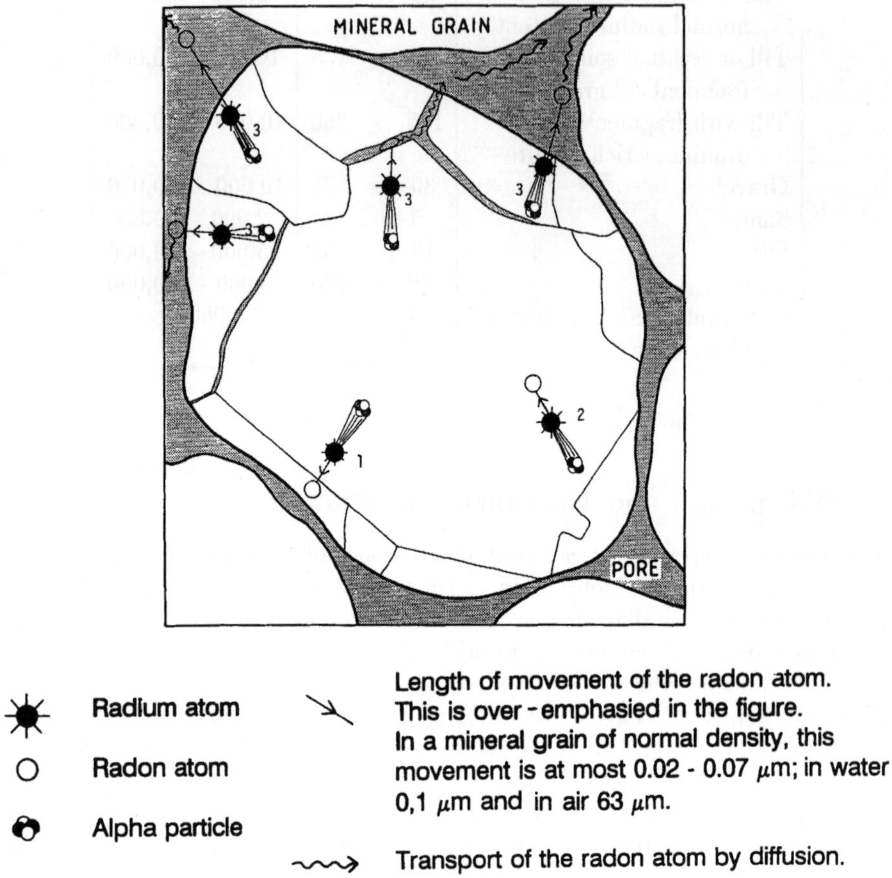

Figure 1.2.4: Principles of radon emanation from a mineral grain, modified after Tanner (1980).
Case 1. When radium decays, a radium atom and an alpha particle are formed. The radon atom is moved into an adjacent crystal by a recoil effect from the ejected alpha particle.
Case 2. The radon atom is moved through the crystal.
Cases 3 and 4. The radon atom is moved from the crystal to a micro-fissure or the air in an adjacent pore. It is assumed that further transport is by diffusion.

If the radium concentration, porosity and emanation for a type of soil are known, the maximum radon concentration that can occur in the pore air can be calculated from eq.

(1.2.1) (Andersson *et al.*, 1983).

Eq. (1.2.1) is used to calculate the maximum radon concentration in the soil's pores when they are completely filled with either air or water[2]

$$C_{max} = Aed(1-p)/p \tag{1.2.1}$$

where C_{max} is the radon activity concentration (Bq m^{-3}) in pore volume with absolutely no ventilation (0 ach); A is specific activity (Bq kg^{-1}), e is emanation[3]; d is compact density (kg m^{-3}), (normal for mineral soils: 2700 kg m^{-3}); and p is porosity, ratio of pore volume to total volume; ach stands for air changes per hour. The value of C_{max} increases as the values of p falls.

Note: This formula is not fully universal. When the concentration of radon atoms per unit of volume in the pore equals that in the surrounding mineral grain, passage by diffusion of radon atoms to the pore ceases if they cannot be transported further. In other words, the concentration of radon gas per unit of volume in the pore cannot be greater than the concentration in the mineral grains. This occurs when

$$p < e \tag{1.2.2}$$

In soil above the water table, radon can be transported from the pore by diffusion, or by soil air or ground water in motion. In soil below the water table, this possibility is strictly limited, as the radon in the water does not have a diffusion length greater than 5 cm (see subsection 1.2.5). In practice, the concentration in soil horizons below the water table will continually maintain the same level, equal to the maximum that can be formed with the radium concentration, porosity and emanation for the soil in question.

Eq. (1.2.1) is used in the example below to calculate the maximum radon concentration in the air of a soil with a specific activity of 10 Bq kg^{-1} of ^{226}Ra.

The soil has 30% porosity, a compact density of 2,700 kg m^{-3} and 25% emanation (normal emanation for till, for instance; see table 1.2.2). Radon concentration C_{max} is then approx. 16,000 Bq m^{-3} per 10 Bq kg^{-1} ^{226}Ra:

$$C_{max} = 10(A)\ 0.25(e)\ 2700(d)\ (1-0.3)(1-p)/0.3(p) \approx\ 16,000\ Bq\ m^{-3} \tag{1.2.3}$$

For a dry soil with a radium activity of approx. 50 Bq kg^{-1}, the maximum radon activity concentration in the pore air (for p = 0.3) is thus about 80,000 Bq m^{-3} with no ventilation (0 ach). In a soil consisting largely of granite rich in uranium and with a radium concentration of 250 Bq kg^{-1}, the maximum radon concentration is 400,000 Bq m^{-3}; and in sand consisting of fragments of alum shale with a radium concentration of 1,250 Bq kg^{-1}, it is about 2 million Bq m^{-3}. Concentrations in air-filled pores will not be as high as in horizons close to the surface as each form of ventilation of the

[2] Radon concentrations are also affected by the extent to which the pore is filled with water (see eq.(1.2.4)).

[3] Fraction of all radon and thoron atoms formed that emanate to the pore space.

soil, and diffusion from it, reduces the radon level in the pores. Thus the normal radon concentration in soil air under Swedish conditions is 10,000-50,000 Bq m^{-3}; but levels are considerably higher in soils with an enhanced radium concentration: in till from granite rich in uranium it is up to 250,000 Bq m^{-3}, and in alum shales up to more than 1 million Bq m^{-3}. If emanation is great, e.g. as from gravel and clay, radon concentrations can be relatively high: 50,000-100,000 Bq m^{-3}, without the radium concentration being above normal for the soil. Table 1.2.2 gives normal concentrations of ^{226}Ra in the soil and ^{222}Rn in the soil air, in various soils. The figures are based on measurements made at a depth of about 1 metre.

The smaller the particle in which the radon atom is formed, the greater the chance of the atom escaping from it. The proportion of radon that emanates from a material is therefore greater the finer grained or more porous it is.

Radon emanation must be seen in relation to the distance the radon atom can be moved by the recoil effect (0.02-0.07 μm) and the grain size of the soil. Examples are:

Type of Soil	Grain Size	
fine clay	< 0.6 μm	(most of the clay particles are often <0.06 μm)
coarse clay	0.6 - 2 μm	
silt	2 - 60 μm	
sand	60 - 2,000 μm	
gravel	2,000 - 60,000 μm	

As can be seen, the clay particles are so small that a large fraction of the radon atoms formed should emanate from them to ambient air or water in the pores when the radium atom disintegrates. This can explain why such high emanations as 40-70% of all radon formed, have been measured from clay.

It is more difficult to explain how emanation from dry sand and gravel can amount to 30%. Something other than the recoil effect is required to explain such a high figure. Ek and Ek (1995) have studied the radium concentration in different particle sizes of gravel and coarse sand. They found two reasons for the enhanced emanation. One is that radium has been absorbed in the fine fractions of the soils (< 0.15 mm), which have 3x to 8x higher radium concentrations than the coarser fractions. Another explanation is that radium atoms have been precipitated on the surface of the grains and fissures in them. Both these ways of occurrence facilitate high emanation. The conditions are results of the selective leaching, transport and precipitation of uranium and its daughter nuclides that takes place in weathering of the mineral grain.

Table 1.2.3 shows radon emanation in various soils and crushed rocks. The figures are taken from tests reported in the literature (e.g. Baretto, 1975; Markkanen and Arvela, 1992; Robé and Labed, 1995) and investigations undertaken by the Swedish Geological Co., the Swedish Geotechnical Institute, and the Swedisch National Testing Institute (Hildingson, 1983).

Thoron emanation from soil has been studied, among others, by Megumi and Mamuro (1974). They found that about 10% of the thoron formed emanates from sand, silt and clay produced by the erosion of granite. As thoron and its daughter nuclides do not readily dissolve in water at a normal pH, and the nuclides have such short half-lives, chemical leaching and precipitation of thorium and its daughter nuclides is of little significance for radon emanation.

Table 1.2.3: Emanation of radon atoms to the pore air, e (expressed as a percentage), in various soils and crushed rocks.

Soil type	e (%)
Gravel	15 - 40
Sand	15 - 30
Clay	30 - 70
Crushed rock (particle size 1 - 8 mm)	5 - 15
Crushed uranium-rich granite (particle size 1 - 8 mm)	15 - 30

One of the factors affecting how large a part of the radon formed emanates from a rock or soil, is the manner in which the uranium or thorium occurs in the rock or rock fragments in the soil. If the uranium is found in the lattice of other minerals, e.g. feldspar, magnetite, apatite, allanite or zircon, radon emanation from the rock is relatively small. If, on the other hand, the uranium forms a uranium mineral such as uraninite (U_3O_8), uranothorite, or coffenite $[U(SiO_4)_{1-x}(OH)_{4x}]$, radon emanation is substantially greater. The latter minerals often occur at the boundaries between larger grains of quartz and feldspar, or between strata in biotite, which facilitates radon emanation. It is also facilitated by the fact that uranium minerals are relatively easily weathered, and many of them tend to be metamictized as a result of their own radioactivity over geological time periods. In granites with a normal uranium concentration of 4-10 ppm U (50-125 Bq kg^{-1} ^{226}Ra), it is most common for the uranium to be included in the lattice of other minerals, while it is more common for the uranium to occur in separate uranium minerals if the concentration is higher.

The chances of emanation from a rock or soil are greater the more porous the material is, which allows a greater diffusion of the radon gas. Emanation in relation to radium concentration is low, for instance, from an uncrushed, unweathered alum shale, which is a dense rock whose pores are filled with kerogen and sometimes oil. Emanation in relation to radium concentration can be large, on the other hand, from a coarse-grained granite. This is confirmed by measurements made by the Swedish Geological Survey from samples of unweathered alum shale and granite rich in uranium.

The measurements were made from freshly crushed material with a grain size of 1.7-7.0 mm. The uranium concentration in the shale was 325 ppm (4,000 Bq kg^{-1} ^{226}Ra) and radon emanation rate 0.76 Bq kg^{-1} h^{-1}, which means that 2% of all radon atoms formed were emanated. The uranium content of the granite was 12-18 ppm (150-220 Bq kg^{-1} ^{226}Ra) and radon emanation rate 0.22-0.45 Bq kg^{-1} h^{-1}, which means 20-27% of the radon atoms were emanated. The results show that radon emanation from the granite is of similar magnitude to that from the alum shale, even though the uranium content is much lower in the granite.

Radon emanation from a rock increases with crushing and weathering because the chances of emanation from the individual mineral grain increase. Emanation from a soil is thus considerably greater than from a coarsely crushed rock, provided they consist of the same type of material.

1.2.4 Radon Concentration, Emanation and Water Content

The concentration of radon gas in the air in soil pores increases with increasing water content in the pores. This is because of the equilibrium relationships between the ratio of radon in air to radon in water, which is dependent on the temperature. It is also due to the fact that the emanation of radon to the pore space increases with moisture.

In a closed volume with equal volumes of air and water, and at temperature of +15°C, the radon concentration is three times greater in the air than in the water. At 0°C the radon concentration in the air is twice that in the water (Allen, 1976; Tanner, 1991). As the number of radon atoms in the pore air in a soil is constant if the emanation in the soil is constant and no radon is supplied to or leaves the pore volume, an increase in the quantity of water in the pore volume means an increase in the radon concentration in the remaining air. If the pores were filled with 75% water and 25% air, for instance, the radon concentration in the pore air would be twice as great as when the pores were entirely filled with air.

A model incorporating temperature effects on ^{222}Rn concentration in the pore air at different water contents is given by Washington and Rose (1990) for depths below which diffusional transport of radon to the surface is negligible, i.e., ca. 1 m, by the following expression

$$C_{max} = \{Aed(1-p)/p\}[F(K_T-1)+1] \tag{1.2.4}$$

where C_{max} is the radon activity concentration (Bq m^{-3}) in a pore volume with absolutely no ventilation; e is the emanation fraction of all radon atoms formed that emanate to the pore space; d is compact density (kg m^{-3}) (normal for mineral soils: 2,700 kg m^{-3}), p is porosity, ratio of pore volume to total volume; F is the water fraction; and K_T is the partition coefficient of ^{222}Rn between units of air and water.

C_{water}/C_{air} values vary from 0.525 at 0°C to 0.226 at 25°C. Variations in the ^{222}Rn concentration of up to a factor of about four can be accounted for as being due to effects of F and K_T. For warm (25°C) and very moist ($F = 0.95$) soil with $A = 30$ Bq kg^{-1}, $e = 0.2$, and $d = 2{,}700$ kg m^{-3}, the ^{222}Rn concentration is calculated to be 78.3 kBq m^{-3}; but when cold (0°C) and dry ($F = 0.05$), the ^{222}Rn activity concentration is calculated to be 21.2 kBq m^{-3}.

The relationships between radon concentrations in pore water and pore air with different amounts of water in the pores and at different temperatures, can also be seen from Fig. 1.2.5. The closer they are to the water table, the more the pores are filled with water. It follows that radon concentration in the pore air increases the shorter the distance to the water table.

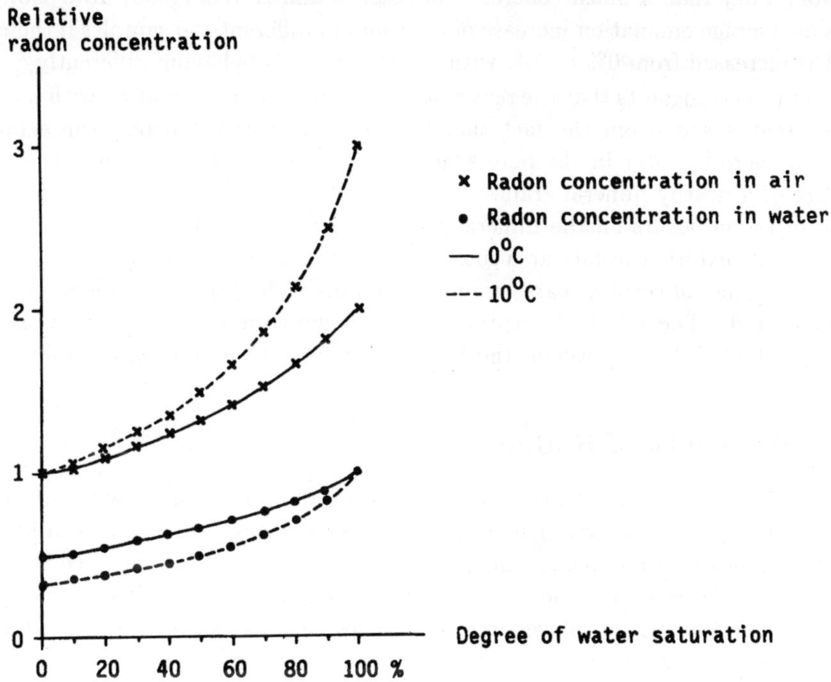

Figure 1.2.5: Radon concentrations in air and water in the pores with varying moisture content in the soil. (Åkerblom *et al.*, 1990).

In a soil horizon below the water table, all pores are filled with water. Therefore all the radon atoms emanating from the soil go into the water. As the diffusion length of radon in water is small - the coefficient is 10^{-9} m^2 s^{-1}, compared to that for air of 10^{-5} m^2 s^{-1} - only a small part of the radon can be moved by diffusion. Normally, there should be equilibrium in a water-filled pore between radon emanating to the pore and radon that disintegrates. This means the concentration in a completely filled pore is equal to the maximum that can be formed in the soil's pores when they are completely filled with water and at the porosity in question. This is so, provided the water does not move more than a few centimetres, from the site where the radon was formed.

Measurements of radon in ground water taken below the water table show results as would be expected from the radium concentration in the soil and the soil's porosity and the emanation rate for the specific type of soil. Using eq. (1.2.1), the figure is about 80,000 Bq m^{-3} if the soil has a radium concentration of 50 Bq kg^{-1}, a porosity of 30%, and there is 25% emanation of all radon atoms formed.

Many researchers have shown that emanation and exhalation from a soil or crushed rock or a concrete slab increase as the pores get filled with water. The radon content of the pore is thus considerably lower in dry soil than in moist soil and radon exhalation lower from a dry than a moist concrete. Markkanen and Arvela (1993) have shown that there is an average emanation increase of 2-3 times in different soil samples if their water content is increased from 0% to 5%, with the various soils behaving differently.

Tanner (1980) suggests that the reason for increasing radon emanation with increasing water content ensues from the fact that the radon atom ejected on disintegration is retarded if there is water in the pore space and thus obstructed from penetrating into the lattice of a nearby mineral grain. Schery and Whittlestone (1989) have shown that sorption of radon occurs on the mineral grains in a dry soil. Of the materials studied, those with high external surface area such as clays had a larger sorption capacity, whereas the sorption values of certain coarse-grained materials such as quartz sands were to small to be measured. The effect of sorption decreases with increasing moisture, and at a moisture content of 10% by weight the sorption was below the limit of detection.

1.2.5 Transport of Radon

Transport of radon through the soil takes place by diffusion and/or with air ambient gases e.q. CO_2 and CH_4 or water moving in the soil horizons. Moving air can be driven by wind, changes in air pressure and percolating rain or melted snow. Differences of temperature in the soil could be a further cause. The outflow of soil air which causes the snow to melt at certain places in gravel ridges such as eskers and above fractures in the bedrock, is assumed to result from the transport of warmer air through the soil or fractures. Owing to the stack effect in ridges, such transport of soil air can be a contributory cause of the high radon concentrations common in buildings on eskers. Soil air can also be transported and conducted to buildings along culverts and cable trenches. Often, the cause of the transport is lower pressure inside the building in relation to the atmosphere outside and in the ground.

Table 1.2.4: Diffusion coefficients, *D*, for ^{222}Rn. Data from UNSCEAR (1982) and various reports.

Medium	D (m^2 s^{-1})
Air	10^{-5}
Boulders - coarse gravel	10^{-5} - $5{\cdot}10^{-6}$
Dry sand	10^{-5}
Moist sand	$2.5{\cdot}10^{-6}$
Till	$5{\cdot}10^{-7}$ - $2.5{\cdot}10^{-7}$
Clayey till	$8{\cdot}10^{-8}$
Water	10^{-9}

The diffusion of radon through the ground is related to permeability, which is dependent on grain-size distribution, degree of compaction and the water content of the soil. Rodgers and Nielson (1991) have described the parameters governing radon transport in detail. Table 1.2.4 lists diffusion coefficients for radon in some soil types (UNSCEAR 1982, etc.).

The table shows that radon diffuses through coarse gravel about as easily as in the air, while there is very little diffusion through a water-saturated clay (ten thousand times less than in the gravel).

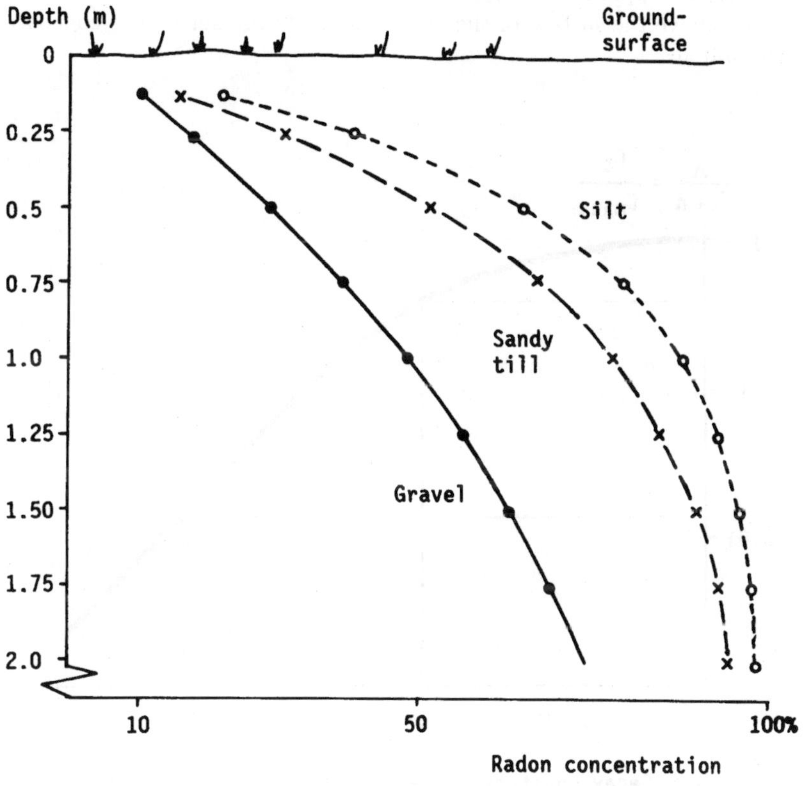

Figure 1.2.6: The effect of diffusion on radon concentrations in soil air at varying depths in gravel, sandy till and silt. Based on theoretical calculations (Åkerblom *et al.*, 1990).

If the soil pores are completely filled with water, the soil has the same diffusion coefficient as water.

As radon has a limited life, the greater part of it will disintegrate after a certain diffusion distance. 90% of the radon emanated from a radon source, e.g. a deep-lying soil horizon, and transported by diffusion, will have decayed after a distance of 5 cm in water, about 2 m in sandy soil with a normal moisture content and 5 m in air (UNSCEAR, 1982).

The transport distance for thoron (^{220}Rn) is insignificant, as its half-life is no greater than 56 seconds.

Diffusion of radon through the soil up to the atmosphere above the surface of the ground is highly important in determining the radon concentration in the soil air at different depths. The curves in Fig.1.2.6 show how radon concentrations in soil air decrease towards the surface as a result of diffusion (the curves are based on theoretical calculations). In gravel, there is equilibrium between radon coming from deeper layers of soil and the radon removed by diffusion, at a depth of about 4 m. In silt and sandy till, this equilibrium is attained at a depth of roughly 2 m. At one metre, the soil-air radon concentration in gravel is about 50% of the maximum at 4 m. For silt and till it is about 80%. At 0.5 m, the approximate figures are 30% in gravel and 70% in till and silt. The effect of winds above the surface of the ground tends to further reduce concentrations of radon in the soil air.

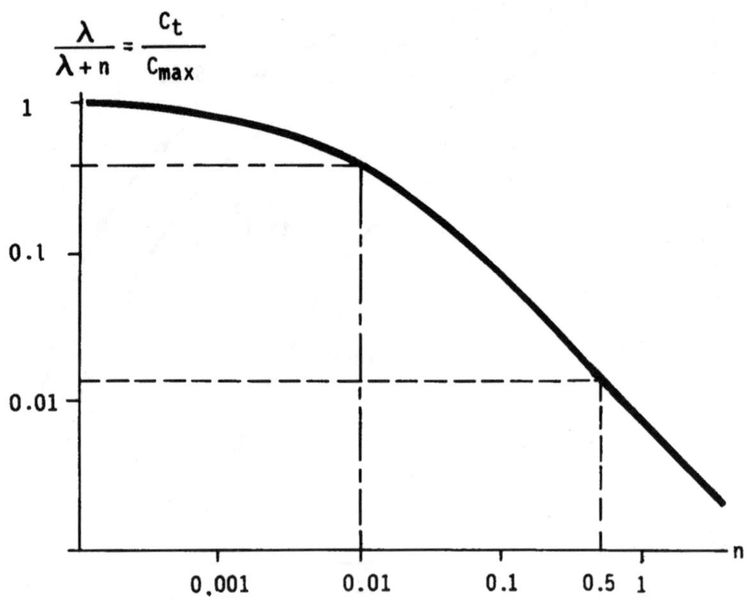

λ = Decay constant of radon-222 = 7.55×10^{-3} (h^{-1})

n = Ventilation (air changes per hour : ach)

C_t = The radon concentration in the soil air (Bq m-3)

C_{max} = The radon concentration at 0 ach (Bq m^{-3})

Figure 1.2.7: The effect of ventilation on radon concentrations.

In sand and gravel, wind can influence the radon concentration down to a depth of about one metre - more if the material has good permeability, e.g. blast stone and coarse gravel. This is particularly so if the soil is exposed to the lateral effect of the wind, as is often the case with an esker, where the wind can blow towards the slope. The result is greater ventilation, resulting in a lower radon concentration in and above the top soil. Fig. 1.2.7 shows how an increase in ventilation affects concentrations. Effects of diffusion and wind should therefore be taken into account in interpreting measurements of radon in soil air.

1.2.6 Radon Concentrations at Different Soil Depths

With regard to radon, erecting a building can be compared to putting a lid on the soil and thereby obstructing diffusion through the surface of the ground and the effect of the wind. Beneath the building, the concentration grows to the maximum from emanated and decaying radon at the radium concentration, porosity and degree of water saturation

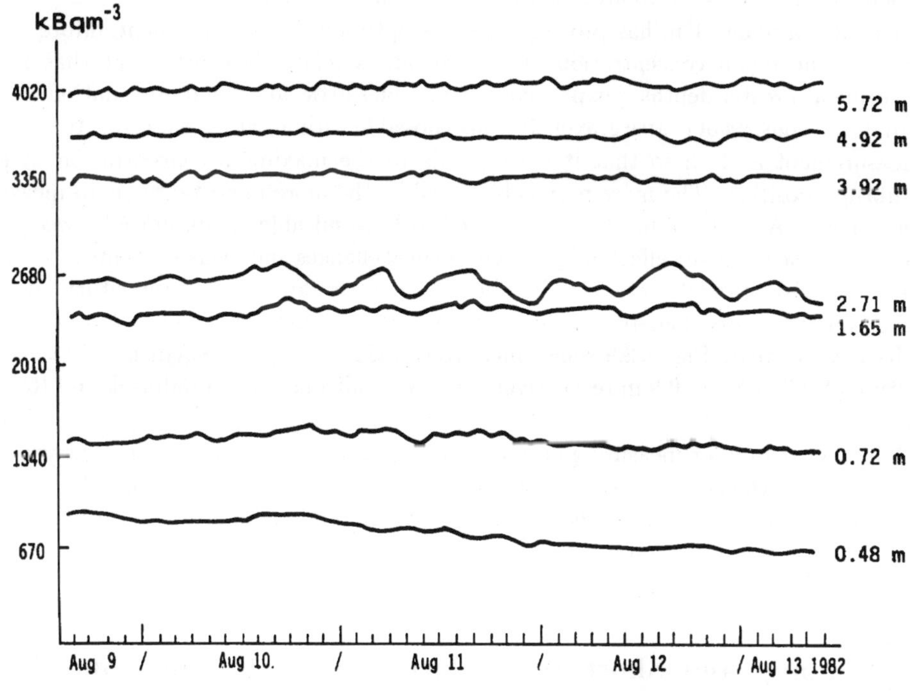

Figure 1.2.8: Radon concentrations at different depths in mill tailings from uranium mining. Porosity 50%. Water content 5-10%. Measurement with α-probe 601 (Bigu *et al.*, 1984).

for the soil in question. It is therefore preferable for measurement to be made at a soil depth where there is equilibrium between radon supplied and removed. For practical and financial reasons this is not always possible, as it is difficult and expensive to dig or drill down to such a depth.

Radon concentrations have been studied at different depths in several research projects (Israelsson *et al.*, 1982; Kraner, 1964). Malmqvist *et al.* (1980) and Hesselbom (1985) have shown that the concentration in the soil air at 0.5 m is in the region of 50% of that with equilibrium between radon supplied to and removed from the soil air, but also that the concentration at so shallow a depth undergoes great change. At about a metre, it is normally approx. 70-80% of the equilibrium position, and more constant. Rose *et al.* (1990) have compiled a large number of results.

Extensive measurements of radon concentrations in soil air were made in a Canadian research project, at depths down to 5.7 m in mill tailings from uranium mining (Bigu *et al.*, 1984). Continuously-registering alpha meters (α-probe 601) were used at the different depths. The tailings consisted of sandy, coarse-silt material and radium concentrations were 3,000-4,750 Bq kg^{-1}. The porosity was 50% and water content 5-10%. The results show how concentrations in the soil air increased with increasing depth, from about 670,000 Bq m^{-3} at 50 cm to around 4 million Bq m^{-3} at 5.7 m (Fig. 1.2.8).

A depth of about 1 m has proved to be the optimum for measurement. More often than not, the radon concentration in the soil air is relatively constant at that level. Measuring at greater depths is expensive and provides little additional information unless the material consists of coarse gravel. It is also possible without any great error, to convert a measurement at 1 m so that it corresponds to the maximum concentration at the equilibrium position. The more permeable a soil is, the more important it is to measure deeper down. A depth of 0.7-1 m has proved to be a suitable compromise between too superficial measurements affected by meteorological changes and measurements at greater depths with risk of the detectors landing below the water table or in water-saturated soil. Measurements at less than 70 cm should be avoided, as uncertainty in interpreting the results is very great. Fig. 1.2.9 shows measurements made by the Swedish Geotechnical Institute at 0.3, 0.5 and 0.9 m respectively, in coarse silt-fine sand (Lindmark and Rosén, 1985).

A further reason for measuring ground radon at depths of around 1 m is that you are then measuring the concentration beneath the rust-coloured B-horizon normally formed in till, sand and silt at a depth of 0-70 cm. Uranium and radium leached from topsoil are enriched in the B-horizon, making the radon concentration higher than in the subsoil below.

1.2.7 Long- and Short-Term Variations of Radon Concentrations in Soil Air

Radon concentrations in soil air vary during the year owing to changes in the permeability and water content of the soil, as well as ventilation of the upper horizons. This variation over time has been studied in several different research projects (Rose *et al.*, 1990).

Fig. 1.2.9 shows results from long-term measurements by the Swedish Geotechnical

Institute (Lindmark and Rosén, 1985). These were undertaken in coarse silt - fine sand during 1982 and 1983 at 0.3, 0.5 and 0.9 m.

The measurements were made using alpha track film (Kodak-LR 115) placed on top of fixed pipes. The radon concentration was highest when the top soil was covered with snow and the pores in the top soil filled with frozen water. Under the frozen top soil the concentration increased to a maximum of 40,000 Bq m^{-3} at a depth of 0.9 m. When the frost in the ground released its hold, ventilation began and radon diffused up to the surface. As can be seen, the ground radon concentration varied and was at its lowest, about 10,000 Bq m^{-3}, just after the frost released its grip, around April/May 1982. In 1983 the pores in the ground were not frozen.

Long-term measurements carried out in sand in northern Sweden by the Swedish

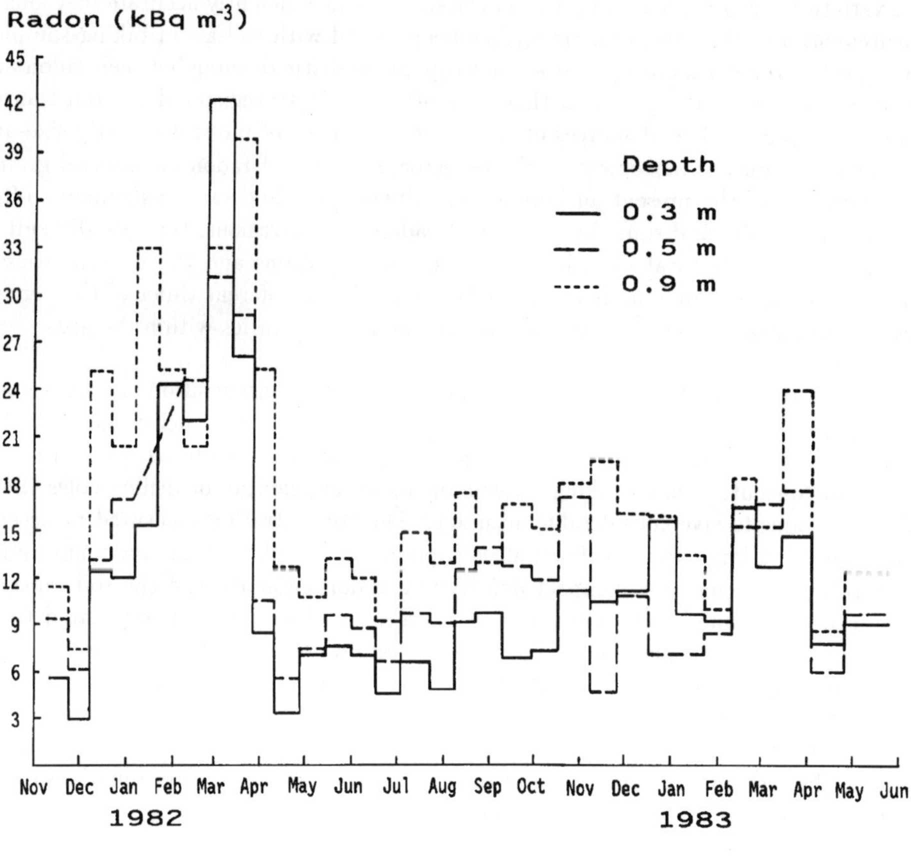

Figure 1.2.9: Radon concentrations at different depths and seasons. Coarse silt. Measurements, using LR 115 alpha track film by the Swedish Geotechnical Institute (Lindmark and Rosén, 1985.)

Geological Co using ROAC (Radon On Activated Charcoal) detectors in drilled holes, gave roughly the same results as those of the Swedish Geotechnical Institute. There was one deviation, however, in that when the frost reached below the depth in question (0.9 m), all radon transport ceased from the surrounding soil and the measured radon concentration in the holes fell to less than 1,000 Bq m^{-3}. This is because the radon could no longer diffuse to the hole, as the pores in the sand were filled with ice.

As conditions change in respect of soil type, moisture content and exposure to wind, radon concentrations will vary even at points close to each other and if measurements are made simultaneously within an area. Variations of a factor of 2 to 3 are not uncommon at points close to each other in a soil that seems to be homogeneous. Rose *et al.* (1990) report variations of up to a factor of 10. However, it cannot be ruled out that the great variations may be caused in several instances by insufficiently accurate methods of measurement and that atmospheric air has been mixed with soil air in taking samples. In using measurement methods that are not capable of distinguishing between radon and thoron, there is also always a risk that part of the activity registered is from thoron. Tanner (1991) gives several sources of error, e.g. the effect of moisture, cavity size and decay of radon during movement to the detector, sorption of radon on mineral grains. The experience of the present authors is that different methods and instruments often give different results and sometimes incorrect radon concentrations, that are difficult to explain. Owing to natural variations in radon concentrations and the uncertainties of measurement, several measurements are often required at different times of the year to assess a normal radon concentration for an area and how it varies within the area.

An alternative method of assessing the maximum radon concentration in a soil's pores with acceptable accuracy is to calculate it from eq. (1.2.1), based on measurements of the soil radium concentration. A portable gamma ray spectrometer can be used to determine the radium concentration by taking measurements in dug cavities or drilled holes. Such field measurements give considerable accuracy. The concentration measured represents the average for a large volume of soil at each point. In calculating the maximum radon concentration, the porosity, compact density and radon emanation of the soil must be determined or assumed. The maximum error in using this method is estimated to be a factor of 2, if all the parameters apart from radium concentration are based on soil determination carried out in the field. The method provides a measure of the soil's radon potential that is independent of moisture conditions. Of course, radium analysis of soil samples can be carried out in a laboratory. The difficulty is in getting representative samples. Therefore, we recommend that the gamma spectrometric determinations of the radium concentrations are done in the field.

Yokel and Tanner (1992) have developed, and given a detailed description of, methods for calculating radon concentration from gamma spectrometer measurements combined with laboratory determination of emanation, porosity and permeability. They recommend this instead of direct radon measurement, which they have found too uncertain for determining radon potential.

1.2.8 Radon Emanation from Fracture and Fault Zones in the Bedrock

Measurements have shown increased concentrations of radon in soil air and ground water above many fracture and fault zones in the bedrock. They have also shown that the concentration changes in conjunction with earthquakes (Birchard and Libby, 1980). These observations indicate that radon could be transported more rapidly than by diffusion or ground water movement in fracture zones. Various theories have been put forward about how this could take place. One is that radon is transported with geogas (e.g. CO_2, CH_4 and N_2) finding its way towards the surface (Hermansson *et al.*, 1991a, b; Kristiansson and Malmqvist, 1982). A reason for increased concentrations over fracture zones is that uranium ions and/or daughter nuclides dissolved in the ground water have been precipitated close to the ground surface, and that radon emanates from the radium formed in their disintegration.

Closer inspection of published radon concentrations in soil air above such zones shows that they normally lie within limits that are natural for the variation occurring in the soil type. However, it is perfectly clear that increased radon concentrations in soil air, beyond those that can be formed by surrounding layers of soil, occur over certain fracture zones. In these cases, they occur together with enhanced levels of CO_2 and often of helium. CO_2 has been shown to act as carrier in transporting radon from the fracture zones to the surface (Hunyadi *et al.*, 1995). Fig. 1.2.10 shows measurements of the radon concentration in ground water above a fracture zone (Lindén *et al.*, 1987). The samples were of ground water taken directly above the surface of the bedrock. There is a clear connection between various gases and radon.

Radon transported by gases must be moved rapidly, as radon has a limited span of life. It has not been established whether transport actually takes place from greater depths, or whether it comes from radium precipitated in the fracture zone close to the ground, or at the water table (see subsection 1.2.9).

It has been fully established that radon can be transported with air from caves and fissures in karst areas. In countries such as Switzerland (Piller and Surbeck, 1989), Ireland (O'Connor *et al.*, 1993), Great Britain and the United States (Pennsylvania), there are problems with radon in buildings on karst. Air is transported in and out of the cavities with changes of pressure and temperature. Even if the radon concentrations in the air are fairly low, large quantities of air can penetrate into a building in direct contact with an underlying cavity. In that way, radon concentrations in the building can be high.

1.2.9 Radon in Ground Water

Radon that emanates from a soil to pores filled with ground water remains in the water-filled pores, as the diffusion length prior to disintegration is small. Only if the ground water is moving rapidly does any significant amount of radon escape from it. Measurements of concentrations in ground water carried out by the Swedish Geological Co. show that in normal cases they are equal to the maximum possible, given the porosity and emanation coefficient of the soil in question.

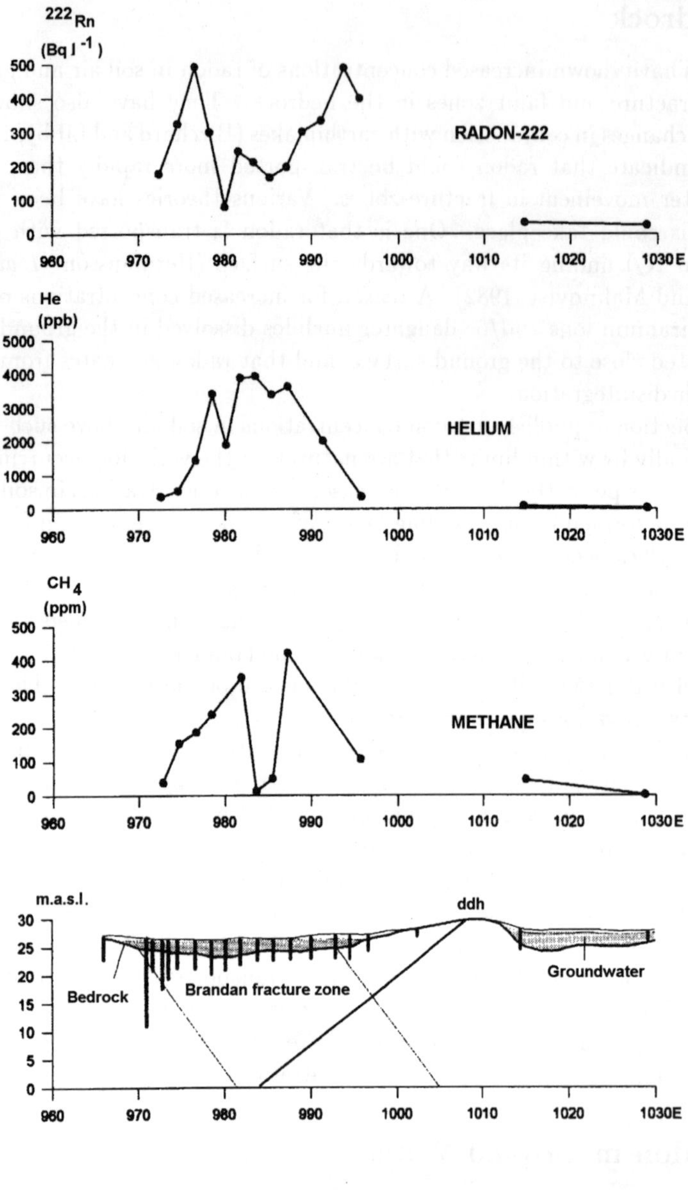

Figure 1.2.10: Radon, helium and methane concentrations in the groundwater directly over the Brandan fracture zone in Finnsjo area, Sweden. The groundwater has been sampled at the bottom of holes drilled through till, down to the bedrock. The gas concentrations at the sampling points 1,015 m E and 1,028 m E represent background values. The Brandan fracture zone is a deep-reaching fault that has been thoroughly investigated within the Swedish nuclear waste program (Lindén *et al.*, 1987). Abbreviations m.a.s.l., and ddh stands for 'metre above sea level', and 'diamond drill hole' respectively.

The situation is different with water in fissures in the bedrock, where the radon concentration is often considerably higher than in the ambient rock. This is exemplified in Fig. 1.2.11, which attempts to illustrate why high radon concentrations are common in water from fissures and fractures in granite.

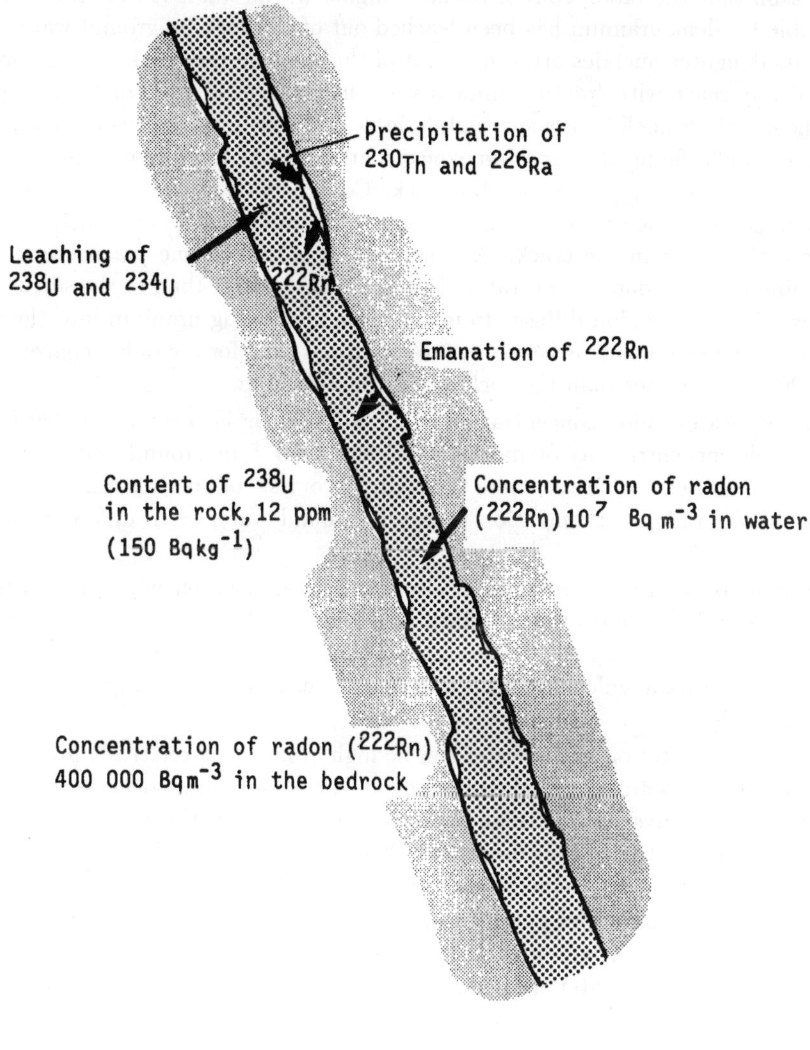

Figure 1.2.11: Radon formation in a fissure filled with water (Åkerblom *et al.*, 1990).

The granite in the example has a uranium concentration of 12 ppm (148 Bq kg^{-1} radium). The ^{226}Ra concentration is assumed to be in equilibrium with this, which in normal cases it is. Hence, 400,000 radon atoms per cubic metre of rock are formed and disintegrate per second in the granite. The radon activity concentration in the granite is thus 400,000 Bq m^{-3}. The concentration in the fissure water can then be up to 10 million Bq m^{-3}, which is not unusual for the conditions in some Swedish granite areas. The explanation for a higher concentration in the fissures and fractures than in the ambient rock cannot be radon diffusion from the granite.

The reason that the radon concentration is higher in the water is that the relatively easily soluble 6-valent uranium has been leached out of the rock by ground water. Uranium and its daughter nuclides precipitate out of the water in the cracks, partly because they chemically react with fracture minerals on the surfaces of the cracks, and partly because the daughter nuclides are sparingly soluble and thus precipitate out. Landström and Tullborg (1990) found that uranium concentrations in the fracture coatings are often 3-20 times higher than in the surrounding rock. Consequently, a coating of radium-rich material builds up on the surfaces of the cracks and fissures, from which radon emanates directly into the water in the crack. Another reason is that in one particular grain of uranium mineral, the radon concentration is many times greater than in the surrounding rock and water. Thus, radon diffuses from the grain containing uranium into the water phase. Since the rock porosity is very low, the conditions exist for the radon concentration to be 10-100 times greater than the rock radon concentration.

The ground water radon concentration in the soil horizon is normally 10-200 Bq l^{-1}. Relatively high concentrations of more than 1,000 Bq l^{-1} in ground water from the bedrock frequently occur in areas where the bedrock consists of granites rich in uranium. Table 1.2.5 shows radon concentration ranges in ground water from different environments.

High radon concentrations sometimes also occur in association with pegmatites, as well as gneisses and vulcanites rich in quartz and feldspar. The concentration in ground water from bedrock is low in areas where uranium concentrations in the rocks are low (e.g. diorite, gabbro and basic vulcanites, and also limestone, sandstone and shale; occasional exceptions occur).

The radon concentration in the water can be high even with a relatively low uranium concentration in the bedrock. On the other hand, a high uranium concentration in the bedrock does not always mean a high radon concentration in the water. By way of example, it can be mentioned that despite the high uranium content in alum shale, high radon concentrations in the ground water are rare in alum-shale areas.

There are many examples of ground water with a high radon concentration causing increased concentrations in buildings, if it is used as household water. Radon escapes from the water when the water is used, e.g. for washing. Extensive studies of radon in ground water as a cause of increased radon concentrations indoors have been carried out in the USA (Hess, 1980; Nazaroff *et al.*, 1987, etc.), Finland (Asikainen and Kahlos, 1980), and elsewhere.

As far as we know, it has not been established that radon transported with ground water and exhaling to the soil air causes high radon concentrations in buildings. However, it is not improbable that concentrations can be high in a building above, or close to, flowing ground water with a high radon concentration. Radon could exhale from the flowing water and then be transported through the soil to the building. Water with a particularly high radon concentration occurs in association with certain types of uranium-rich granites and uranium mineralization.

The first study to discover whether radon transported by ground water can be the cause of increased radon concentrations in buildings was carried out in the uranium-mining town of Elliot Lake in Canada, commissioned by the Canadian Atomic Control Board. The background to the study was that increased concentrations were found in many buildings in the town even though they stood on ground with a normal concentration of uranium (radium). A conceivable cause was assumed to be that radon was transported with the ground water from uranium mineralizations and uranium ores in the bedrock on which the town is built. The study included hydrological and geochemical investigations and went on for a year. It showed that the ground water in the town had normal concentrations of both radium and radon and that radon from the ground water could not be the cause of the high concentrations in the buildings (DSMA ATCON Ltd., 1981).

Table 1.2.5: Radon and radium activity concentration in Swedish ground water. Normal ranges and maximum concentrations.

	^{222}Rn (Bq l^{-1})	^{226}Ra (Bq l^{-1})
Lake and sea water	<2	0.0005 - 0.007
Dug wells:		
normal	10 - 300	0.001 - 0.09
with uranium-rich granite	40 - 400	
Drilled wells:		
sedimentary rock	10 - 50	
crystalline rock	70 - 500	0.01 - 0.25
uranium-rich granite	500 - 4,000 (max 57,000)	
uranium-rich pegmatite	max (15,000 - 28,000)	max (0.35 - 2.5)
Ore deposits:		
Kiruna, iron-ore mine	50 - 300 max 600	0.1 - 6
Uranium ores	2,000 - >100,000	

It is likely that, in normal circumstances, radon escaping from ground water in the soil does not give rise to higher concentrations in buildings. The reason is partly that concentrations in the soil air are normally as high as or higher than in the ground water, partly that the water content in the soil horizon immediately above the water table is so great that diffusion through the soil is prevented. Only if the radon concentration in the ground water is much higher than in the soil air, and if the ground water is in motion or flows out at the surface, is it likely that sufficient radon will be released to affect the concentration in a building.

1.2.10 Concluding Remarks

Radon follows the normal laws of nature. ^{222}Rn is formed by the decay of ^{226}Ra; it is the concentration of radium which governs how many radon atoms are formed. How many of the produced radon atoms leave, i.e. emanate from, the mineral grain or matter and enter the pore spaces depends on: where the radium atoms are situated in the grain, the texture and size of the grain, the permeability of the grains, temperature and pressure. The radon concentration in the pore spaces is determined by: sorption of radon gas on the grain surface, water content, temperature and pore size, and transport of radon by diffusion or by pressure-driven flow of air or water from, or to, the pore space. Diffusion or active movement of air or water is affected by permeability; the combined effect of moisture, grain size and packing. Air pressure, wind-speed, and percolating ground water have a large influence on the radon concentration down to a depth of at least one metre. Thus, the radon concentration in a pore has a relatively large variation, which is greater closer to the ground surface. This has to be borne in mind when interpreting results on radon concentration in soil gas measurements. Results from radon measurements which exceed radon activity concentration by more than 50% above that which could be emanated from the radium in the surrounding soils should be regarded with caution. Radon may have been transported from an underlying soil layer or bedrock, which is richer in radium; but the results may also have been due to measurement errors. The last is usually the most probable cause.

References

Åkerblom G. (1986) Investigation and mapping of radon risk areas (Paper presented at the International Symposium on Geological Mapping in the service of Environmental Planning: Subcommission for Maps on Environmental Geology, Trondheim, Norway, 1986). In *Geology for Environmental Planning* (ed. F.C. Wolff), Geological Survey of Norway, pp. 96-106.

Åkerblom G. (1994) Ground radon - monitoring procedures in Sweden. *Geoscientist* **4**, 4, 21-27.

Åkerblom G. and Wilson C. (1982a) Radon gas - a radiation hazard from radioactive bedrock and building materials. *Bulletin of the International Association of Engineering Geology* **23**, 51-61.

Åkerblom G. and Wilson C. (1982b) Radon - geological aspects of an environmental problem (Papers presented at the International Meeting on Radon - Radon Progeny Measurements, Montgomery, Alabama, August 27-28, 1981). *Geological Survey of Sweden, Reports and Communications* **30**, pp. 1-47.

Åkerblom G., Clavensjö B. and Anderson P. (1983) Soil gas radon - a source for indoor radon daughters. *Radiat. Prot. Dosim.* **7**, 1-4, 49-54.

Åkerblom G., Pettersson B. and Rosén B. (1990) *Handbook on Investigation of the*

Radon Situation in Areas Before Building. The Swedish Council for Building Research and the Swedish National Board for Physical Planning and Building, Report R88:1988, Revised edition 1990, pp. 1-160 (In Swedish).

Allen J. W. (1976) Development of a portable radon detection system. *Bendix fieldengineering corp., Grand Junction*, Colorado, GBJX-50(76).

Andersson P. (1987) Radon chamber for soil gas detectors. Swedish Radiation Protection Institute, SSI-Report 87-01, pp. 1-86.

Andersson P., Clavensjö B. and Åkerblom G. (1983) The effect of the ground on the concentration of radon and gamma radiation indoors. Swedish Council for Building research, Report R9:1983, pp. 1-442. (In Swedish).

Asikainen M. and Kahlos H. (1980) Natural radioactivity of drinking water in Finland. *Health Phys.* **39**, 77-83.

Baretto P. M. C. *et al.* (1975) Physical characteristics of ^{222}Rn emanation from rocks, soils and minerals: is relation to temperature and alpha dose. In *Natural Radiation Environment* (Symposium 1972), Houston.

Bigu J. *et al.* (1984) Study of radon gas concentration, surface radon flux and other radiation variables from uranium mine tailings areas. *Uranium* **1**, 257-277.

Birchard C. F. and Libby W. F. (1980) Soil radon changes preceding and following four magnitude 4.3-4.7 earthquakes on the San Jacinto fault in southern California. *J. of Geophys. Res.* **85**, B6, 3001-3106.

DSMA ATCON Ltd (Dilworth, Secord, Meagher and Associated Limited) (1981) Ground water investigation. Development program for radiation reduction. Atomic Energy Control Board, Canada, Report 12.

Ek J. and Ek B. M. (1995) Radium and uranium concentrations in two eskers with enhanced radon emission (Paper presented at the International Symposium on the Natural Radiation Environment (NRE IV)), *Environment International*, in press.

Fleischer R. L. (1980) Isotopic disequilibrium of uranium: Alpha-recoil damage and preferential solution effects. *Science* **207**, 979-981.

Fleischer R. L. (1982) Alpha-recoil damage and solution effects in minerals: Isotopic disequilibrium and radon release. *Geochim. Cosmochim.* Acta **46**, 2191-2201.

Gundersen L. C. S. *et al.* (1991) Preliminary radon potential map of the United States. In *Proc. of the 1991 EPA International Symposium on Radon and Radon Reduction Technology*, Vol. **5**, paper IX-4, pp. 1-24.

Hesselbom Å. (1985) Radon in soil gas - A study of methods and instruments for determining radon concentrations in the ground. *Geological Survey of Sweden*, Serie C, No 803, pp. 1-58.

Hermansson H. P. Cyssler J. Åkerblom G. and Lindén A. (1991a) Geogas - A carrier or a tracer? *Swedish National Board for Spent Nuclear Fuel*, SKN Report 51, pp. 1-66.

Hermansson H. P., Sjöblom R. and Åkerblom G. (1991b) Geogas in crystalline bedrock. *Swedish National Board for Spent Nuclear Fuel*, SKN Report 52, pp. 1-32.

Hess C. T.*et al.* (1980) Radon-222 in portable water supplies in Maine. Maine 04469.

Hildingson O. (1983) Radon emitted from gravel and aggregate; Testing method and consistencies for indoor environment. Swedish National Testing Institute. Technical

Report 1983:28, pp. 1-45 (In Swedish).

ICRP (1993) Protection against ^{222}Rn at home and at work. International Commission on Radiological Protection publication 65. Annals of the ICRP **23**, 2, 1-45.

Hunyadi I., Csige I., Hakl J., Hertelendi E. and Vasarhelyi A. (1995) Radon enriched deep earth-gas upflow in a seismically active inhabited area. *Proc. 3rd International Colloquium on Rare Gas Geochemistry*, Amritsar, 1995, in press.

Israelsson S. Knudsen E. and Ungethöm E. (1982) Natural radioactivity in soil gas: Exhalation of natural radioactivity from the ground surface. *The Meteorological Institute at Uppsala University*, Report No 30, Uppsala.

Kigoshi K. (I971) Alpha-recoil ^{234}Th. Dissolution into water and the ^{234}U/^{238}U disequilibrium in nature. *Science* **173**, 47-48.

Kogan R. M., Nazarov I. M. and Fridman Sh. D. (1971) *Gamma Spectrometry of Natural Environments and Formations.* Atomizdat, Moscow (1969), Israel program for scientific translations, Jerusalem, pp. 1-337.

Kraner H. W. *et al.* (1964) Measurements of the effects of atmospheric variables on ^{222}Rn flux and soil gas concentration. In *The Natural Radiation Environment.* (eds. J.A.S. Adams and W. M. Lowder), University Press, Chicago, pp. 191-190.

Kristiansson K. and Malmqvist L. (1982) Evidence for non-diffusive transport of ^{222}Rn in the ground and a new physical model for the transport. *Geophysics* **47**, 1444-1452.

Kristiansson K. and Malmqvist L. (1984) The depth - dependence of the concentration of ^{222}Rn in soil gas near the surface and its implication for exploration. *Geoexploration* **22**, 17-41.

Landström O. and Tullborg E. L. (1990) The influence of fracture mineral/groundwater interaction on the mobility of U, Th, REE and other trace elements. Swedish Nuclear Fuel and Waste Management Company, Technical Report 90-37, pp. 1-71.

Lindén A., Mellander H., Åkerblom G. and Jönsson G. (1987a) Concentrations of radon, methane and helium in ground water over a faultzone in the Finnsjö area. Swedish Geological Co. Report IRAP 87020, pp. 1-42 (In Swedish).

Lindén A., Mellander H., Åkerblom G. and Jönsson G. (1987b) Concentrations of radon, methane and helium over a fracture zone in the Finnsjö area. Swedish Geological Co. Report IRAP 87020, pp. 1-37 (In Swedish).

Lindmark A. and Rosén B. (1985) Radon in soil gas - Exhalation tests and in situ measurements. *The Science of the Total Environment* **45**, 397-404.

Markkanen M. and Arvela H. (1992) Radon emanation from soils. *Radiat. Prot. Dosim.* **45**, 1-4, 269-272.

Megumi K. and Mamuri T. (1974) Emanation and exhalation of radon and thoron gases from soil particles. *J. Geophysical Res.* **79**, 23, 3357-3360.

Malmqvist L.*et al.* (1980) Radon measurements with alpha track film. A new technique for geological surveys, exploration and for mapping of radiation. The Swedish Board for Technical Development, STU Report 77-3590 (In Swedish).

Minell H. (1983) Gamma radiation and contents of K, U and Th in Swedish soils. *Geological Survey of Sweden*, Report IRAP 83051 (In Swedish).

Minell H. (1990) Natural radioactivity from different ground types. In *Quaternary Economic Geology in the Nordic Countries* (eds. L. K. Kauranne and L. K. Königsson), Uppsala, Vol. **29**, pp. 73-78.

Nazaroff W. W. Doyle S. M. Nero A. V. and Sexto R. G. (1987) Portable water as a source of airborne ^{222}Rn in U.S. dwellings: A review and assessment. *Health Phys.* **52**, 3, 281-295.

O'Connor P. J., Gallangher V. Madden J. S. *et al.* (1993) Assessment of the geological factors influencing the occurrence of radon hazard areas in a karstic region. *Geological Survey of Ireland*, Report Series RS93/2.

Piller G. and Surbeck H. (1989) Radon and karst. In *The Radiology of Natural and Artificial Radionuclides* (ed. W. Feldt), *Fackverband för Strahlenschutze*, V. Publication series: Progress in Radiation Protection. FS- 89-48-T.

Reimer G. M. and Tanner A. B. (1991) Radon in the geological environment. In *Encyclopedia of Earth System Science* (ed. W. A. Nierenberg), Vol. **3**. Academic Press Inc., San Diego, pp. 707-712.

Robé M. C. and Labed V. (1995) Explaining the variation in soil radon concentrations: a study of the influence of some intrinsic properties of a rock matrix on the radon emission factor. In *Gas Chemistry* (ed. C. Dubois), *Science Reviews*, 535-542.

Rogers V. C. and Nielson K. K. (1991) Multiphase radon generation and transport in porous materials. *Health Phys.* **60**, 6, 807-815.

Rose A. W., Hutter A. R. and Washington J. W. (1990) Sampling variability of radon in soil gases. *J. Geochem. Explorat.* **38**, 173-191.

Schery S. D. and Wittlestone S. (1989) Desorption of radon at the earth's surface. *J. Geophys. Res.* **94**, D15, 18,297-18,303.

Tanner A. B. (1964) Radon migration in the ground: A review. In *The Natural Radiation Environment* (eds. J. A. S. Adams and W. M. Lowder), University Press, Chicago, pp. 191-190.

Tanner A. B. (1980) Radon migration in the ground: A supplementary review. In *The Natural Radiation Environment III.* (eds. T. F. Gesell and W. M. Lowder), National Technical Information service, Springfield, VA. CONF-780422, pp. 5-56.

Tanner A. B. (1991) Error in measuring radon in soil gas by means of passive detectors. *Nucl. Geophys.* **5**, 1-2, 25-30.

UNSCEAR (1982) Ionizing radiation: Sources and biological effects Annex D. *United Nations Scientific Committee on the Effect of Atomic Radiation. 1982 Report to the General Assembly, with annexes.* United Nations publications, Sales No.:E.82.IX.8.

Wilkening M. H. (1980) Radon transport processes below the earth's surface. In *The Natural Radiation Environment III.* (eds. T. F. Gesell and W. M. Lowder). National Technical Information service, Springfield, VA. CONF-780422, pp. 90-104.

Washington J. W. and Rose A. W. (1990) Regional and temporal relations of radon in soil gas to soil temperature and moisture. *Geophys. Res. Lett.* **17**, 6, 829-823.

Yokel and Tanner A. B. (1992) Site Exploration for Radon Source Potential. United States Dept. of Commerce Technology Adm., Nat. Inst. of Standards and Technology, NISTIR 5135, pp. 1-61.

M. M. Monnin[1] **and J. L. Seidel**[1]

1.3 Radon Measurement Techniques

1.3.1 Introduction

Radon is the only alpha-emitting radioactive gas. It has three natural isotopes, ^{222}Rn, ^{220}Rn or thoron (Tn), and ^{219}Rn or actinon (An). Actinon has a very short half life and is of little interest within the scope of this book. The same applies to a lesser extent to thoron. Accordingly, in the following text, the word "radon" will refer to ^{222}Rn exclusively, unless otherwise stated.

As described in section 1.1, when radon decays it produces "radon daughters" or "radon progeny". The detection and measurement of radon can thus be performed directly on radon itself (called also "radon alone" measurements) or indirectly on radon and/or its daughters. Radon and several of its daughters decay away by emitting α-particles. ^{214}Pb, ^{214}Bi, ^{210}Pb and ^{210}Bi are also β emitters and γ-rays accompany their β-emissions. Accordingly, radon detection and measurement can be performed through α, β or γ radiation detection. Therefore the understanding of the detection and measurement techniques requires a basic knowledge of the interaction of those radiations with matter. Alpha-particles are dealt with in section 2.1. Information on β and γ emissions will be given when needed below.

1.3.2 Radon Measurements in the Laboratory

Let us now examine the various experimental and technical possibilities that are offered when the question of measuring radon concentration in the environment arises. There are many techniques available, and they demonstrate the ingenuity of a large number of investigators. The scientists involved in radon measurements mainly do so within two large fields of investigation: health physics and geosciences. In the first case, the work is motivated by the possibility that cancers of different types might be induced by exposure to radon. The number of publications related to this topic is increasing exponentially with time. In the second case, radon measurements are performed for different applications. These range from mineral prospecting (see section 4.4.) to geothermal studies and prediction of volcanic eruptions and earthquakes (see Chapter 4). Accordingly, reviewing the available experimental techniques calls for extensive reading in various types of journals and patents. In the following text not all authors will be cited, but only those that are felt to be most demonstrative of a given technique from the pedagogical viewpoint.

[1] Université Montpellier II, Montpellier, France

The first selection that can be made is to determine whether to measure radon itself or its daughter products. In both cases, alpha and gamma radioactivities are the detectable phenomena, either independently or simultaneously. For some of the daughters, β radioactivity can also be considered. One of the following techniques can now be used for detection:

- nuclear emulsion
- adsorption
- solid scintillation
- liquid scintillation
- gamma spectrometry
- beta monitoring
- solid state nuclear track detectors
- electrometer or electroscope
- ionization chambers
- surface barrier detectors
- thermoluminescent phosphors
- electret
- collection.

Many authors are using one, two or even three of the above techniques. Mathematically, more than 300 possible combinations are available to the scientist, engineer and inventor's imagination. It is therefore no wonder to find that, during the last decade alone, more than a hundred basic or priority patents have been deposited throughout the world regarding radon measurements. Faced with this huge amount of literature and versatility on the topic, instead of going patiently and systematically through all the possibilities, we will essentially describe the most representative of all the feasible techniques and merely mention the various developments, unless a very astute version has been developed. In this case, details will also be given.

Practically all of the techniques that are used for field measurement (next subsection) can be used in the laboratory. Therefore, let us focus in this subsection on measurement techniques that are not considered of common use for field measurement and that are best performed in the laboratory environment only.

In the laboratory one relies basically on the general techniques for the determination of radioactivity and counting. The main difference or characteristic is the adaptation of these techniques to the gaseous nature of radon.

Scintillation counting in the gaseous phase has been extensively used for the purpose. Lucas (1957) was probably the first to publish anything about it. It is one of the oldest and most reliable techniques. Since then many variations on the theme can be found in the literature (e.g. *Radon, Radium and other Radioactivity in Groundwater*, B. Graves, ed., Lewis Publishers, Michigan, 1987). The equipment is in general a glass vessel internally covered with a scintillating material such as ZnS. The tiny light flashes produced by alphas hitting the phosphor are seen by a light amplifier. Those have been photomultipliers for a long time. Now they sometimes are replaced by microchannel plates.

Sensitivity depends mainly on the length of the counting period. It improves slightly by increasing the volume of the flask. A sensitivity of a few Bq m^{-3} can be achieved. It can be further increased by electrically collecting ^{218}Po atoms directly on the detector (Bochicchio and Risica, 1990).

As a variation, coincidence α and β-particle scintillation counting can be performed (Centro de Investigaciones Energeticas, 1995). Scintillation can also be observed after filtering the radon progeny and placing the resulting α-emitter-rich filter in front of the scintillation screen (Bovi *et al.*, 1993; Fuji Photo Corp., 1993; Phelan and Robert, 1969; Powell, 1979). A more sophisticated version of the Lucas cell is the multiple scintillation chamber device which gives, as a function of time, a better quality response (Demel, 1988). Continuous gas scintillation counting is performed in an apparatus produced by Madnick and Sherwood (1990). Their equipment comprises several light-tight chambers in which scintillation counting of radon and radon daughters is performed. This provides an inexpensive radon gas detector suitable for continuous radon monitoring in homes. Finally, the concentration of airborne radon products is measured as a function of time using Bressan and Larson (1979) equipment. Radon daughters are collected on a moving strip of filter. The particle-bearing area of the filter tape is periodically advanced away from the collection position to a detecting position. The β radiation emitted from the particle-bearing area is detected by a plastic scintillator plus photomultiplier detector.

Scintillation counting can also be performed for radon level assessment in the liquid phase, once it has been extracted. One good example of radon extraction by solvent is the case of radon determination in ground water. Radon is quite soluble in water and can be transported over large distances by this means. However, radon is even more soluble in toluene. As a consequence, it is fairly easy to extract radon from water by mixing it with toluene. Since toluene and water are not miscible liquids, the separation of radon-rich toluene fraction is easy. The amount of radon contained in the toluene is then determined by liquid scintillation, either using the toluene itself as the scintillating liquid or by adding known amounts of liquid scintillator such as PPO and POPOP (Fukui, 1985; Navratil, 1995).

Ionization chambers (LeConte Cathey, 1978) and proportional counters (Melvin *et al.*, 1978) are the most widely used radiation measurement techniques applied to radon determination by sampling it from the atmosphere. These techniques have been extensively described in nuclear technology books. Since measuring radon, when compared to measuring any other radioactive element, does not involve drastic changes in the way one uses the technique, we would like to invite the interested reader to refer to those specialized articles. Their special outdoor versions of these devices are described below.

Spectroscopy can, in general, be used for indirect radon determination. In all cases, radon daughters are the emitting material. Alpha spectroscopy is more generally the so-called α- gross counting (Bochicchio and Risica, 1990). With these methods the total number of α-particles emitted by the filter in a selected interval of time is counted. The α-gross methods - especially those taking longer than 30 min - are sensitive to the presence of long-lived thoron daughters (^{212}Pb and ^{212}Bi). This effect can be corrected by taking more counts and solving several sets of linear equations. Many hours are required to complete the measurement.

Gamma spectroscopy is the second most used spectroscopy method. It allows, in principle, discrimination between products of the U and Th series. It is based on the interaction of γ-rays with matter. It is known that a single monoenergetic γ-ray emitted by a radionuclide produces a spectrum, in which several different peaks appear, that can be described as follows. The photoelectric peak, asymetric in shape; the short term Compton continuum due to the escape of photons from multiple Compton scattering; the Compton continuum, the single escape peak produced when the energy of the incoming photon exceeds 1.022 MeV; the annihilation peak produced by 511 keV photons; the back-scattering peak and the X-ray peaks produced by the ionization of the inner shells of the detector atoms (Sgorbati, 1989). Several examples of equipment based on γ-spectroscopy are available. In the Charueau *et al.* (1989) equipment, the radon gas is trapped homogeneously within the instrument and left to decay until its decay products form a γ-source. Stieff (1992) uses a gamma-compensated radon measurement system together with an electret. The additional electret comes very handy in correcting for high backgrounds. Other versions of γ-spectrometers, specially designed for radon daughter measurements, can be found in Danilov *et al.* (1972) and Asahi (1987).

Let us not leave the laboratory without taking a look at some rather unusual techniques. The first one c.f., Riedel (1995), intends to evaluate the internal radiation exposure of a patient. The equipment determines the patient's pulse rate which monitors, as well as the respiration air/pulse function, the radon progeny measurement in the subject's breath. The variation of the data with time gives an indication of the internal dose. Kuzhevskii *et al.* (1995) have proposed the most unusual device of those described in these pages. It achieves radon determination by passing the test substance (air or water) through a cell, fitted with beryllium gratings in which alphas from radon induce a nuclear reaction with beryllium, thus producing neutrons that are measured by an outer neutron counter. The authors claim that the apparatus provides highly accurate measurements on a continuous basis.

1.3.3 Radon Concentration Measurements in the Field

Radon concentration measurements *in situ* require detectors to be put in place and left for a sufficient length of time, determined after previous testing. A measurement can be continuous or discrete by carrying out grab sampling. It can be passive when the radon concentration is measured under natural conditions, radon entering the detection volume by mere diffusion; or active, which involves the pumping of gas into or through a detecting instrument.

1.3.3.1 Passive devices

In the case of direct measurement of the radiation, the detection sensor is usually placed inside a container that has an opening to let radon enter it. The container is meant to protect the detector and to make room around the detector for a sensitive volume large enough so as to have as many alpha particles produced and detected in as short time as possible. The simplest of such containers can be a simple inverted cardboard cup (Fleischer and Mogro-Campero, 1978) in different versions (Malmqvist and Kristianson, 1980b), or a more sophisticated vessel. Since alpha particles from radon and its progeny have a range of the order of 5 cm in the air, it is indeed easily understood that a free space of this order in front of the detector will ensure maximum efficiency.

The measurement of radon concentration is in principle a simple affair. However, some consideration must be given to the "thoron problem". The two isotopes of radon (radon i.e. ^{222}Rn and thoron i.e. ^{220}Rn) generally occur together, and the ratio of the relative amounts of the two isotopes varies according to the amounts of uranium and thorium present in the vicinity. Then both radon isotopes would theoretically be detected unless some sort of discriminating mechanism can be applied (Hambleton-Jones and Smit, 1980). There are essentially two ways to solve the problem: the first possibility consists in preventing the thoron entering the detecting device; the second is to discriminate between thoron and radon response once both isotopes have entered the detecting device.

In the first case, thoron is kept off the detecting device by taking advantage of the difference between its half-life (55.6 s) and that of radon (3. 83 d) and of the fact that they have the same coefficient of diffusion in a given medium, since they both are isotopes of the same element. Accordingly, the basic idea is to put a diffusion barrier in their way in such a manner that it takes too long for thoron to pass through it before decaying away, while radon can go through without appreciable loss. The simplest diffusion barrier can be just a certain length of a column of air (Seidel, 1982). It has been shown that travelling 30 cm in the air by diffusion is enough for thoron to be reduced down to a negligible fraction of its initial amount. Thus, some set-ups consist of a 30 to 35 cm long pipe instead of a mere inverted cup. The bottom end of the pipe is open to the environment while the upper end is plugged by the detector or by a detector-holder, the detector being placed inside the pipe and always facing down. This type of arrangement offers the most rugged way to dispose of the thoron problem. A slightly different version is presented in devices where radon and thoron are made to circulate in a tortuous air channel before reaching the SSNTD chamber (Steck, 1992; Venkov *et al.,* 1983). This geometry provides the further advantage that tracks are produced at a certain constant angle. Other diffusion barriers are made of thin membranes of different materials (Fleischer *et al.,* 1980; Jha *et al.,* 1982; Pohl-Rüling *et al.,* 1980; Smith and Kanter, 1978; Terradex, 1976). The use of such membranes allows a smaller set-up, and also ensures a moisture-free atmosphere within the volume of detection, provided the chosen membrane is itself waterproof (e.g. Felice *et al.,* 1981; Hassib and Piesch, 1979; Likes *et al.,* 1979; Smith and Kanter, 1978; Terradex, 1976).

The second way to deal with radon and thoron by differentiating between them depends on the detecting sensor itself. It will be described in later sections.

The various detectors that can be used for radon concentration measurement are essentially solid state nuclear track detectors, electret detectors, thermolumiscent phosphors, solid state electronic detectors such as photodiodes or silicon barrier chips, and tiny electrometers and ion chambers.

1.3.3.1.1 Solid state nuclear track detectors. Solid State Nuclear Track Detectors (SSNTDs) have been used for a long time for radon measurements. The functioning of SSNTDs is described elsewhere in this book and a comprehensive treatment of the question in general can be found under the pen of Fleischer, Price and Walker (1975) or of Durrani and Bull (1987). Let us only note here that every detectable α-particle produces in a SSNTD a single trail of damage which, after chemical enlargement, turns into a narrow channel and is made visible under the microscope. In this text, the term "track" will refer to such a revealed channel, etched along the path of an alpha particle.

SSNTDs exhibit different sensitivities: some of them in particular are sensitive to alpha particles in the energy range of the particles emited by radon. Mostly, these SSNTDs are cellulose esters (nitrate and acetate), and polycarbonates like bis-phenol-A polycarbonate and CR-39. On the other hand, however sensitive they may be to heavily charged particles, SSNTDs are largely insensitive to beta and gamma rays. In other words, β- and γ-rays do not produce etchable individual tracks. (However, pre-irradiation with non-track-forming radiation has also been found to increase sensitivity and etching speed (US Atomic Energy Com., 1968). SSNTDs also have the advantage to be mostly unaffected by humidity, low temperatures, moderate heating and light. They present unique characteristics for long-term integrating measurements of radon gas for large-scale surveys (Tommasino, 1990). They of course do not require an energy source to be operated since their detecting property is an intrinsic quality of the material they are made of. They are consequently very well adapted for long-term measurements in the field and have received particular attention. More than half of the devices one can find in the literature operate with SSNTDs.

The ways in which SSNTDs are used for radon and/or radon daughter detection differ mostly in the geometry of the arrangement; the use or not of diffusion membranes; and the use (or not) of filters to keep daughters out of the device (or to detect the latter especifically, once they are stuck on the filter surface).

Let us review the most favoured arrangements. Besides the very simple inverted cup containing a single strip of detector, one may find devices in which the detectors form a double active layer placed in the centre of a single radon chamber. This helps to reduce the amount of film required and allows the detector to be compactly packaged (Yoder, 1988).

In other cases, the detector is placed vertically (e.g. Wheeler, 1987). When radon alone measurements are wished, airborne radon daughters are kept off the inner chamber of the device by using filters (see below) and by devising a special geometry for the chamber. Essentially, the size of the chamber is such that α-particles emitted form the daughters that will form out of the pure radon that has entered the chamber cannot reach the detector (e.g. General Electric, 1968) or it has a specific hemispherical form (e.g. Port *et al.*, 1990).

Some devices have two measuring chambers, separated or not by a filter and/or a membrane. Good examples of such designs can be found in Hassib and Piesch (1981) or Guhr and Sarenio (1992).

The use of filters to pick up radon daughters is well examplified in various devices such as those of Pai (1989a), Alter and Oswald (1989), or Port *et al.* (1990). Special mention should be made of a design proposed by Pai (1989b). In this device, radon daughters ^{218}Po and ^{214}Po are collected on a filter. In front of the filter are placed two collimators leading to two different SSNT detectors, having different sensitivities. By this means the differential response of the two radon progenies is feasible.

An original concept was proposed by Leuschner *et al.* (1993). The detector itself is part of a self operating electrochemical etching system (see section 2.3). The device provides an inexpensive, easily operable radon daughter dosemeter.

A way to improve the detecting ability of SSNTDs is to apply to the detecting surface a negative voltage of approximately 300 V (Pretzsch *et al.,* 1990). A four-fold improvement is obtained in terms of accuracy, for instance, with respect to the more traditional way, where no negative voltage is applied.

The so-called "plate out effect" will be discussed in a further section. This effect is such that conventional track detectors can produce erroneous results due to the presence of static electrical charges on the detector. This can be avoided by covering the upper surface of the detector with a very thin conducting film that is either grounded or provided with a constant voltage from a power supply (Boliden, 1985)

Finally, a strong drawback in the use of SSNTDs is that they only integrate the received flux of particles. They do not provide time-dependent response. This shortcoming is overcome with the device of Ishankuliev *et al.* (1993), where the detector is moved step by step by means of an electrical motor. The same applies to the device proposed by Monnin and Seidel (1982). The latter have produced a different time-dependent SSNTD equipment. It is made of 20 disks that are put on top of each other. A track detector is fixed at the bottom of each disk. The pile is suspended at the upper part of a vertical pipe. A clock driven motor release the disks at fixed time periods. While it is at the bottom of the pile, the lower detector is exposed to the radon atmosphere. When released it falls down and comes in contact with the flat bottom of the detection volume and cease to be exposed to the Rn atmosphere. The next detector is exposed; then in turn it falls down, comes in contact with the upper part of the previous disk holding a detector, and so on. When the detectors are in contact with the previous disk-holder, the α-particles are no longer recorded except for those few coming from the Rn that has sneaked in between the two flat surfaces and has emitted particles at a low angle. Of course, experimental corrections must be made; but this device offers a reliable way to make time-dependent measurements with SSNTDs.

The apparent complexity of geometrical arrangement used with SSNTDs is counterbalanced by their utmost simplicity of use. For instance, Henshaw and Campling (1990) have proposed a SSNTD detector sheet with a perfect 1 square centimetre grid engraved onto its surface, treated with a standard recipe for etching, providing constant and large enough tracks to be counted on a screen when the detector is placed in a conventional slide projector.

With the exception of this particular case, after exposure in the detector holding device, SSNTDs are recovered and subsequently processed according to various recipes that can be found abundantly in the literature (e.g. Durrani and Bull, 1987; Monnin,

1980a,b, 1990 and Tommasino, section 2.3, Cross and Tommasino, 1971; Somogyi *et al.,* 1978). However, it is worth mentioning an arrangement where the sensitive layer is placed next to a polymeric dispersant layer containing an indicator which changes colour after the etchant has reached through the tracks themselves (Patel, 1987).

When thoron is not blocked before entering the device, it is necessary to split the device response in two parts, that to thoron, and that to radon. This can be achieved by playing with the difference in the range of the alpha particles emitted by the two isotopes (Urban and Kiefer, 1984). In such a case, the inner geometry of the device is such that α-tracks from Tn are automatically separated from Rn tracks. Separating Tn and Rn responses can also be achieved by performing track geometry analysis on the SSNTD itself (e.g. Henshaw, 1992). Lucas (1990) offers a different discriminating alternative. His apparatus is formed of two parallel chambers. They both contain a detector partly covered with a thin α-particle filter. Upon exposure to the gas to be sampled the α-particles emitted by the thoron and radon are detected in the target material. However, as the mixture decays with different α-particle energies, the filter limits the range of the particle, thus enabling the analysis of the relative concentration of the two isotopes.

In some cases, SSNTDs are used in conjunction with another type of detector. For instance, SSNTDs can be paired with electrets (see below). In this way a prompt response can be obtained together with an energy differenciation (e.g. Pretzsch *et al.,* 1985).

It is not possible to leave this section devoted to solid track detectors without mentioning, from the historical viewpoint, that photographic emulsions were also track detectors of their own kind and were also used for radon detection. As an example one could see the device patented by Evans (1966).

1.3.3.1.2 Alpha card. According to its manufacturer Alphanuclear, Canada, the alpha card system is a passive radon detecting method which provides a means for sensitive measurement of radon in soil-gas. The alpha card is a 4.5 cm $\times$5 cm card with a 2.5 cm diameter hole pierced in the center of it. This hole is covered with a thin membrane. The card is deposited in the ground, very much like SSNTD, suspended in an inverted cup for instance. It is left in place for a given length of time, from 12 hours up to several days. After entering the detection volume the radon decays, and its daughter products make an active deposit on the thin membrane (plate-out effect). The alpha card is then recovered from the field and read. Its activity in terms of alpha particles is measured using silicon detectors sensitive to alpha particles only. The reading device contains two silicon detectors facing one another, between which the card is inserted. Since the collecting membrane is very thin, alpha particles issuing from both sides of the card impinge upon either one of the silicon detector. In this manner a rather good counting efficiency is achieved. The alpha card reader is a small, portable, battery operated equipment that can be used in the field. The normal reading time is of the order of 5 minutes.

The thoron problem is solved in this technique by taking advantage of the fact that thoron daughter products have long half-lives compared to those of radon daughters. Therefore something like 5 hours after the alpha card has been recovered, the alpha activity due to Rn daughters has dropped to virtually zero. A second reading achieved afterwards provides an indirect measure of the thoron concentration when the appropriate calibration has been made.

Very similar is the collector proposed by Card and Bell (1981). A collector made of silver, copper, brass, aluminium or even platinum and gold (corrosion-proof materials) is placed in a housing where it collects the decay products of radon. After a suitable time, the collector is removed and α, β or γ-ray radiation from the decay products is measured directly.

1.3.3.1.3 Electret detector. A recent technique offered to investigators in the field of radon concentration measurement is the electret radon monitor (Kotrappa *et al.*, 1988). Let us note here that an electret is a piece of dielectric material that exhibits an almost permanent electrical charge if not otherwise perturbed. This charge produces a strong electrostatic field which is able to collect ions of the opposite sign and the total charge of the electret decreases. It was Marvin (1955) who first suggested to use this property to make gamma radiation detectors out of the electrets. The idea has not been practical until the recent development of high dielectric fluorocarbon polymers that have made electrets become reliable electronic components even under high temperature and humidity conditions.

An electret radon monitor is basically made of a steel can, on the inside top of which the electret dosimeter is fixed. At the bottom of the can a small inlet allows the radon gas to enter the assembly through a filter. Decaying away, radon produces ionizing particles that in turn produce ions within the volume of the device. These ions are collected and the total charge of the electret changes. After some time has elapsed the surface potential of the electret is measured by the shutter method (Kotrappa *et al.*, 1982, 1984; Gupta *et al.*, 1985) using a battery-operated, lightweight piece of equipment. This method is non-destructive and the electrets can be measured several times without having their charge value appreciably altered. After proper calibration, the electret dosimeter can deliver a dose response curve. It has been found that the sensitivity and dynamic range of the detector depend upon the volume of the chamber of the device and of the thickness of the sensitive material. As for SSNTDs, various geometrical arrangements have been devised. For instance, the measurement can be activated timewise by having a moving piece translating the electret in the housing (Ramsey and Stieff, 1990). Detection of radon alone is achieved by using a 4π-geometry chamber, on the wall of which another electret is placed for daughter detection (Gast, 1992). In some other instances, a positively charged electrode is used to increase the electret collection (Pretzsch, 1984). This counter electrode can be coated with a SSNTD or with a TL phosphor for completeness of detection and measurement.

In another method, both radon and radon daughters are counted. The device is made of a container with a flat bottom and an upwardly and outwardly extending truncated conical side wall. A cover overlies its upper edge. This cover has a disc-shaped plate with holes in a circular arrangement. A filter is placed on the disc. A negatively charged electret is held against the underside of the central portion of the disc. A positively charged electret is placed at the bottom of the housing. Radon enters the housing, while airborne daughters are filtered. Inside the housing, radon decays. Its positively charged products are attracted and counted by the first top electret, while negative ions are counted by the bottom one (Dempsey and Stieff, 1990).

As pointed out by Kotrappa *et al.* (1984) the electret dosimeter offers several advantages, among which worth mentionning are its ability to store information over relatively long periods of time, the fact that it is independent of the humidity in its environment, and its ease of reading. On the other hand, among its disadvantages with respect to other detectors one may mention the fact that its response curve does not efficiently cover the very low or very high doses, and also that it is sensitive to normal gamma radiation background which in some cases may induce a significant error or even preclude its use. This latter objection can be overcome by a special arrangement in which three detectors are used. One is placed outside and two inside two different chambers completely sealed except for an equalisation aperture. By comparing the response of the three electrets one gets the radon concentration accurately without distortion introduced by γ-radiation presence (Dorfi *et al.,* 1992). In our view, as they stand at present, the electret detectors could offer an encouraging alternative for radon measurement, particularly in the case of geophysical studies.

1.3.3.1.4 Adsorption technique. This technique is a radon measurement system based upon the adsorption of the gas on an adsorbing medium, such as charcoal, and then on the measurement of the gamma activity of the trapped daughters ^{214}Pb and ^{214}Bi.

When the adsorbing medium is activated charcoal, the technique is better known under the name ROAC (Radon On Activated Charcoal) technique. Basically, the tool for the adsorption of the radon is a plastic can containing a few grams of activated charcoal. The charcoal is kept in place even when the can is placed upside down, by a fibrous mat or by a membrane such as silicon rubber or PTFE. The plastic can is placed in the ground, with its lid open, and left in place for a few days; typically from 4 to 12 days. The can is then retrieved and the lid closed. It is then brought to the field lab where gamma activity is determined by means of a NaI scintillation counter. The radioactivity of the entrapped radon decreases according to the laws of radioactivity, provided that the can, the lid and the lid seal are gas-tight. After proper calibration, one can determine the radon concentration at the test point from the decay activity curves. The thoron problem is solved in principle by taking advantage of the decay curves of both isotopes and their daughters. As a matter of fact, normally after 3 to 5 days, all the thoron and its daughters contained in the cup should have disappeared and the count rate should represent radon only. The type of results one gets in high thoron environments seem to depend on the quantity of activated charcoal used. Using some 400 g of charcoal, Megumi and Mamuro (1977) found it possible to determine the ratio between thoron and radon emanation from a soil. On the other hand, by using 8 g of product Hambleton-Jones and Smit (1980) found that their charcoal cartridges were unable to adsorb significant amounts of thoron and they concluded that their ROAC set-up could be "used with impunity in high thorium environments".

Another way of determining the amount of radon is to desorb it into a scintillation liquid and to count the α and β activity by liquid scintillation (Perlman, 1990). A variant of this is to introduce the liquid scintillating desorbent directly into the charcoal container. The radon is desorbed *in situ*. Without removing the adsorbent the detector container is introduced into the scintillation counter. Of course the detector container has light transmitting walls (Grodzins, 1993).

1.3.3.1.5 Thermoluminescent detectors. When an ionizing particle travels within a crystal it ejects some electrons from their normal position into the conduction band, leaving a hole in the valence band. The electrons are then trapped into defects of the crystal and stored there in a very stable position. If, later on, the crystal is heated, the energy given to the electrons allows them to escape from their trap. They then recombine with the holes in the valence band, or with recombination centres, by means of a radiative process: light is emitted. (Horowitz, 1984; McKeever, 1985). Based upon this principle, several Thermoluminescent Dosemeters (TLDs) have been developed.

In the case of radon detection, one relies mostly on the alpha activity of the radon daughters. Radon is allowed to enter a detection volume containing the TLD. A metallic plate is placed at a short distance in front of the TLD. This plate can either be electrically charged for better collection efficiency, or not. Radon daughters deposit on the plate and ultimately decay, thus producing energy storage in the TLD. After proper exposure to the radon rich atmosphere, the TLD is recovered and "read" in a TLD apparatus. This apparatus is essentially a tiny oven where the temperature is raised under a nitrogen atmosphere at a very well-controlled rate. The amount of light emitted by the TLD falls on the photocathode of a photomultiplier and is converted into an electrical signal. TLD spectra are recorded. Each peak corresponds to a particular trap. The data analysis is not always very simple, depending on the type of phosphor used as TLD, and it usually calls for curves deconvolution. Generally speaking, the amplitude and/or the area below a peak is proportionnal to the irradiation dose. In many commercially available devices the data analysis is made automatically.

Various type of detectors containing TLD sensors have been used for field measurements, and in the sphere of different applications. Hochman and Ypma (1988) are using them for detecting oil gas reservoirs. Harley and Maiello (1987) have developed a detection system, which incidentally also holds an electret detector for γ-background measurement, having three TLDs. One is a pre-irradiated TLD which will give information on the natural fading of the signal registered in the two others. The first generates a signal responsive to the radioactivity produced by the radon that has diffused into the detection volume; and the second generates a cross measurement of the γ-field. In some cases the housing of the TLD sensor is made such that the whole detector can withstand the temperature at which the light is released from the TLD when reading it (Seidel, 1982). See also Felice *et al.* (1981).

1.3.3.1.6 Solid state electronic detectors. Electronic detectors are not widely used for radon measurements in the soil because they usually require appreciable amounts of energy not easily available in the field, and also because they are rather expensive and fragile. However, some types of equipment have been put on the market for temporary measurements. One may mention, for instance, the Alphameter manufactured by Alpha Nuclear Co, Canada. The active part of the equipment is a Si(Li) detector associated with a counter unit, both operated on batteries. The counting is directly displayed on the instrument. Some similar pieces of equipment are also operable in the field, e.g. Alphaguard by Genitron Intruments, Germany; or the RMS-4501, manufactured by ECIL, India (Singh, 1992); or the Radon gas meter TN-RN2000 manufactured by Thomson & Nielsen Electronics, Canada; or the Atmos by Gamma Data, Sweden; or the Rad7 manufactured by Niton, USA; or the Prassi and Mod.4S models, by Silena Corp., Italy.

Using a cheap photodiode (Hamamatsu S3590-02), we have developed a very simple equipment, called Clipperton, that can be operated on pen-size batteries for three months. The detector does not require a polarization voltage, and the amplifying electronics require very low consumption. Data are stored in memory cells of a microcomputer. Up to 4,000 different data can be stored at intervals of time previously fixed. Data can then be transferred into memory cards for further analysis on a larger computer. This unit is able to work down to 100 metres below the water level, and has actually been working for months on a routine basis at a depth of 30 m. In addition to being usable in air as well as in the ground, this equipment is particularly suited for hydrology and hydrogeology studies. Of very similar properties is the Dataqua marketed by the Dataqua Co, Hungary. If the Clipperton performs radon-only measurements, the Dataqua simultaneously achieves radon and radon products counting, thanks to a collection disk situated in front of the silicon detector. A special version of the Dataqua allows water level, temperature and conductivity measurements to be made in the meantime (see also Yamamoto *et al.*, 1987). Based also on a silicon chip (400 mm^2) is the Barasol, marketed by the Algade Co., France, for ground measurements.

For other electronic devices, see Kotlyarov and Krisvashcheev (1994), who have produced a detector especially for indoor measurements; or Matthies (1994), who has built a measuring instrument with a hemispherical bell chamber fitted with a solid state alpha detector; or Simon *et al.* (1987), who use a solid state diode, whose depleted layer is located near to the test medium, and two connected amplifiers in order to reduce background and unwanted noise; or Sciocchetti and Scacco (1981), who carry out both α and β counting with their equipment; the same double counting is performed with Berker and Martz (1988) equipment; or the device of Hurst (1989), which allows continuous monitoring of air-conditioning system for dwellings; or that of Wolfert (1978), which, using a desiccant, can determine radon levels near the earth's surface.

Having more specific characteristics are the following set-ups. Branker (1980) combines a solid state photodiode with a solid scintillating ZnS layer and a thin metal layer to shield the diode from natural light. Thomson (1991) uses several active cells for counting radon daughters. Based on light measurement with solid state detectors is the device of Decuir (1990). It is a cubic-shaped chamber with a photon-blocking system from the outside, but with inlet and outlet ports which allow gas or liquid flow from the environment to be monitored. The radioactivity is detected and counted by an alpha sensor. Dumbeck (1989), in an original way, uses an ionization detector along with a second one which incorporates a microcurie continuous-radiation source. Thus, by comparing the two response levels the equipment is set to give an alarm when the activity of the sampled air reaches 160 Bq m^{-3}.

The main advantage of electronic equipments is that they allow dynamic studies, i.e. the study of short-term variations in the radon concentration.

1.3.3.1.7 Ion chambers and electrometers. Allowing dynamic studies are also the devices that are based on the measurement of ion pairs created along the path of the alpha particles in the air. The detection is mostly carried out with electrometers and ion chambers. As an over-simplified description of the detecting part, one can say that, in the first case, one measures the quantity of electrical charges that have been created, without applying a constant voltage. In the second case, one collects the created charges that have been attracted between two electrodes in an electrical field.

Again, as with the SSNTDs, the multiplicity of design and conception is very large. Shape, size, voltage, external polarized grid, type of amplifying set-up, field strength, etc., are amongst the many parameters that are found to change from one model to another. For instance, Kershner and Burgess (1987) can achieve radon-alone or radon-and-daughters measurements by applying, or not, a negative voltage. Others have been able to miniaturize the device by adding many of the improvements found separately (Aloca, 1993a). The last-mentioned device has two ionization chambers working together. A higher sensitivity is achieved, which allows real-time measurement of radon concentration (Aloca, 1993b).

Regarding this topic, see also Aloca (1993c); Jia (1985); Carlsson (1989); Davidson and Vandrish (1995); Klein (1989); Klein *et al.* (1995).

1.3.3.2 Active devices

In contrast to what we have been examining up to now, most of the following techniques are not based on equipment in which the radon gas enters under natural drive. It is usually either pumped from the environment or extracted by means of a gas or liquid. In addition, the active part for detecting radon, or its daughter products, is an electrical or electronic device in general.

Most of the apparatuses described above could be turned into an active device by using a pump or a fan. SSNTD and a pump go together in the apparatus patented by Gellev *et al.* (1993). However, other examples can be found involving different detectors. The ROAC technique coupled with a pump is examplified by Scheible's patented device (Scheible, 1987). The radon-charged air is forced to pass through a passageway around which the adsorbent is placed. The subsequent analysis of the adsorbent is carried out in the laboratory. Ilmasti (1987) has devised a detector where radon is pumped from the ground soil. The gas flow passes through the electronic meter to the outside of the equipment. The radon concentration is thus continuously measured. Direct scintillation can also be used for measuring the radon level by means of forcing air through the detector (US Secretary of Interior, 1976).

See also Allander *et al.* (1994), Commissariat à l'Energie Atomique (1972), Diamondis (1994), Fimian *et al.* (1988), Gast (1992b), Horn (1993), Nindel (1990), Shepherd (1980),

Tetley *et al.* (1988, 1990) and Wilhelm (1988).

A different way to induce the flow of radon-rich air is to play with thermal pumping. A heating element is a part of the device patented by Kutelev *et al.* (1995) in which the sensitive part is an electrometer. This adds a second advantage: it avoids moisture effects on the reading.

The second type of active measurement is radon sampling or grab sampling. After sampling, the air or the water or the solvent that contains radon is brought to the laboratory for measurement. Sampling techniques are numerous, but they all require one to be particularly cautious not to contaminate the sample. Sampling can be achieved by the suction action of pumps, syringes, pistons, empty vacuum vessels, water sampling, degassing on the spot, etc. Many examples are found in the literature, which the reader will find inspiring. However, we would like to stress here the fact that blanks must be made regularly to check the quality of one's sampling technique, whatever this technique might be. For the measurements see the above sections.

1.3.4 Short- and Long-Term Measurements

The discussion whether short-term or long-term radon measurement should be preferred is different in nature, depending on the information sought. To make it simple let us divide the topic in two: (a) radon-related health problems; (b) geology and geodynamics studies.

In the first instance, and regarding home radon level evaluation, instances exist when short-term measurements were performed and recommended. The objective was to get an almost immediate answer. It was additionally recommended to perform the measurement in the basement in winter, i.e. when the radon readings are expected to be at their highest. The main argument was that "false positives are preferable to false negatives". One of the authors (Nyberg, 1989) further added that he was proposing a two-step strategy: short-term evaluation – and in cases of values which would cause concern, a long-term assesment to be performed subsequently. This view was strongly opposed by others (Fleischer, 1989; Stewart and Kneale, 1989). These latter authors mostly advocated that the reliability of short-term measurements was far from acceptable. The dispersion of data due to many well known factors, such as seasonal variations, ventilation, moisture content in the air, outdoor/indoor temperature gradient, wind speed and so on was too great; and in addition the fact that basement readings are on average three times higher than in living quarters makes it difficult to believe that a short-term measurement would provide any valubale information. The main recommendation that was made is: "Health-physics dosimetry should be relevant to the radiation exposure of individuals. Normally, radon causes effects only after years of exposure. It follows that meaningful readings can only be made over a reasonably long time, such as a year, with negligible harm caused to most individuals by the delay" (Fleischer, 1989).

Regarding Earth Sciences problems the question is different. Radon carries information of a different nature. Radon mapping is performed for the location of permanent features of the Earth's crust, or of its contents. We are referring here to ore-body location (uranium, oil, steam or hot water). In this case, radon mapping over a long time (at least one climatic cycle) is needed because of the seasonal and nyctemeral variations of the radon response to environmental stress. Averaging of readings leads to a better picture of the situation. In a somewhat intermediate state is the question of location of faults. Long-term mapping can provide a good answer if one has the chance to locate the detectors properly. A brief survey is probably preferable beforehand and could save much time in the end. The analysis of air circulation in fracturated natural media like caves (Somogyi *et al.,* 1989) calls for medium-term measurements. As a matter of fact, the variations that one observes are more of the order of weeks or months than of days or hours. In such cases, measurements performed by track detectors are quite adequate. When it comes to time-dependent studies, the situation is different. For instance, in hydrological or hydrogeological studies it is now obvious to us that short i.e. hourly or briefer measurements are needed. Strong variations of the radon content in waters are observable. When it comes to the use of radon measurements with the aim of evaluating the near occurrence of an earthquake or a volcanic eruption, both long and short-term measurements have been carried out. Long-term measurements have provided the first evidence that radon could be a precursor of such catastrophic events. However, it now seems that short-term measurements provide more reliable data; but it still needs to be definitely proved.

1.3.5 Concluding Remarks

Determinations of the concentration of radon and radon daughters is a field of growing interest for various reasons. Accordingly, many techniques have been developed for this purpose, and the newcomer may have to make a difficult choice among the available possibilities. However, his choice will be guided by the very nature of his interest in radon measurements; whether he is, for instance, interested in time-dependent variations or in space-dependent variations of the concentration values. In order to help him and for the sake of completeness, a number of bibliographical references have been included in the literature list.

References

Abdelkazek M. (1984) An integrated radon flux monitor. *Nucl. Instr. Meth.* **224**, 233-237.

Abe S., Ochifuji K. and Yokoyama S. (1988) An electrostatic integrating ^{222}Rn monitor with cellulose nitrate film for environmental monitoring. *Health Phys.* **54**, 2, 139-148.

Akber R. A., Khan H. A., Ahmad I. and Jamil K. (1980) Some important considerations in the use of solid state nuclear track detectors for radon gas concentration measurements. *Nucl. Instr. Meth.* **173**, 183-189.

Allander K. S., Bounds J. A. and Macarthur D.W. (1994) High sensitivity radon detector for environmental monitoring. *U.S. Patent 940,125* from *92US-864,746,920,407.*

Aloca Co. Ltd (1993a) Measurement equipment to detect background alpha radiation.

Japanese Patent 049,404,930,310.

Aloca Co. Ltd. (1993b) Radioactive substance measurement equipment. *Japanese Patent 049,317,930,310.*

Aloca Co. Ltd. (1993c) Measurement equipment to detect radioactive substances. *Japanese Patent 049,299,930,310.*

Alpha NUCLEAR, technical data. 6380 B Viscount Road, Mississauga, Ont., Canada.

Alter H. W. (1980) Disposable alpha particle detector especially for radon gas detection. *U.S. Patent 139,382.*

Alter H. W. and Oswald R. A. (1989) Radon gas detector for personnel and environment monitoring. *U.S. Patent 435,197,821,019.*

Alter H. W. and Price P. B. (1968) Detecting system for alpha radiation from gases. *U.S. Patent 764,686,680,717.*

Asahi K. (1987) Evaluation of ground foundation using device to detect gamma rays in spectrum containing radon, thoron and potassium gamma rays. *Japanese Patent 311,736,871,209.*

Baranyi I., Futo K., Gerzson I., Lazar J., Varhegyi A. and Vecsey K. I. (1992) Forecasting earthquake dangers. *Hungarian Patent 921,028.*

Becker G. K. and Martz D. E. (1988) Alpha-beta continuous air monitoring. *U.S. Patent 211,647,880,627.*

Bigu J. and Kaldenbach R. (1983) Design and performance of a microprocessor-based automated continuous monitor with alpha spectroscopy capabilities for the determination of radon and thoron and their short-lived decay products. Int. Report 0087, CANMET, Energy, Mines and Resources, Elliot Lake, Ont., Canada, 1983, p. 26.

Bigu J., Raz R., Golden K. and Dominguez P. (1984) Design and development of a computer-based continuous monitor for the determination of the short-lived decay products of radon and thoron. *Nucl. Instr. Meth.* **225**, 399-406.

Bluszcz A. and Pazdur A. (1979) Some physical problems of very low ^{222}Rn concentration measurements with the proportional counter. *Nucl. Instr. Meth.* **167**, 299-303.

Bochiccio F. and Risica S. (1990) Active radon and radon daughter monitors. In *Proc. Int. Workshop on Radon Monitoring in Radioprotection, Environmental Radioactivity and Earth Science* (eds. L. Tommasino *et al.*), Trieste, 1989, World Scientific, Singapore, pp. 110-121.

Boliden A. B. (1985) Detection and determination of substance emmitting alpha rays such as radon and its decay products. *European Patent 52,636,851,002.*

Bovi M., Baldassini P. and Porcu I. (1993) Radon monitoring diffusion device. *Italian Patent 3,588,190,900,704.*

Brancker R. W. S. (1980) Solid state alpha radiation detector. *Canadian Patent 1,078,974, 800,604.*

Bressan D. J. and Larson R. E. (1979) Atmospheric radon monitor for airborne particulate matter. *U.S. Patent 4,140,912,790,220.*

Burgess E. T. and Kerschner C. J. (1989) Airborne alpha radiation detector. *U.S. Patent 890,502.*

Card J. and Bell K. (1981) Detecting radon gas especially in uranium ore prospection. *U.S. Patent 4,297,574,811,027.*

Carlsson R. (1989) Radon gas measurement chamber. *Swedish Patent 000,513,890,215.*

CEA (1972) Detecteur portable de particules alpha. *French Patent 025,550,720,713.*

Centro de investigaciones energéticas (1995) Equipo para mediciones continuas de concentraciones de radon in aire. *Spanish Patent 2,073,990,950,816.*

Charuau J., Guelin M., Legac J. and Le Gac J. (1989) Mesure de la concentration des différents isotopes du radon. *French Patent 007,109,890,530.*

Chyi L. L. and Quick T. J. (1995) Exhaust system to control air contaminant in building. *U.S. Patent 5,438,324,950,801.*

Clements W. E. and Wilkening M. H. (1974) Atmospheric pressure effects on Rn-222 transport accross the air-crust interface. *J. Geophys. Res.* **79**, 33, 5025-5029.

Commissariat â l'Energie Atomique (1972) Détecteur portable de particules alpha. *French Patent 025,550,720,713.*

Cross W. G. and Tommasino L. (1970) A rapid reading technique for nuclear particle damage tracks in thin foils. *Radiat. Effects* **5**, 85-91.

Danilov V. S., Matveev A. V. and Filimonov V. V. (1972) Gamma ray spectrometer with additional resolving amplifier compensating the uranium error. *S.U. (former) Patent 793,697,720,605.*

Dancereau R. E. and Hutchinson J. A. (1995) Radon collector and bubbler for sampling liquid for testing. *U.S. Patent 5,381,699,950,117* from *92US-958,047,921,006.*

Davidson L. A. and Vandrish G. E. (1995) Integrated electroscope and condenser ion chamber. *Canadian Patent 2,116,252,950,824.*

Decuir J. (1990) Radon monitoring device. *European Patent 357,697,900,314.*

Demel F. (1988) Radon activity measuring apparatus. *Czeckoslovak (former) Patent 8,604,867,880,615.*

Dempsey J. C. and Stieff L. R. (1990) Compact monitor for radon in air. *U.S. Patent 4,926,053,900,515* from *88US-289,438,881,223.*

Dempsey J., Kotrappa P. and Dempsey J. C. (1989) Cumulative radioactive radon gas monitor for domestic use. *U.S. Patent 890,801* from *86US-936,240,861,201.*

Diamondis P. J. (1994) Forced airflow radon detector with electrically conductive inlet filter. *U.S. Patent 940,607* from *91US-679,911,910,403.*

Dorfi K., West J.E. and Sessler G.M. (1990) Discriminating electret radioactivity detector system and method for measuring radon concentration.

Dorfi K., West J. F. and Sessler G. M. (1992) Discriminating electret detector of radon gas concentration. *European Patent 477,308,920,401.*

DSMA Atcon Ltd. (1983) Review of existing instrumentation and evaluation of possibilities for research and development of instrumentation to determine future levels of radon at a proposed building site. Res. Rept. INFO-0096, Atomic Energy Control Board, Ottawa, Canada, p. 29.

Dumbeck R. F. (1989) Radon detector monitoring air quality in dwellings. *U.S. Patent 4,868,546,890,919.*

Durrani S. A. and Bull R. K. (1987) *Solid State Nuclear Track Detection: Principles, Methods and Applications.* Pergamon Press, Oxford.

Evans R. D. (1966) Radiation detector. *U.S. Patent 539,691,660,404.*

Felice P. E., Gonzalez J. L. and Seidel J. G. (1981) Thermoluminescent dosimeter for alpha radiation detection. *U.S. Patent 874,976,780,203.*

Fimian K. S., McGreevy J. E. and Tucker R. W. (1988) Radon detection system. *U.S. Patent 148,729,880,126.*

Fimian B. P., Fimian K. S., McGreevy J. E. and Tucker R. W. (1993) Radon detection system with means to measure another parameter. *European Patent 930,630.*

Fleischer R. L., Price P. B. and Walker R. M. (1975) *Nuclear Tracks in Solids: Principles and Applications.* University of California Press, Berkley.

Fleischer R. L. and Mogro-Campero A. (1978) Mapping of integrated radon; emanation for detection of long distance migration of gases within the earth: Techniques and principles. *J. Geophys. Res.* **83**, B-7, 3539-3549.

Fleischer R. L. and Mogro-Campero A. (1980) Techniques and principles for mapping of integrated radon emanation within the earth. *Nat. Rad. Env.* III, 57-71.

Fleischer R. L. (1980) Radon flux from the earth: Methods of measurement by the nuclear track technique. *J. Geophys. Res.* **85**, 7553-7056.

Fleischer R. L. (1989) Reply to radiation cancer risk series. *Health Phys.* **57**, 5, 842.

Fuji Photo Film Co. Ltd. (1993) Detection of radon in air. *Japanese Patent 930,427.*

Fukui M. (1985) Continuous monitoring of Rn-222 concentration in unconfined water. *J. Hydrol.* **82**, 371-380.

Gast H. (1992a) Measuring chamber for determining alpha activity of radon gas in carrier gas especially air. *German Patent 200,308,920,109.*

Gast H. (1992b) Flow density determination of radon gas passing from gound into atmosphere. *German Patent 200,307,920,109.*

Gellev M. G., Penchev O. Y., Pressianov D. S., Pentschev O. J. and Geilev M. G. (1993) Radon and thoron daughter product monitoring device. *U.S. Patent 5,225,673,930, 706* from *91BG-093,584,910,104.*

General Electric Co. (1967) Radon and its emitting descendents detector. *U.S. Patent 654,631,670,719.*

General Electric Co. (1968) Radon gas detector. *U.S. Patent 787,431,681,227.*

Grodzins L. (1987) Radon detection. *U.S. Patent 94,192,870,908.*

Grodzins L. (1993) Radon concentration measuring method for air in building. *European Patent 306,949,930,224* from *87US-094,192,870,908.*

Guhr A. and Sarenio O. (1992) Radon decay product content measurement in air. *Australian Patent 9,183,142,920,317* from *90DD-343,507,900,817.*

Gupta P. C., Kotrappa P. and Dua S.K. (1985) Electret personal dosemeter, *Radiat. Prot. Dosim.* **11**, 107.

Hambleton-Jones B. B. and Smit M. C. B. (1980) *ROAC - A New Dimension in Radon Prospecting.* Atomic Energy Board, Pretoria.

Harley N. H. and Maiello M. L. (1987) Passive measurement of environmental radioactivity. *U.S. Patent 021,450,870,304.*

Hassib G. and Piesch E. (1979) Passive dosimeter for radon detection. *European Patent 21,081,810,107* from *79DE-926,491,790,630.*

Henshaw D. L. (1992) Detector for alpha radioactive radon and thoron daughters in air. *British Patent 921,021-22,544,990* from *91GB-007,541,910,410.*

Henshaw D. L. and Campling G. C. (1990) Radon detector. *British Patent 22,28,610,900, 829* from *88GB-029,414,881,215.*

Hochman M. B. and Ypma P. J. (1988) Location of oil and gas reservoirs. *Australian Patent 01,988,788,000.*

Hopke P. K. and Hurst G. S. (1992) System for real time determination of radon health-risk parameters. *European Patent 503,568,920,916.*

Horiuchi K. and Murakami Y. (1983) A new method for the determination of radon in soil air by the open "vial" and integral counting with a liquid scintillation counter. *J. Radioanal Chem.* **80**, 1-2, 153-163.

Horn W. (1993) Process and apparatus for measurement of radon content in ground air. *German Patent 310,096,930,325.*

Horowitz Y. S. (1984) *TL and TL dosimetry.* CRC Press, Boca Raton, Fl, USA.

Hurst G. S. (1989) Radon monitor and control system. *U.S. Patent 364,209,890,609.*

Ilmasti V. (1986) Portable detector for measuring radon content in air. *Finish Patent 004,055, 861,007.*

Ilmasti V. (1987) Device for measuring radon gas concentration. *Finish Patent 002,717, 870,618.*

Ishankuliev D., Matveev I. V. and Panomarenko V. A. (1993) Subsoil radon concentration registration device. *S.U. (former) Patent 1,796,066,930,215* from *91SU-947,389,910,624.*

Israelson S. (1980) Meteorological influences on atmospheric radioactivity and its effects on the electrical environment. *Nat. Rad. Env. III*, 210.

Jha G., Raghavaya M. and Padmanabhan N. (1982) Radon permeability of some membranes. *Health Phys.* **42**, 723-725.

Jia W. and Xu B. (1985) Unit for radon and daughter isotope detection in soil. *Canadian Patent 107,800,851,023.*

Kershner C. J. and Burgess E. T. (1987) Airborne alpha radiation detector. *U.S. Patent 039,392,870,416* and *88US-197,496,880,815.*

Klein D. (1989) Un appareil de mesure de terrain du radon 222 en continu. *Revue Générale Nucléaire* **2**, 134-138.

Klein D., Barillon R., Demongeot S., Tomasella E. and Chambaudet A. (1995) *Radiat. Meas.* **25**, 1-4, 553-556.

Knauss K., (1980) Radon-gas extraction and counting system for analysing radon and radium in groundwater in seismically active areas. *UCRL-53088*, Lawrence Livermore Laboratory.

Kotlyarov A. A. and Krivashcheev S. V. (1994) Measuring chamber of volume activity of radon in surrounding air samples. *Russian Patent 2,008,694,940,228.*

Kotrappa P., Dua S. K., Pimpale N. S., Gupta P. G., Nambi K. S. V., Bhagwat A. M. and Soman S. D. (1982), Passive measurement of radon and thoron using TLD or

SSNTD on electrets. *Health Phys.* **43**, 3, 399-404.

Kotrappa P., Dua S. K., Gupta P. C., Pimpale N. S. and Khan A. H. (1984) Measurements of potential alpha energy concentration of radon and thoron daughters using an electret dosemeter. *Radiat. Prot. Dosim.* **5**, 49-51.

Kotrappa P., Dempsey J. C., Hickey J. R. and Stieff L. R. (1988) An electret passive environmetal ^{222}Rn monitor based on ionization measurement. *Health Phys.* **54**, 1, 47-56.

Kristiansson K. and Malmqvist L. (1980) A new model mechanism for the transportation of radon through the ground. In *S.E.G. Ann. Int. Meeting*, Tech. papers, **5**, 2535-2566.

Kotlyarov A. A. and Krivashcheev S.V. (1994) Measuring chamber of volume activity of radon in surrounding air samples. *Russian Patent 940,228* from *92RU-005,688,921,110.*

Kutelev A. S. (1994) Capacitor type ionization chamber for measuring gaseous type radioactive nuclides. *Russian Patent 940,430* from *92SU-5,044,407,920,527.*

Kutelev A. S., Lapshin V. I. and Shavrin N. Y. (1995) Radiometer for measuring volume activity of airborne gaseous nuclides and radioactive aerosols. *Russian Patent 950,510* from *93SU-0,0303,930,119.*

Kuzhevskii B. M., Maduev V. L. and Nechaev O. Y. (1995) Radon concentration determination and measurement in substances. *Russian Patent 950,520* from *92SU-5,038,222,920,224.*

Leconte Cathey (1978) Continuous radon concentration monitoring. *Rev. Sci. Inst.* **49**, 12.

Leuschner A. H., Rolle R. and Strydom R. (1993) Portable personnal radon daughter dosemeter for use in mines. *British Patent 2,261,106,930,505* from *91ZA-008,601,911,029.*

Liehu A. E., Makelainen I., Valkiainen and Winqvist K. (1982) Calibration of LR-115 and CR-39 for radon and comparative measurements in indoor air in Finland. In *Proc. 11th Int. Conf. on Solid State Nuclear Track Detectors* (eds. P.H. Fowler and V. M. Clapham), Bristol, 1981, Pergamon Press, Oxford, pp. 557-560.

Likes R. S., Mogro-Campero A. and Fleischer R. L. (1979) Moisture insensitive monitoring of Radon. *Nucl. Instr. Meth.* **159**, 395-400.

Linborg L. (1983) Radon residual beta particle detection. *Swedish Patent 004,544,830,822.*

Lucas H. F. (1957) Improved low level alpha-scintillation counter for radon. *Rev. Sci. Instr.* **28**, 9, 680-683.

Lucas H. F. (1990) Apparatus to determine thoron and radon concentration in mixture. *U.S. Patent 4,975,574,901,204.*

Madnick P. A. and Sherwood R. W. (1990) Radon gas detector for continuously monitoring radiation levels. *U.S. Patent 4,894,535,900,116.*

Malmqvist E. L. and Kristiansson K. (1980a) Monitoring radon ground gas. *Swedish Patent 007,338,801,020.*

Malmqvist E. L. and Kristiansson K. (1980b) Measuring radon below ground level. *Swedish Patent 007,337,801,020.*

Marvin H. B. (1955) How to measure radiation with electrets. *Nucleonics* **13**, 82-87.

Matthies D. (1994) Radon concentration measuring instrument. *German Patent 940,908* from *93DE-307,015,930,305.*

Mc Keever S. (1985) *Thermoluminescence of Solids.* Cambridge University Press, Cambridge.

Megumi K. and Mamuro T. (1977) Method of measuring radon and thoron exhalation from the ground. *J. Geophys. Res.* **77**, 17, 3052-3056.

Melvin J. D., Shapiro M. H., Copping N. A. (1978) An automated radon-thoron monitor for eathquakes prediction research. *Nucl. Instr. Meth.* **153**, 239-251.

Mendenhall M. H., Shapiro M. H., Melvin J. D. and Tombrello T. A. (1981) Preliminary spectral analysis of near-real-time radon data. *Geophys. Res. Lett.* **8**, 5, 449-452.

Mertens K. and Spicka M. (1995) Measuring radioactive contamination of component. *German Patent 4,334,320,950,413* from *93DE-334,320,931,008.*

Monnin M. (1980a) Visualisation of latent damage trails, *Nucl. Instr. Meth.* **173**, 1-14.

Monnin M. (1980b) Methods of automatic scanning of SSNTD, *Nucl. Instr. Meth.* **173**, 63-72.

Monnin M. (1990) Physical basis for radon emission and measurements techniques. In *Proc. Int. Workshop on Radon Monitoring in Radioprotection, Environmental Radioactivity and Earth Science* (eds. L. Tommasino *et al.*), Trieste, 1989, World Scientific, Singapore, pp. 199-230.

Monnin M. and Seidel J. L. (1982) An automatic device for Rn activity measurements. In *Proc. 11th Int. Conf. on Solid State Nuclear Track Detectors* (eds. P.H. Fowler and V. M. Clapham), Bristol, 1981, Pergamon Press, Oxford, pp. 513-517.

Monnin M. and Seidel J.L. (1988) Sur une hypothétique émission intense de radon avant un événement géophysique majeur: une analyse théorique. *C.R. Acad. Sci. Paris,* **307**, série II, 1363-1368.

Nakahara H., Kudo H., Akiba F. and Murakami Y. (1980) Some basic studies on the absolute determination of radon concentration in the air by a cellulose nitrate track detector. *Nucl. Instr. Meth.* **171**, 171-179.

Navratil O. and Sury J. (1995) Determining volume activity of radon from soil and atmospheric air. *Czechoslovak (former) Patent 9,302,500,951,018.*

Nindel R., Miersch G. and Raepke S. (1990) Pressure chamber measuring gas exhalation. *German (former Democratic Republic) Patent 900,228.*

Nyberg P. C. (1989) Protocol for indoor radon measurement: A reply. *Health Phys.* **57**, 5, 843.

Pai H. L. (1989a) Radon dosimeter to survey in and outdoor and underground contamination. *U.S. Patent 48,611,985,890,829.*

Pai H. L. (1989b) Active radon daughter dosimeter used in mines. *U.S. Patent 4,864,143, 890,905* from *88US - 274,681,881,121.*

Patel G. (1987) High energy radiation monitor. *U.S. Patent 074,592,870,717.*

Perlman D. (1990) Method and apparatus for radon detection. *European Patent 359,770, 900,328.*

Phelan L. G. and Robert F. (1969) Portable radon sampling apparatus. *Canadian Patent 04,7259,690,329.*

Pohl-Ruling J., Steinhausler F. and Pohl E. (1980) Investigation on the suitability of various materials as ^{222}Rn diffusion barriers. *Health Phys.* **39**, 299-301.

Pörstendorfer J., Schraub A. and Wicke A. (1980) Methods for continous registration of radon, thoron and their decay products indoors and outdoors. *N.R.E. III*, pp. 1293-1307.

Port E. A., Bitner J. M., Lazarus I. P., Liaboe T. W. and Toohey R. E. (1990) Radon measuring device. *U.S. Patent 4,948,970,900,726.*

Port E. A., Liaboe T. W. and Sebesta E. E. (1990) Radon alpha track detector device. *U.S. Patent 4,980,550,901,225.*

Powell B. (1979) Sampler for detecting radon. *British Patent 019,214,790,601.*

Pretzsch G. (1984) Radon detector comprising two electrodes and electret. *German (former Democratic Republic) Patent 265,673,8,407,226.*

Pretzsch G., Borner E., Lehmann R. and Sarenio O. (1985) Radon diffusion chamber with passive integrating detector. *German (former Democratic Republic) Patent 276,690,850,528.*

Pretzsch G., Borner E., Lehmann R. and Sarenio O. (1990) Radon detector with a polymer track detector in a diffusion chamber. *German (former Democratic Republic) Patent 279,594,900,613.*

Ramsey P. M. and Stieff L. R. (1992) Programmable control exposure radon measurement system. *U.S. Patent 920,421.*

Riedel M. (1995) Measurement of individual internal radiation exposure resulting from radon and its progeny inhalation. *German Patent 4,341,523,950,608.*

Scheible J.N. (1987) Radon collection apparatus. *U.S. Patent 091,589,870,831.*

Schiocchetti G. and Scacco F. (1981) Determination of daughter products of radon and thoron in air. *Italian Patent 049,036,810,803.*

Seidel J. L. (1982) Thermoluminescent dosimeter for monitoring subsurface radiation. *U.S. Patent 941,995,780,913.*

Seidel J. L. (1982) Radon emanométrie appliquée à la géophysique interne. *Thesis n°679D, University of Clermont-II* (France).

Seidel J. L. and Monnin M. (1982) Some Radon activity measurements of geophysical significance. In *Proc. 11th Int. Conf. on Solid State Nuclear Track Detectors* (eds. P.H. Fowler and V. M. Clapham), Bristol, 1981, Pergamon Press, Oxford, pp. 517-523.

Sgorbati G. (1989) Gamma spectroscopy of natural radionuclides. In *Proc. Int. Workshop on Radon Monitoring in Radioprotection, Environmental Radioactivity and Earth Science* (eds. L. Tommasino *et al.*), Trieste, 1989, World Scientific, Singapore, pp. 189-198.

Shapiro M. H., Melvin J. D., Copping N. A. and Whitcong J. H. (1980) Automated radon-thoron monitoring for earthquake prediction research. *Nat. Rad. Env. III*, 137-153.

Shepherd R.A. (1980) Portable radiation dosimeter. *U.S. Patent 166,063,800,707.*

Simon W. E., Powers T. L. and Ernsberger G. W. (1987) Inexpensive radon detector. *U.S. Patent 046,933,870,505.*

Singh B. (1992) Earthquake prediction studies in Kangra valley and Punjab. *Thesis, Guru Nanak Dev University*, Amritsar (India).

Simon W. E., Powers T. L. and Ernsberger G. W. (1987) Inexpensive radon detector. *U.S. Patent 046,933,870,505.*

Smith R. C. and Kanter I. E. (1978) Locating underground depositis of uranium. *U.S. Patent 868,779,780,112.*

Somogyi G., Hunyadi I. and Varga Z. (1978) Spark counting of radiograms recorded on stripable cellulose nitrate LR 115 film. *Nucl. Tracks. Detect.* **2**, 4, 191-197.

Somogyi G., Hunyadi I., Hakl J. (1989) Historical review of one decade radon measurements in Hungarian caves performed by SSNTD. In *Proc. 10th Congress of Speleology*, Budapest.

Stewart A. and Kneale G. (1989) Protocol for indoor Rn measurements. *Health Phys.* **57**, 5, 842.

Steck D. J. (1992) Track registration detector for radon concentrations. *U.S. Patent 920,721.*

Stieff L. R. (1992) Gamma radiation compensated radon measurement system. *U.S. Patent 920,707.*

Tanaka F., Maki Y. and Nojiri T. (1982) Automatic monitoring of ^{222}Rn daughter concentration in soil air by an electrode collector. *Nucl. Instr. Meth.* **203**, 587-593.

Tanner A. B. (1964) Radon migration in the ground: A review. *Nat. Rad. Env. I*, 161-268.

Tanner A. B. (1978) Radon migration in the ground: A supplementary review. *Nat. Rad. Env. III*, 5-56.

Terradex Corp. (1976) Detecting radon gas in uranium prospecting. *U.S. Patent 715,225, 760,818.*

Tetley W. C., Cummmings B. A. and Westcott D. R. (1988) Radon progeny detector. *U.S. Patent 184,758,880,422.*

Tetley W. C., Cummmings B. A. and Westcott D. R. (1990) Radon detector for radon daughter products. *U.S. Patent 4,963,730,901,016.*

Thomson I. (1991) Radon detector. *U.S. Patent 4,983,843,910,108.*

Tidjani A., Monnin M. and Seidel J. L. (1990) Enhancement of radon signals in geophysical studies with the track technique. *Pure and Appl. Geophys.* **132**, 3, 505-520.

Tommasino L. (1990) Radon Monitoring by alpha track detection. In *Proc. Int. Workshop on Radon Monitoring in Radioprotection, Environmental Radioactivity and Earth Science* (eds. L. Tommasino *et al.*), Trieste, 1981, World Scientific, Singapore, pp. 123-132.

Urban M. and Kiefer H. (1984) Passive dosimeter with open diffusion chamber. *German Patent 404,301,840,208.*

Urban M. and Piesch E. (1981) Low level environmental radon dosimetry with a passive track etch detector device. *Radiat. Prot. Dosim.* **1**, 2, 97-109.

US Atomic Energy Com. (1968) Personnel dosimeter for radon and daughter products. *U.S. Patent 749,725,680,802.*

US Atomic Energy Com. (1969) Integrating thermoluminescent dosimeter. *U.S. Patent 798,284,690,211.*

US Sec. of Interior (1976) Monitoring Radon gas in air. *U.S. Patent 653,314,760,129.*

US Sec of Interior (1976) Radiation dosimeter for use in mines. *U.S. Patent 670,213,760, 325.*

US Sec. of Navy (1977) Atmospheric radon monitor fir airborne particulate matter. *U.S. Patent 818,709,770,725.*

Venkov V. A., Kuvshinnik E. I. and Sozanska E. E. (1983) Measuring instrument for radioactive emanation in air. *SU-588817 830915* from *S.U. (former) Patent 377,734, 760,628.*

Ward Alter H. and Fleischer R. L. (1981) Passive integrative Radon monitor for environmental monitoring. *Health Phys.* **40**, 693-702.

Wilkening M. H. (1980) Radon transport mechanisms below the earth surface. *Nat. Rad. Env. III*, 90-104.

Wheeler R. V. (1987) Detector for radon gas and its decay products. *U.S. Patent 012,266.*

Wheeler R. V. (1988) Alpha particle radiation detector element holder. *U.S. Patent 218,640,880,713.*

Wilhelm I. (1988) Apparatus determining radon contamination in atmosphere. *Czechoslovak (former) Patent 000,421,880,122.*

Wolfert L. G. (1978) Alpha particle counter for radon gas detector. *U.S. Patent 4,104,523, 780,801.*

Yamamoto K., Fujii Y., Icotooka Y. and Katayama T. (1987) Highly stable silicon pin photodiode. *Nucl. Instr. Meth.* **A253**, 542-547.

Yoder R. C. (1988) Radon gas detector with sensing film. *U.S. Patent 211,516,880,624.*

2.

ETCHED PARTICLE TRACK METHODS

S. A. Durrani[1]

2.1 Alpha-Particle Etched Track Detectors

2.1.1 Background

Today, plastic track detectors provide the commonest method of radon measurement, both in the home and in the field, all over the world. The reasons are easy to understand. These detectors are small (typically 1 cm x 1 cm, $\sim$ 100-500 μm thick); easy to use (requiring no electronics); inexpensive (a sheet yielding around 400 detectors costing perhaps US $40); and simple to process and read (a thermostatic etching bath, simple chemical reagents, and an optical microscope being the minimum requirements). In addition, they are 'integrating devices' - i.e. if the signal recording rate is low, they can be simply left *in situ* for a long enough time to accumulate sufficient counts. Quite often this latter property in fact offers the extra advantage of giving an average reading in case of a signal that varies with time - thus smoothing out any erratic variations.

Track-etch plastic (polymeric) detectors are 'passive devices' in the sense described in subsubsection 1.3.3.1 above (see section 1.3 Radon Measurement Techniques). While this property is useful in some senses, it also constitutes one of the main limitations of such detectors: namely that the results cannot be examined in real time or 'on-line' - without elaborate instrumentation (which would militate against their 'simplicity', etc., extolled above). Another drawback is that counting the etched tracks under a microscope 'manually' (or, put more accurately, by human eye: ocularly) is a tedious and time-consuming matter - although a number of semi-automatic or quasi-automatic methods have been developed for this purpose (see subsection 2.1.3.5 below, and also section 2.3 Track Registration: Etching and Counting Methods for Nuclear Tracks).

As explained in detail elsewhere (see, e.g., subsection 1.1.2 in section 1.1 Radon: Overview of Properties, Origin and Transport), radon (whether ^{222}Rn, or ^{220}Rn, sometimes called 'thoron') decays through a number of short-lived steps by emitting α- and β-particles, ending finally in stable lead isotopes. It is the α-particles emitted by the radon gas as well as by its short-lived solid 'daughter products' (or more correctly 'progeny') that are recorded by the plastic detectors. By counting the α-particle tracks recorded over a given time, and assuming a state of secular equilibrium between ^{222}Rn (half-life $\tau_{1/2}$ = 3.825 d) and all its daughters - or using some measured or assumed value of the *equilibrium factor* (q.v.), viz. ratio of the daughter activity to the parent activity - one is able, in principle, to calculate the radon concentration (in Bq m^{-3}, etc.) in a closed system. Many problems arise if one wishes to determine absolute values - but we defer their consideration to the section on Calibration and Standardisation of Etched Track De-

[1]University of Birmingham, Birmingham, UK

tectors (section 2.4). The contribution from thoron (^{220}Rn) can usually be discriminated against by various means (see subsection 2.1.5 below).

2.1.2 Track Formation Mechanisms in Dielectric Media

A detailed consideration of this topic is beyond the scope of the present monograph, where our main emphasis is on applications. For an in-depth coverage of the theoretical aspects of the Solid State Nuclear Track Detection (SSNTD) technique, the reader is referred to standard texts such as *Nuclear Tracks in Solids* (Fleischer, Price and Walker; University of California Press, Berkeley, 1975) or *Solid State Nuclear Track Detection* (Durrani and Bull; Pergamon Press, Oxford, 1987). For our present purposes it should be sufficient to summarize the main characteristics of the method as follows.

1. Heavy, charged particles (from protons upwards), but not β-particles - nor γ-rays, etc., - can leave latent quasi-continuous trails of damage in dielectric media (but not in most metals or semiconductors), which are relatively long-lived at normal (and sub-zero ^{o}C) temperatures. These 'latent tracks' can survive even at high temperatures, depending on the medium and the intensity of the ionization (i.e. the linear density of displaced electrons): the higher the latter, the more stable (or temperature-resistant) the track or part thereof close to the *'Bragg peak'* (q.v.), which occurs towards the end of the charged particle's trajectory. The latent tracks can only be seen by Transmission Electron Microscopy (TEM), being only some tens of nm in diameter, or less.

2. The latent tracks can be etched with the help of a suitable etchant (often a simple alkali or acid, e.g. NaOH or KOH; HF or HNO_3), thereby not only 'fixing' them - i.e. making them permanent - but also enlarging them sufficiently so that they become visible under an optical microscope (with diameters of a μm, or more).

3. Different dielectrics have different 'registration thresholds', such that only those particles which, in a given part of their trajectory as they slow down, have linear rates of energy loss (dE/dx), or primary ionization rate J, above a 'critical value', or threshold, will be registered in a given dielectric medium (see Fig. 2.3.1 on p. 130). Thus polymers (plastics) have far lower registration thresholds than glasses and mineral crystals, and can therefore often record tracks of such low-Z ions as α-particles (Z being the atomic number). The widely used polymer CR-39 (a polyallyl diglycol carbonate) can record even proton tracks. Crystalline media (minerals), by contrast, can only record high-Z ions (say, with $Z > 10$ - and that only towards the very end of their range in the case of very energetic particles such as cosmic rays). This property is often a useful discriminatory tool in the case of a mixed radiation field. Differential annealing (before the etching process) can also help by preferentially eliminating the latent tracks (or the less stable parts thereof) produced by low-LET particles, i.e. particles with low Linear Energy Transfer rates (see Durrani and Bull, 1987).

2.1.2.1 Charged-particle tracks in polymers

Polymeric plastics are the most widely used nuclear track detectors in all fields other than geology (fission track dating, ancient cosmic-ray tracks in meteorites, etc.) or high-temperature environments (e.g. in nuclear reactor cores). Thus they are much used in radon studies, nuclear physics, accelerator-based research, neutron monitoring, space research, etc. Their simplicity, good geometry because of their small size, and ability to integrate the response over long periods are amongst the reasons for their popularity.

As stated above, plastics are in fact the most sensitive of all known nuclear track detectors - for some of them can register low-charge particles down to protons. The last remark is true of CR-39 and also of many cellulose nitrates - at least for low energy protons; while all the cellulose nitrates can record α-particles (within a certain energy window depending on the etching conditions; see Fig. 2.3.1). Some plastics, however (e.g. polycarbonates such as Lexan, Makrofol), normally have a practical charge threshold of around Z=6 - although they can record α-particles by adopting special techniques, e.g. ultraviolet exposure, addition of ethyl alcohol to the etchant, or use of energy-degrading foils. As stated above, none of the known polymers can register β-particles. (*N.B.* Most of the plastics mentioned above are commercially available, under quality-controlled conditions - see some manufacturers' names and adresses in the footnote below[2]).

The CR-39 plastic can also respond to fast neutrons by recording the (n,p) recoil tracks as well as tracks from (n,α) reactions with the constituents of the polymer itself (H, C and O). Such tracks are sometimes referred to as 'intrinsic' or 'internal' tracks (since the charged particle originates within the detector) as opposed to 'extrinsic' or 'external' tracks from an external source of charged particles. An example of the latter is tracks originating from an external 'converter screen', say lithium borate ($Li_2B_4O_7$), exposed to thermal neutrons, which yield α-particles from the (n,α) reaction on the boron (or on the lithium), which then impinge on the CR-39 detector placed in contact with the converter screen.

As mentioned at the beginning of this subsection (2.1.2), a detailed description of track formation mechanisms in polymers is beyond the scope of the present book; the theoretical framework has, in any case, not yet been fully worked out nor universally agreed upon. Speaking broadly, the generally accepted view is that the passage of heavily charged particles results in the scission of some of the polymeric chains (see Fig. 2.1.1). This also produces highly reactive species such as free radicals, which may be quite long-lived. When the plastic is subsequently treated with an appropriate etchant (e.g. alkali hydroxides, such as aqueous solutions of NaOH or KOH, sometimes with the addition of alcohol), the area of scission is dissolved and the resulting hole can become enlarged by factors of 10^2-10^3. The damage trails thus become quasi-continuous as well

---

[2]CR-39 (Polyallyldiglycol carbonate): (i) Page Mouldings (Pershore) Ltd, Pershore, Worcs., UK; (ii) American Acrylics and Plastics, Stratford, CT, USA; (iii) Tastrak, c/o H H Wills Physics Laboratory, Bristol, UK

Lexan (Bisphenol-A polycarbonate): General Electric Co., Schenectady, NY, USA

Makrofol (Bisphenol-A polycarbonate): BAYER AG, Leverkusen, Germany

LR 115; CN 85 (Cellulose nitrate): Kodak Pathé, Vincenne, France.

as permanently fixed as 'tracks' (so that the application of heat cannot anneal them, unlike the latent trails which can be thermally annealed out). The etched tracks are also, through the enlargement, rendered visible under an ordinary optical microscope. The amount of damage, and hence the etchability and size, etc., of the etched track, depend on the rate of Linear Energy Transfer (LET) of the charged particle over its trajectory. The total energy lost by the particle in the medium also plays a role in determining the size of the 'etch pit' in the detector under any given etching conditions (see section 2.1.5).

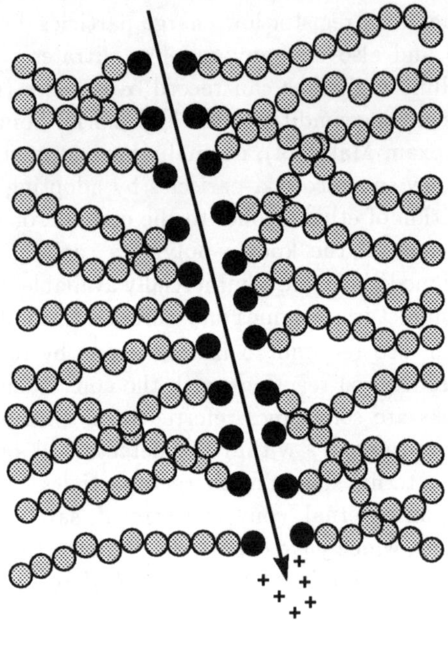

Figure 2.1.1: Schematic diagram of chain scission in polymers caused by the passage of heavily charged particles. A thermal spike along the trajectory of the charged particle causes localized melting; and this, together with excitation caused by the ionization, leads to chain-breaking and production of new chain ends. Chemically reactive sites are formed at the points of scission, which may be enlarged when an etchant such as NaOH enters the damaged region by dissolving the disordered and degraded portions.

2.1.2.2 Etching characteristics of tracks in polymers

The shape of the 'etch pit' and the diameter (or the major/minor axis) of the 'etch-pit opening' depend basically on two parameters: (i) the etching characteristics of the given (plastic) medium, signified by its 'bulk etching velocity' V_B for the etchant used under the given etching conditions such as the type, molarity and temperature of the etchant, and (ii) the characteristics of the charged-particle interaction with the medium, signified by the 'track etching velocity' V_T along the trajectory of the particle. The ratio $V = V_T/V_B$ is an important parameter that is very useful in determining the registration properties of charged particles in dielectric media (both crystalline and polymeric). This ratio also gives us parameters such as 'etching efficiency' and 'critical angle of etching' (θ_c), etc.

The latter concept can be explained by Fig. 2.1.2. Essentially, it is the lower limit for the angle with the detector surface that the tracks can have if they are to be revealed by etching.

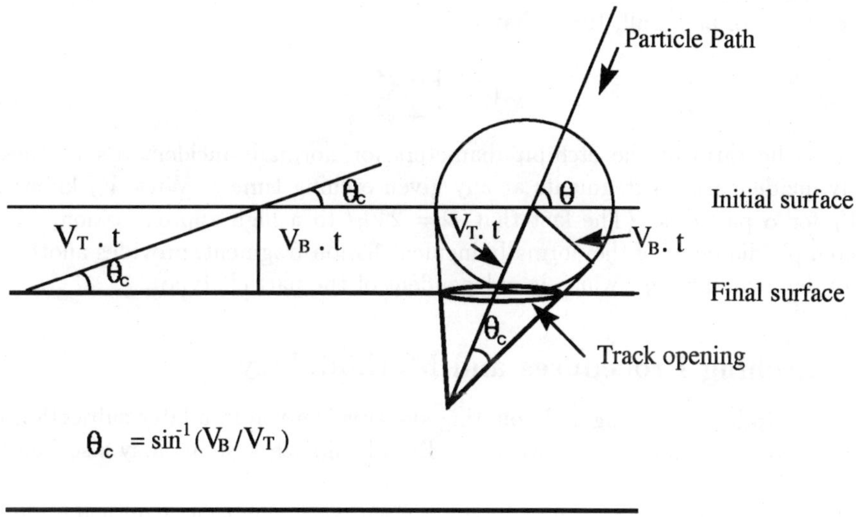

Figure 2.1.2: Critical angle of etching. When an irradiated detector is treated with an appropriate etchant, the velocity of etching along the latent track (V_T) is larger than the bulk velocity of etching (V_B) elsewhere in the medium. In the figure shown, the ratio $V_T/V_B = 3$. There is an angle θ_c for each medium and a given heavy ion such that, by the time that the etchant travels a distance $V_B t$ vertically into the body of the detector, it reaches the end of the range of the particle proceeding along that 'dip angle' θ_c at the same instant - i.e. $V_B t/V_T t = sin\theta_c$. Only tracks making dip angles with the detector surface, such that $\theta > \theta_c$, will thus leave observable track openings. The half-cone angle of all such etch pits is also $\theta_c = \sin^{-1} (V_B/V_T)$.

It can also be shown geometrically (see Durrani and Bull, 1987) that the probability of observation of tracks (i.e. revelation of their etch pits) resulting from a 'thick' source of charged particles incident on a detector placed in contact with it (where 'thick' means of thickness $>R$, the range of the particles in the source material) is given by $cos^2\theta_c$, where $\theta_c = sin^{-1}(V_B/V_T)$ for the given detector. This probability of observation is usually referred to as the 'etching efficiency'.

The value of the track-etch velocity V_T in a given medium is a function of the type and energy of the charged particle. For high-Z (i.e. heavily-charged) particles, such as fission fragments, V_T can be many times V_B (the general velocity of etching in the medium). In such a case, the track has a long, needle-like, quasi-cylindrical shape, with a very small etch-pit opening compared to its length (the diameter D after an etching time t being $\approx 2V_Bt$, and the length $\approx V_Tt$, or the range of the particle, whichever is least). This is especially the case in crystalline media (but also for fission fragments and heavy-ion beams from accelerators incident on plastics). For α-particles, however, $V = V_T/V_B$ is not very high - e.g. in the CR-39 plastic, the value of V may only be 2 to 3. To measure V_B in a plastic is, of course, very easy (e.g. the thickness h of the removed layer from *each* face of the plastic upon etching for time t is $h = V_Bt$); but the measurement of V_T for low-V particles such as α's is not straight-forward. In such a case it can be shown (see, e.g., Durrani and Bull, 1987) that

$$V = \frac{1+x^2}{1-x^2} \tag{2.1.1}$$

where x is the ratio of the etch-pit diameters for normally-incident α's to those for normally-incident fission fragments at any given etching time t. With V_B known, this gives V_T for α-particles. (The fact that $D = 2V_Bt$ to a high approximation, where D is the etch-pit diameter of the normally incident fission fragment, provides another easy means of determining V_B - which is independent of the particle type).

2.1.3 Etching Procedures and Methodology

A fuller description of etching and counting systems is given in a later subsection (2.3), so we shall give here only a brief overview. Details of such systems may also be seen in Durrani and Bull (1987).

Of the etching methods employed for subsequent track-counting by optical microscopy - whether used 'manually' or automatically/semi-automatically - the most common are: (a) chemical etching, and (b) electrochemical etching. As examples of non-microscopic methods, we shall briefly refer only to: (c) spark counting, and (d) scintillator-filled etch-pit counting methods. Only plastic detectors are covered in the following descriptions.

2.1.3.1 Chemical etching

This is the most basic method in SSNTD. Chemical etching (CE) is usually carried out in a thermostatically-controlled bath at temperatures ranging from 40°C to 70°C (exceptionally, up to $\approx 90^o$), and the commonest etchant is an aqueous solution of NaOH (or KOH) at a molarity of from 2 to 6 M (e.g. for 6 M NaOH, 6 x 40 (the gram molecular weight of NaOH) = 240 g of NaOH is made into a litre of aqueous solution, using distilled water). Typical etching times range from 2 to 6 h. As the molarity of the etching reagent, and/or the etching time, and/or the etching temperature, increase, so does the size of the resulting etch pit. In certain applications, ethyl alcohol may be added to the solution in a certain proportion (e.g. a PEW solution, which is made of potassium hydroxide, ethyl alcohol, and water); the alcohol helps to enhance the registration sensitivity of some plastics (polycarbonates) - although it lowers that of others (CR-39). The plastic detectors are normally suspended by means of wires/strings attached to small springs (which can grip several detector foils simultaneously between the coils), etc. Alternatively, a suite of detectors may be suspended vertically in frames, which hold the individual foils at a fixed distance apart. It is important that there is room for the etchant to circulate, and that the currents are not impeded. The detectors are immersed in the etching solution, with or without stirring/rotation, inside beakers (with their tops covered), which are placed in a constant-temperature water bath. At the end of etching, the detectors are removed, washed in running water, and preferably placed in a small ultrasonic bath of distilled water for a few minutes to remove the etching residue from the etch pits. After drying, the detectors are ready to be counted under an optical microscope. The etched track diameters are typically a few μm in size, although they may grow to 50 μm or more after prolonged etching.

2.1.3.2 Electrochemical etching

Here, the purpose of electrochemical etching (ECE) is to further enlarge the etch pits by factors of $\sim 10^2$ - so that they can be counted by means of low-magnification devices (e.g. a microfiche reader, a slide projector, or a low-magnification ($\sim$10x) microscope). ECE tracks are, in fact, easily visible to the naked eye.

Figure 2.1.3: Typical electrochemically etched (ECE) track-spots produced in CR-39 by α-particles from the decay of radon and its short-lived daughters. Some of the ECE spots have failed to develop fully because of local electric field shielding effects.

The basic mechanism is to apply a high (AC) electric field across the plastic detector which acts as a barrier between two parts or compartments of an electrolyte (say an aqueous solution of NaOH) - each compartment containing one 'pole' of the electric field (see Durrani and Bull (1987) for a diagram of an ECE chamber). A typical electric field is 50 kV cm^{-1} (peak-to-peak; $= 50/\sqrt{2}$ rms voltage) at 2 kHz. For a plastic detector, say 200 μm thick, an applied voltage of only 1000 V is sufficient to give a field of 50 kV cm^{-1}. Such a field (especially at high frequencies - e.g. 2-50 kHz) results in an electric breakdown - leading to sparking or 'treeing' - originating at the end-point, or the 'tip', of the etched track in the dielectric medium, i.e. the plastic. This phenomenon is analogous to 'Lichtenberg figures' resulting from electric stress at the tip of an embedded conductor in an insulator when a high electric field is applied (see Fig. 2.1.3).

Electrochemical etching often consists of two or even three sequential steps, viz. (i) a period of CE (which lets a track develop into a needle-like or conical shape), (ii) ECE proper (when 'treeing' takes place), and - possibly - (iii) a post-ECE chemical-etching step (which simply enlarges the ECE 'etch-spots'). The final size of an ECE etch-spot may well be around 100-200 μm. (If the track density is too high, say greater than a few times 10^3 cm^{-2}, the average size of the etch spots becomes small (because of mutual repulsion of electrical charges), and also some of the tracks may totally or partially fail to spark on account of being (partially) 'shielded' from the electric field (see details in Durrani and Bull, 1987).

2.1.3.3 Spark counting

This is one of the (semi-) automatic counting techniques, but it can be only used to count relatively small ($\leq 10^3$ cm^{-2}) track densities. The technique was invented by Cross and Tommasino (1970). A very thin (≈ 20 μm) plastic detector foil, irradiated with α-particles (or heavy ions/fission fragments) is first etched to make 'through holes' along the charged particle tracks. A 'pre-sparking' step may then be interposed to punch out such holes that were originally not quite etched through - by applying an electric field across the thickness of the foil. Finally, the holes are counted by an ingenious spark-counting device. This is done by placing an aluminized Mylar plastic film on top of the etched detector foil and applying a high electric field across the thickness of the foil (say 500 V across 20 μm = 250 kV cm^{-1}). The aluminized layer (with the metal face in contact with the detector foil) provides a conducting path, and short-circuits the field, containing a capacitance. In doing so, the electric current burns off the aluminium directly above the hole, so that the spark cannot go through that particular hole again. The spark thus jumps through all the holes one by one in a random sequence. A scaler incorporated into the circuit automatically counts the sparks as they occur. The efficiency of hole-counting by this method is usually $\sim$95% $\pm$ 5% compared to optical counting. The Mylar film retains a vivid replica of the counted holes. Further details of the system may be found in Durrani and Bull (1987).

2.1.3.4 Scintillator-filled etch-pit counting

This is a semi-automatic method that allows very fast 'counting' of treated etch pits, and was introduced by Harvey and Weeks (1982). The irradiated plastic detector is first etched to produce reasonably sized etch pits (~10-100 μm in diameter by CE or CE+ECE). A quick-drying paste of scintillating material (e.g. ZnS: Ag powder of grain size ≤ 2 μm, made with the help of an automobile anti-freeze fluid, etc.) is first smeared across the etched top-surface of the detector. The scintillator paste gets into the etch pits, and the excess material is wiped off the detector surface. The etch pits are then counted by placing the detector - with the etched surface uppermost - above the end-window of a Photo-Multiplier Tube (PMT) within a light-tight assembly. A strong (strip-form) source of α-particles ($\sim$ 100 μCi) is then placed above the detector (close to, and facing, the etched surface - all within the light-tight assembly). The scintillations produced by the α-particles falling on the paste-filled etch pits go through the body of the detector foil - although the incident α's themselves cannot get through the (>100 μm thick) detector - and are detected, amplified and automatically counted by the PMT and its associated electronics system. Calibration of the scintillation count rate (which can be 10^3-10^5 s^{-1}) against the number of etch pits (determined microscopically), establishing the linearity of the PMT response vs the track (or etch-pit) density, and possibly determining the calibration factor, requires auxiliary experiments. For routine use, however, it is sufficient to calibrate the scintillation count rate against that given by a standard detector under the same processing conditions. *NB* It should be noted that detector foils that have been 'read' in the above manner can no longer be etched any further (e.g. in order to enlarge the etch pits) *for re-use* by this method: for the readout α-source will have formed a huge number of new latent tracks on the top surface of the detector, which would produce a plethora of extraneous etch pits if the detector were re-etched for re-use.

2.1.3.5 Fully automatic track-counting systems

A number of fully automatic track-counting systems are currently available on the market, while others have been developed in-house in various laboratories (Dreute *et al.*, 1986; Fews, 1992; Molnar *et al.*, 1984; Skvarč, 1993). Examples of commercially available systems are Quantimet 900; Magiscan 2; ELBEK Automatic Track Measuring and Counting System (Siegen, Germany), etc., while 'home-made' systems range from DIGITRACK (ATOMKI, Debrecen, Hungary) to TRACOS (J. Stefan Institute, Ljubljana, Slovenia). Such image analysis systems use CCD (Charge-Coupled Device) cameras and recognize tracks on the basis of their grey level and shape.

With the help of appropriate software, these systems can carry out tasks such as measuring spatial density of etched tracks, determining their two-dimensional coordinates, areas, statistical distribution, etc. Because of their speed, such automatic systems are popular for routine work. However, where heuristic and in-depth research is the basic objective, examination of the tracks by human eye, using an optical microscope, still has the edge.

2.1.4 Track Counting Methods and Statistics

In this subsection we shall concentrate on 'manual' (or ocular) counting of tracks, using an optical microscope. Automatic counting of tracks is based on similar considerations, but such systems have built-in criteria for track recognition, overlapping tracks, and background contributions, based on grey-levels, pixels, etc., as well as for error calculation. We shall also ignore details such as methods of measuring the diameters of etch pits (or, in ECE, etch spots), which are more relevant to basic research or track geometry studies, etc.

For radon work, the main requirement generally is simply to count etched tracks on a detector so as to yield statistically reliable values. Only relative (rather than absolute) values are normally required, and the track densities are seldom too high to present problems of significant track overlapping. Etch-pit sizes and shapes will, of course, vary: vertically incident α-particles will form circular etch pits; while the majority of etch pits will be elliptical, resulting from α-particles incident on the detector surface at shallower 'dip angles' (angle of the track with respect to the surface of the detector) - down to the critical angle of etching θ_c, but no shallower. Also, since the α-particle may originate from any emission point in the air up to the full α-range in air (i.e. from 0 to say $\sim$ 4 cm), the final etch-pit size will range from $\sim$ 0 (or, strictly speaking, $\approx$ 0.3 μm, which is the practical resolution limit for an optical microscope) to several μm in diameter (or major axis), depending on the etching conditions. It is usual to choose any reasonable lower size-limit convenient to the observer, and then *consistently* ignore any smaller etch pits. It is also necessary to distinguish between genuine track etch-pits and background 'defects' and artefacts, which are always present on the surface of a plastic sheet and which also become enlarged upon etching. Any scratches or 'string of pearls' type artefacts are easily discounted. The genuine track etch pit (in CE) may be identified by slowly moving the fine focus of the microscope up and down and looking for a bright point of internally reflected light at the bottom tip of the etch-pit cone. A genuine chemically etched pit should also have a well-defined geometric shape (with a circular or elliptical mouth-opening at the detector surface). In electrochemical etching, the tracks form Lichtenberg-like figures, with craggy edges, and shapes that look a little like the petals of a carnation flower (Fig. 2.1.3). It is assumed that a vast majority of the defects, etc. - which lack a sharp needle-like damage-trail penetrating the body of the detector - fail to produce 'treeing' and the resulting characteristic etch-spot shape. The observer chooses an arbitrary lower limit for the ECE etch-spot size, and then consistently employs that criterion for all track counting.

There is always a background of genuine track pits on any detector surface, whether resulting (more rarely) from any radioactive impurities in the plastic or from cosmic rays/fast neutrons incident on it, or (more commonly) from exposure to ambient radon (or radioactive dust) from the time of manufacture to its deliberate exposure to radon for experimental/measurement purposes. For this reason, one should, preferably, use plastic sheets or individual detectors which have a thin, peelable, layer of protective foil on the surfaces. Thus, Pershore Mouldings Ltd (UK) used to supply sheets of CR-39 that had a very thin ($\sim$ 30 μm) top layer of greenish plastic attached to each face of the sheet, which could be peeled off just prior to use. More recently, they (and their successors, Page Mouldings (Pershore) Ltd) supply CR-39 sheets that are carefully enclosed in individual plastic covers and the whole package wrapped in thin aluminium foil and black plastic, etc. This layer (or the protective wrapping) prevents α-particles from radon and other α-emitting impurities in the ambient air from reaching the detector surface and producing an undesirable background. It is also advisable to store the well-wrapped detector sheets in a refrigerator - or, better still, in a freezer - until required for use, so as to minimize undesirable effects of aging, unnecessary exposure to oxygen and light, etc. Some laboratories store the detector sheets in an inert (dry nitrogen) atmosphere, prior to use.

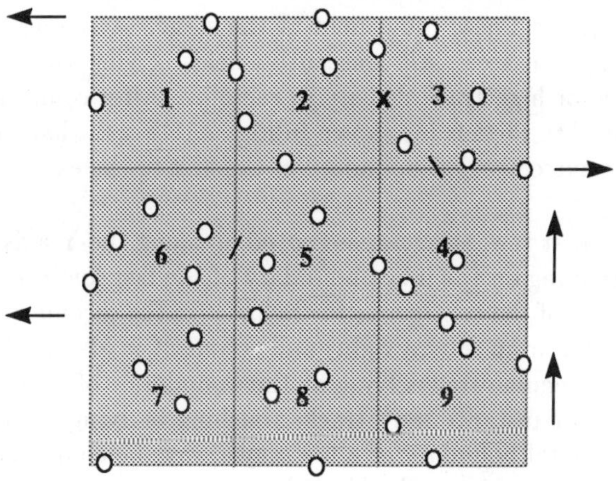

Figure 2.1.4: Systematic scanning by an optical microscope fitted with a movable stage. In this example, the stage is first moved from right to left along the x- axis for counting tracks in the consecutive fields of view (fov's) designated as 1,2,3 ... . After having counted all the tracks in the first fov, the stage is moved in such a way that some fiducial mark on, say, the extreme right-hand boundary (on a graticule visible in the eyepiece) now falls on the extreme left-hand boundary, and so on. At the end of the first line, the stage is moved upwards along the y-axis, again using some fiducial mark to ensure exact coverage of all consecutive fov's; the stage is now moved from left to right to count fields 4,5,6 ...; etc. To ensure that no tracks are missed or counted twice, a convention is adopted, e.g. to count a track if it touches or falls on the left vertical and the bottom horizontal (graticule) boundary of a fov, but to ignore those on the other two boundaries (which would be counted on subsequent viewing of the adjacent fields). Similar conventions apply to automatic scanning. A lower limit to the size of tracks taken to be genuine is set from the beginning by the observer, and smaller etch pits are then consistently ignored. Conventions on overlapping tracks, shape, size and sharpness of a genuine track, etc., are also adopted in both manual (ocular) and automatic scanning.

In any case, it is necessary to count background tracks (in unexposed but etched detectors) on sufficiently large areas of detectors to give good statistics. Background track densities of 10-40 tracks cm^{-2} are common. If genuine (or, more strictly, total) track densities are, say, 1000 tracks cm^{-2}, then the background tracks would contribute (assuming that one has counted a total of 1000 tracks as well as 40 background tracks) a negligible error. [$(1000 \pm \sqrt{1000})$ - $(40 \pm \sqrt{40}) = 960 \pm \sqrt{1040}$, i.e. a total error of 3.36% *vs* an error of 3.16% in the total tracks and $\sim$ 16% in the background tracks]. If the genuine track density is comparable to the background track density, then, of course, it becomes necessary to count the background more thoroughly, i.e. over large areas, to ensure good statistics.

It is assumed that a moving-stage research microscope is available to the experimentalists. In such a case it is necessary, at the outset, to adopt an unbiased routine for track counting. For example, a linear movement of the stage, say along the X-axis, may be adopted, over the exposed area. *All* consecutive fields of view (fov's) are then counted, regardless of nil, low, or high track density in each fov. After counting the tracks in one fov, the stage is moved to the left (as shown in Fig. 2.1.4) such that some fiducial feature moves from, say, the extreme right boundary of the first to the extreme left of the second fov (see Fig. 2.1.4).

At the end of the first row, the stage is moved up along the Y-axis in such a way that some fiducial feature lying on the lower boundary of the last fov in the first row now lies on the upper boundary of the contiguous fov below it in the second row; the stage is then moved along the X-axis from left ro right, and so on. It is also normal to decide, from the beginning, which genuine tracks lying at the boundaries of a fov are to be counted, and which ignored. For instance, any tracks touching or lying astride the left vertical and the bottom horizontal boundary of a fov may always be counted, while those on the other two boundaries always ignored. These latter tracks would then be counted in the horizontally and vertically adjacent fov's, respectively.

Poisson statistics are assumed to apply to track counting - in common with other nuclear events (where, of a large number of atoms available, only a very small fraction undergo some observable reaction). In these statistics, the error is given by the square root of the *events actually counted.* An elementary but extremely important factor here - which is sometimes overlooked - is that the error is given by the *actual* events counted and *not* by any calculated frequency or density. For example, if one counts just 400 tracks, the standard deviation error is 5% (i.e. $400 \pm \sqrt{400}$), in whatever way one expresses the result. If, for instance, the number of counted tracks (N=400) was in a field of view which was A=10^{-3} cm^2 in area, the track density ρ (tracks cm^{-2}) is given by

$$\rho = \frac{N \pm \sqrt{N}}{A} = \frac{400 \pm 20}{10^{-3}} = (4 \cdot 10^5) \pm 5\% \text{ (tracks cm}^{-2}) \tag{2.1.2}$$

i.e. $4 \cdot 10^5 \pm 2 \cdot 10^4$tracks cm^{-2}.

It is absolutely wrong to express the value as

$$\left(\frac{400}{10^{-3}} = 4 \cdot 10^5\right) \pm \sqrt{4 \cdot 10^5}(\text{tracks cm}^{-2}) \tag{2.1.3}$$

i.e. $4 \cdot 10^5 \pm 632$ tracks cm^{-2} which would purport to contain a $(\sqrt{4 \cdot 10^5}/4 \cdot 10^5) \cdot 100\% = \pm 0.16\%$ error instead of the $\pm$ 5% error. A few hundred tracks, counted cumulatively over a number of fov's, are usually considered to give reasonable statistics.

Incidentally, the absolute area of the fov corresponding to any magnification on the objective lens of the microscope is determined by using a graticule provided by the manufacturer. The magnification provided by the eyepiece lens on the microscope contributes to the total magnification of the 'virtual image', but does not improve the 'resolution', which is determined by the objective lens. For producing a real image, e.g. to make a photomicrograph through a camera attachment, or to project the image on a TV screen, a modified 'eyepiece' or projection lens is used. High-magnification (up to $\sim$ 163 x) objective lenses that are dry (i.e. do not require immersion in oil on the detector surface) are now available on the market - although they are quite expensive.

To avoid excessive overlap of tracks, and also to make visual counting (whether directly by eye, or on a TV screen, or on a microfiche projection) not too stressful - or even well-nigh impossible - it is necessary to ensure, during exposure or irradiation, that the track density (estimated *before* the exposure/irradiation) is not higher than a few times 10^4 cm^{-2}. Under normal etching conditions a density of 10^6 cm^{-2} becomes virtually impossible to count, and even 10^5 cm^{-2} may be uncomfortable. Too low a track density also makes track counting both tedious (since a large number of fov's become necessary to count for good statistics) – unless automatic track counting equipment is used – and imprecise, since the genuine track density begins to approach the background value. In the case of electrochemical etching, a further consideration to bear in mind is that, if the etch-spot density exceeds values of $\sim 10^3$ cm^{-2}, partial 'shadowing' by the field of nearby sparks suppresses the development of some of the potential etch spots (Al-Najjar *et al.*, 1979). Not only do all of the eventual spots thus become undersized, but the actual numbers of etch spots (which can only result from successful sparking) fall as well. Hence the value of the counted track density in a high-density exposure would appear to be (proportionately) smaller than it should be compared to a low-density exposure.

2.1.5 Energy Dependence of Alpha-Particle Tracks in Polymers

Alpha particles emitted in radioactive decay have, at the point of emission, a precise energy and hence a fairly precise range in a given medium (except for a very small amount of 'straggling' resulting from the statistical nature of the ionization process during their trajectory). The etching process, however, introduces a statistical variation in the size of the etch pit (say the diameter of the etch-pit opening in the case of a vertically incident α-particle). This results in the production of a histogram (for both chemical and electrochemical etching) when the etch-pit diameter is plotted against the observed frequency, such that the 'resolution' ($\Delta D/D$, where ΔD is the full width at half maximum) is often 10% or worse - even for a nominally fixed energy of α-particles which are incident at a fixed angle. It is also observed that, in general, the mean diameter of the etch pits for *fully etched tracks* increases with the energy of the incident particle under the same etching conditions. If, however, the etching covers only a short part of the particle's

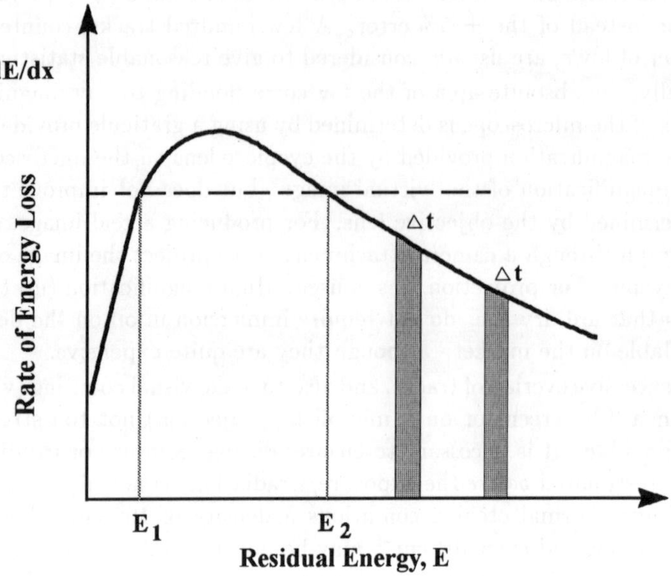

Figure 2.1.5: A 'Bragg curve' showing the linear rate of energy loss (dE/dx) as a function of residual energy (E) of a heavy ion or an alpha particle (the ion is travelling from the right to the left along the X- axis as it loses energy). At high energies, dE/dx is small (because the ion has less time per unit distance traversed to interact with and ionize the atoms in the medium); it then goes through a peak; after which a decline sets in as the effective charge, Z_{eff}, of the heavy ion gradually decreases through the capture of the medium's electrons. This behaviour roughly represents also the registration (or track production) properties of the charged particle - although a more accurate indicator is the rate of primary ionization, J: such that tracks are registered in a medium only when the value of J exceeds a critical threshold, J_{crit}.
This curve can also be used to give a phenomenological or qualitative indication of the dependence of the diameter of a track on the etching procedure and the energy of a particle. Thus, if the etching is carried out for short, but equal, intervals of time Δt - for particles at different residual energies up to the Bragg peak - then the area under the curve corresponding to that time interval is *smaller* at high energies than at lower energies, so that the track diameter (or etch pit opening) mirrors that behaviour. If, however, the etching is done, in all cases, to the end of the track (where $dE/dx = 0$) - i.e. for different lengths of time - the total area under the curve is *greater* in the case of high-energy particles than when they have slowed down to low residual energies.
The 'degeneracy' mentioned in the text can be seen corresponding to residual energies E_1 and E_2, both of which have the same dE/dx value. However, in the case of E_2 the residual range is much larger than for E_1. This difference can be recognized by prolonged etching, and exploited for energy spectrometry.

range, a high-energy particle will generally produce a smaller pit than its counterpart which had slowed down before entering the detector surface (for the same duration of etching). This can be understood by looking at Fig. 2.1.5 - where the dE/dx values (as also the qualitatively similar ionization rates, J) are low at high energies (generally - except at the very end of a heavy ion's range) and are higher at low energies. Thus, when tracks are *fully etched* to the end of the respective range, the energy deposition in the medium will correspond to the total area under the curve (i.e. $\sim \int(dE/dx) \cdot dx$) in Fig. 2.1.5. - which would be larger for a high-energy particle than for a low-energy one. If, however, a short etching time, Δt, is used in all cases, the area under the curve at the high-energy side would be smaller than on the low-energy side for the *same (short) etching time* (cf. Fig. 2.1.5). We make the reasonable assumption that the track diameter (or the etch-pit opening) is proportional to, or qualitatively represents, the total energy deposited in the medium over the etched portion. Wong and Tommasino (1982) have reported an inverse relation between the mean diameter and the α-energy in the case of ECE, based on the electrochemical etching of tracks produced in a polycarbonate detector by a mixed source of α-particles (^{239}Pu with E_α = 5.16 MeV; ^{241}Am: 5.49 MeV; and ^{244}Cm: 5.81 MeV). All these α-particles (incident in 2π-geometry) had been degraded in their energy to 2.6-3.6 MeV by placing a 20 μm thick PC foil in their path. The reason for the inverse relation given by the authors was that 'since tracks of particles of lower energy will be etched first [which presumably means: at a greater etch rate V_T], they will develop a greater diameter'. In any case, all three peaks of the histogram had wide spreads (with $\Delta D/D \sim$ 30 to 50%), and there were large overlaps - partly because of differences in the angles of incidence of the α-particles - between the front and back 'tails', of the histogram peak. It is therefore difficult to assign a diameter, falling in the region of overlap, to one or the other peak (and hence to the corresponding α-energy). It may, however, be of interest to point out that some ingenious ways do exist of overcoming this 'degeneracy' caused by the fact that there are (an infinite set of) two equal values of dE/dx on either side of the Bragg peak (leading to equal values of the etch-pit diameter for short etching times, Δt); these, however, correspond to different values of the α-particle residual range in each case (see Fig. 2.1.5). Continued etching can, thus, show whether a given diameter in the histogram corresponds to the high- or the low-energy side of the Bragg peak (for details see Abdel-Naby *et al.*, 1986). However, so far as routine measurements are concerned, dependence on the diameters to distinguish between α-particles of different energies (unless these energies are far apart, and the α-beams have been collimated) is not a reliable or a viable proposition.

A further, and major, complication in discriminating between tracks of different α-particles is introduced in radon measurements. Suppose the range of α-particles (from radon or one of its progeny) in air is R. Then α's originating at all distances 0-R from the surface of the detector, possessing all (residual) energies from 0 to E, where E is the full energy of the α-particle at the point of emission, will be able to reach the detector. They will register etchable tracks at dip angles ranging from 90° to θ_c with the detector surface, where θ_c is the 'critical angle of etching' (subsection 2.1.2.2). This will also mean that the size of the diameter (or major axis) of the etch pits will vary from ~0 to a maximum value dependent on the total energy of the α-particle and the etching conditions. Moreover, one has to take into account not only the α-particle accompanying the decay of the ^{222}Rn itself (E_α = 5.49 MeV), but also the α-particles emitted by its short-lived (solid) daughter products, viz. ^{218}Po (E_α = 6.00 MeV) and ^{214}Po (7.69 MeV). In addition, if there is a sufficient concentration of 'thoron' (a descendant of ^{232}Th, i.e. ^{220}Rn, $\tau_{1/2}$ = 55.6 s) present in the environment, then one has also to take into account the α's emitted by ^{220}Rn (E_α = 6.29 MeV) as well as by its short-lived (solid) decay products: ^{216}Po (6.78 MeV), and ^{212}Bi (36%: 6.05 MeV) plus ^{212}Po (64%: 8.78 MeV).

Thus, there is a whole spectrum of α-particles shooting around, of which only those issuing from the direct decay of ^{222}Rn itself have an energy below 6 MeV (viz. 5.49 MeV), while all the others - whether from ^{220}Rn or any of the progeny of radon (^{222}Rn) and thoron (^{220}Rn) - have energies (equal to or) above 6.00 MeV. Full α-energy from any of the solid daughters of radon or thoron is received by the detector (whether an SSNTD polymer or a semiconductor detector), in particular, from the "plate-out" component - i.e. the solid decay products that get deposited on the surface of the detector. (Thus a semiconductor detector feeding its signal to a multichannel analyser will show distinct, prominent peaks corresponding to the full-energy α-particles from the active adsorbed layer deposited on its top face and corresponding to each of the solid radioactive decay products.)

2.1.5.1 Energy spectrometry and discrimination

From the above description, it is clear that a number of α-particles are involved in the serial decay of the radon isotopes (^{222}Rn: 5.49; 6.00; 7.69 MeV; ^{220}Rn: 6.29; 6.78; 6.05/8.78 MeV) - even if one completely ignores any contribution from ^{219}Rn (a descendant of the minor uranium isotope ^{235}U (natural abundance, 0.7%), which has a very short half-life: 3.96 s). As explained in the preceding subsection, the α-energy received at the detector surface from the surrounding air will range from E_{max} to 0 for each type of α-particle. Discrimination between the various sources of the α-particles reaching the detector is, therefore, a very difficult task. In the case of a semiconductor detector, discrimination between α-energies can be easily achieved by the use of a multichannel analyser, where different energies peak at different channels. However, even there, a thick-source geometry in conjunction with a multiplicity of α-sources will result in a good deal of overlap between the spectra. Distinct peaks will be seen corresponding to the full-energy α's issuing from the various plated out solid daughter products of radon (and thoron), together with overlapping 'tails' of the energy-degraded α's arriving from various distances in the surrounding air. Only the highest-energy α's (say 7.69 MeV from ^{214}Po out of ^{222}Rn, or 8.78 MeV from ^{212}Po out of ^{220}Rn) will show an unadulterated peak corresponding to the plated-out component.

The only way that a passive detector, such as an SSNTD sheet, can reliably discriminate between α-particles of different discrete energies (all incident at the same angle, say vertically) is by interposing degrader foils of various thicknesses. These foils may be metallic (e.g. thin Al), but are often plastic (e.g. a polycarbonate such as Makrofol KG) of well-controlled thicknesses. The thickness to be interposed is chosen by working out the range of α-particles in the degrader material (and several layers of a thin plastic may be used additively). For example, if one wants to discriminate between α-particles of 6.00 MeV (and lower energies) and 7.69 MeV, one chooses a degrader foil of thickness (slightly) greater than $R_{\alpha,1}$ but smaller than $R_{\alpha,2}$, where the two ranges correspond to the two alpha energies ($E_{\alpha,1} = 6.00$ MeV, and $E_{\alpha,2} = 7.69$ MeV). It must be remembered that only those α-particles of initial energy 7.69 MeV will get through which possess residual energies of more than 6.00 MeV at the time of arrival at the degrader surface after having traversed a certain distance in the air. A series of degrader foils of different total thickness (or a wedge-shaped degrader assembly) may need to be employed if a number of α-energies need to be discriminated. (In using active, e.g. semiconductor, detection devices, the same principle is employed where a series of filters of different thicknesses are used to discriminate between the various radon and thoron daughters). It must, however, be remembered that, analogously to a 'discriminator bias' in an electronic set-up, one cuts off the lower parts of even those signals that have peak energies above the 'threshold' - so that even those α-particles that get through the degrader will have only short ranges in the detector underneath, corresponding to the residual energy remaining after the traversal of the degrading foil.

A second basis for the discrimination of different sources of α-emission and their progeny is their half-lives. The half-life of radon (^{222}Rn), viz 3.825 d, is very much longer ($\sim$ 6000 x) than the half-life of thoron (^{220}Rn), viz. 55.6 s. Hence, if the two gases proceed from the same point by diffusion in the air, the introduction of filters or other constrictions - or, alternatively, sufficiently long pathways between the source and the detector - would ensure that most of the thoron has decayed by the time the gases reach the detection point. This is the approach adopted in, e.g., radon measurements in ground soil, when filters are employed to cover the mouths of inverted collector cups containing CR-39 detectors attached to the underside of their 'tops', or radon measurements inside dwellings, when plastic detectors are stationed at varying heights above the ground (which is the main source of the exhalation of radon and thoron): at heights of greater than $\sim$ 25 cm, a vast majority of the surviving radon is ^{222}Rn. (With a diffusion coefficient D of ~ 0.1 cm^2 s^{-1} in air for all radon isotopes, the "diffusion length" $(L = \sqrt{D\tau})$, where τ is the *mean life in seconds*, is found to be $\sim$ 218 cm for ^{222}Rn and 2.8 cm for ^{220}Rn. The intensity of the gas falls by a factor of e after each distance L).

The less reliable and more problematic method of energy discrimination by etch pit diameter measurements has been discussed above, and need not be repeated here. It may be possible, in an idealized case, where α-energies are discrete and well-separated (without overlapping 'tails' resulting from a thick-source geometry), to discriminate between different α-particles by simultaneously measuring (i) the surface track-size parameters (area of etch pit, minor axis), and (ii) 'grey level' (or average brightness) corresponding to a single track, using a microcomputer-controlled image analysis system. Such 'nuclear spectroscopy' on a CR-39 detector, irradiated by α-particles of three different energies (3.0, 3.8, and 5.1 MeV), has been reported by Ilić *et al.* (1993).

A radically different approach to separating α-particles of widely differing energies is to employ such SSNT detectors, simultaneously or in succession, that have different registration properties. For example, cellulose nitrate (CN) plastic detectors - of which commercial varieties are available as CN 85 and LR 115 (manufactured by Kodak Pathé, France) - have an upper limit to the registration of α-particles at energies of roughly 4 to 6 MeV (depending on the etching conditions); whereas CR-39 has, for all practical purposes, no such upper limit. Thus, if CR-39 is exposed to ^{222}Rn and its short-lived daughter products, it will register all three α-particles (5.49 MeV from ^{222}Rn; 6.00 MeV from ^{218}Po; and 7.69 MeV from ^{214}Po). LR 115 type II, with a sensitive layer of 12 μm - or CN 85, if etched for a short time - will, on the other hand, only register the ^{222}Rn α, but fail to register the full-energy α's from the decay of radon daughters ^{218}Po and ^{214}Po settling on the surface of the detector as a plate-out. (These latter (CN) detectors will, however, record α's from these two daughter products if the particles have been sufficiently degraded in energy by the time they reach the detector surface after travelling some distance through the air). It must, of course, be remembered that if one continues to etch into the body of the CN 85 detector one would eventually reach a point - after having eaten through a sufficiently thick 'removed layer' - when the residual energy of the incident α-particle falls below the etchability ceiling for the plastic, and it does, thus, begin to be revealed.

A comparison of the track densities on the CR-39 and the CN detectors under standard etching conditions can, thus, in principle, indicate the contribution of the radon daughter-product α's. Since the critical angle of etching θ_c (and hence the etching or revelation efficiency $cos^2\theta_c$) depends on the energy of the α-particles and also on the etching conditions; and also taking into account the fact that there is a whole spectrum of α-particle energies involved in the decay of radon and its short-lived daughters, coupled with the thick- source geometry: therefore only broad values for θ_c can be quoted for CR-39 and for the cellulose nitrate detectors such as CN 85. In all cases, the values of θ_c range from $\sim 10^o$ to 30° (Tidjani, 1990; Zamani *et al.*, 1986). The corresponding values of $\cos^2\theta_c$ are: 0.97 (for $\theta_c = 10^o$); 0.88 ($\theta_c = 20^o$); and 0.75 ($\theta_c = 30^o$). The LR 115 (cellulose nitrate) detector has, in addition, certain etching limitations since its (red) sensitive layer is only $\sim$12 μm thick ('type II' film) and $\sim$ 6 μm ('type I' film).

2.1.5.2 Calculation of radon dose from track density measurements

In radon dosimetry, it is usual to calculate the radon activity concentration in the surrounding air (in terms of Bq m^{-3}), since most regulatory reference levels are specified in these units. If it is required to estimate doses, a conversion factor may be used, such as that given by ICRP (1993; Publication 65), which yields an effective dose of 3 mSv per year for a mean radon concentration in dwellings of 200 Bq m^{-3}.

Most dosimeter devices, made available, say, by the UK National Radiological Protection Board (or the Karlsruhe Nuclear Research Centre in Germany), simply let the surrounding air diffuse slowly into a small hemispheroidal chamber (diameter 5 cm (or 3 cm); maximum height of dome 2 cm (or 1.8 cm), respectively). The slow diffusion through a filter paper or through the gaps of a snap-fitting lid prevents the entry of the solid decay-products of radon and thoron, and reduces or eliminates the ingress of thoron gas. The radon in the chamber decays through its progeny, which rapidly plate out on the detector surface or the walls of the dosimeter. These dosimeters are usually calibrated by the issuing authority with the help of a radon chamber containing a known absolute concentration of ^{222}Rn. Thus the observed track density on the (CR-39) plastic within an experimental dosimeter can be converted into the activity concentration of ^{222}Rn in Bq m^{-3} in the air, knowing the duration of the experimental exposure.

Before concluding this section, we wish to draw the reader's attention to some model calculations given in the Appendix: "Some model calculations concerning radon exposure of a detector".

2.1.6 Concluding Remarks

Although radon can be separated from thoron, it will be seen from the preceding subsections that it is quite difficult to separate the individual contributions to the recorded track density made by each of the different members of the radon (^{222}Rn) and thoron (^{220}Rn) decay series. This is because the 'thick source' geometry applies, such that the α-particles arriving at the surface of the detectors from points in the surrounding air located from 0 up to R, the range of a given α-particle in air, posses energies from 0 up to E_{max}, the full energy of the α-particle. Thus the total track length, or the diameter/major axis of the etch-pit opening, is no sure indicator of the source of any given α-particle. Methods such as use of degrader foils or filters can be employed, but only to eliminate tracks by α-particles below a certain energy - in other words, only the energies above a given ceiling and up to E_{max} will enable an α-particle to get through the degrader/filter to register a track (with characteristics corresponding to the residual energy of the α after it has traversed the degrader/filter).

Differing types of track detectors (e.g. LR 115, CR-39) may be utilized with different values of the upper threshold (or ceiling) to the α-particle registration energy. Thus cellulose detector would, under normal etching conditions, fail to register any α-particles with energy above $\sim$ 4-6 MeV, and so offer a means of discrimination. Another approach is to use large differences in half-life, e.g. $\tau_{1/2}$ = 3.825 d and 55.6 s, respectively, for ^{222}Rn and ^{220}Rn ('thoron'), as a means of stopping most of the thoron from getting into a can or other dosimeter, by introducing a filter or some other delaying mechanism in the path of a radon plus thoron mixture: if it takes more than a few minutes for thoron to diffuse through such a barrier, most of it will have decayed in the meantime. Distance (in air) from the source may also, by itself, be sufficient to achieve the same end: for, after diffusing over a distance of 15-20 cm in air, most of the ^{220}Rn would be dead.

In general, in most dosimetry work, calibration against a known, standard source is the only practical way of measuring the absolute activity level of radon. Thus the UK National Radiological Protection Board (and corresponding regulatory bodies in many other industrialized countries) would provide calibration for their own dosimeters - or perform a similar service for their customers - in terms of tracks cm^{-2} vs exposure (Bq m^{-3} h) to radon and all its daughter products for normal values of the equilibrium factor, *F*. It is necessary to allow sufficiently long exposure of the track detector to achieve good statistics in track counting. *If* the steady-state level of ^{222}Rn activity concentration is being built up from $\approx$ 0, through the decay of its parent ^{226}Ra, *and* the track detector is *in situ* from the beginning, a (negative) correction of $\approx \tau$ (= 5.5 d, the mean life of ^{222}Rn) needs to be made to the actual exposure time in order to equate it to the 'effective' exposure time at the steady-state level throughout the period of exposure (or measurement) - see Appendix.

Appendix: Some model calculations concerning radon exposure of a detector

Here, we shall derive *approximate* values for the track density accumulated by a track detector exposed to radon gas in the ambient air; for a more thorough treatment, see Fleischer (1984). Suppose the 'active thickness' of air from which ^{222}Rn α-particle tracks can be received by a detector (say CR-39) is R cm (the range of 5.49 MeV α's in air). Then the volume of the active cylinder of air, upon a base of 1 cm^2, will be $V = R$ cm^3 = R$\cdot 10^{-6}$ m^3 (see Fig. 2.1.6).

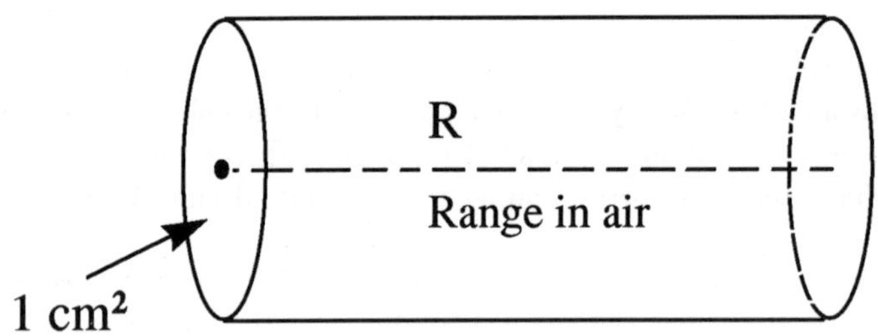

Figure 2.1.6: 'Thick-source' geometry of an α-emitter. It is assumed that the source (say, radon) is uniformly distributed in the air, in which the α-range is R(cm). Then the maximum distance from which an emitted α can reach the detector (taken to be 1 cm^2 in area) is R cm. The energy of the α-particles arriving at the base of the cylinder (of height R) will then vary from E_{max} (the full energy), in the case of particles contiguous to the detector at the base, to 0 for α's emitted at the top of the cylinder of air. Of the α-particles emitted in the decay of the source-nuclei contained in the cylinder, only 1/4 will reach the detector. The factor 1/4 comes from integrating the detectable emission over the thick-source geometry, and is, strictly speaking, the value obtained when a hemisphere of radius R is considered, where the 'height' of detectable α-particles is governed by the angle subtended on the detector (see details in Durrani and Bull, 1987, p. 66), and only 1/2 of all the emitted α's travel in the direction of the detector. A minimum angle with the detector surface (θ_c) also needs to be taken into account for the revelation of tracks by etching.
Many of the short-lived daughter products of ^{222}Rn will end up by being plated out on the surfaces of the actual can or dosimeter as well as on the detector itself. For our approximate calculations, these solid products are also assumed to remain suspended in the cylinder of air.

We assume the detector to be at the bottom of a can which has dimensions larger than R; for a smaller dosimeter, a proportionately rather larger track density will obtain than that calculated below. Also, strictly speaking, one should consider a hemisphere of radius 4 cm from which the tracks are received, rather than a cylinder as assumed by us.

Since thick-source geometry applies, therefore only 1/4 (rather than 1/2, as expected in the 'forward' hemisphere) of the α-particles emitted in all directions by the decay of the radon atoms in the active volume (R$\cdot 10^{-6}$ m^3) will be received by the detector (of which a fraction $\cos^2\theta_c$ will be revealed by etching, θ_c being the critical angle of etching for the 5.49 MeV α's).

Suppose also that the value of radon activity concentration in the ambient air is 50 Bq m^{-3} (roughly the average in the US homes). Then the decay rate per second (and hence the α-particles emitted per second) is 50 x R$\cdot 10^{-6}$. If, for convenience, we take R = 4 cm in air, the active volume of the cylinder of air will be $4 \cdot 10^{-6}$ m^3, giving an α-activity of $200 \cdot 10^{-6}$ Bq, of which 1/4, i.e. $50 \cdot 10^{-6}$ α's per second, will reach the detector base of 1 cm^2. The corresponding etched track density will therefore be $50 \cdot 10^{-6} \cdot 0.93$ cm^{-2} s^{-1} (assuming a mean value of $\theta_c = 15^o$ for CR-39 for the α-particles, so that $cos^2\theta_c = 0.93$). If the detector is exposed for 3 months (90 d = $7.776 \cdot 10^6$ s), the etched track density will be = 362 tracks cm^{-2}. These are tracks from the decay of ^{222}Rn itself. In addition, if the daughter products, ^{218}Po and ^{214}Po, are in full secular equilibrium (i.e. $F = 1$), then one would get 2 further α's from these daughter products per radon decay, leading to a track density of 1,085 tracks cm^{-2}; or if $F = 0.5$, then one would get only 1 α from the daughter products per ^{222}Rn decay - leading to 723 tracks cm^{-2} over 90 days from a ^{222}Rn activity level of 50 Bq m^{-3}. (This corresponds to a track density of 0.241 tracks cm^{-2} per (Bq d) of exposure to radon and its daughters at $F = 1$ and $cos^2\theta_c = 0.93$). So the track densities expected from average Rn concentration levels in the USA homes, resulting from a 3-month exposure, are not very high - i.e. around 720 to 1,100 tracks cm^{-2} (or around 290 to 430 tracks cm^{-2} corresponding to the average Rn concentration of 20 Bq m^{-3} in the UK homes). Counting is usually done under 25 or 40 x objective magnification (or, in special cases of very high track densities, at 63 x). If, under 25x, one field of view (fov) is $2 \cdot 10^{-3}$ cm^2, then each fov will, on average, only have at most $(1.1 \cdot 10^3) \cdot (2 \cdot 10^{-3}) = 2.2$ tracks (corresponding to 50 Bq m^{-3} for 90 d at $F = 1$), or as few as $290 \cdot (2 \cdot 10^{-3}) = 0.58$ track per field of view (corresponding to 20 Bq m^{-3} for 90 d at $F = 0.5$). A large number of fields of view will, thus, need to be counted to accumulate reliable statistics (normally 400 $\pm$ 20 tracks are regarded as acceptable, yielding a $\pm$ 5% accuracy).

It should be pointed out that the above are approximative calculations. For example, the solid, short-lived α-emitting daughter-products of ^{222}Rn (viz. ^{218}Po and ^{214}Po) will largely plate out on the walls of the can and the surface of the detector. However, since the products result from the radon in the air - and while some wall surfaces may be at distances of $>R$ from the detector and so fail to contribute tracks to the detector, whereas those plated out on the detector itself will contribute tracks *in toto* - we may, as a first approximation, consider all the progeny to remain suspended within the sensitive volume of air (i.e. in the 4 cm high cylinder on a 1 cm^2 base).

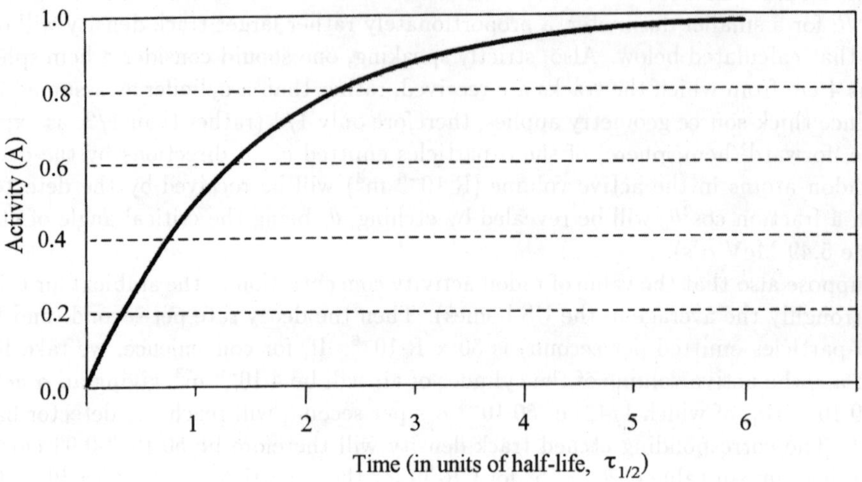

Figure 2.1.7: Growth of activity (A) of a radiogenic nuclide with time (t). The Activity follows the well-known equation $A = A_o\ (1 - e^{-\lambda t})$, where λ is the decay constant of the nuclide concerned, and A_o is the final value of the activity when $t \to \infty$ or is $>> 1/\lambda$. The figure corresponds to the case when the initial activity of the nuclide in question is 0 and steadily grows, in a closed system, through the decay of its parent (which is assumed to be abundant and long-lived: e.g. ^{226}Ra with a half-life of 1,600 yr, cf. the half-life ($\tau_{1/2}$) of 3.825 d for the daughter ^{222}Rn).
The time is plotted in units of $\tau_{1/2}$ of the *daughter*; such that, for instance, after one half-life, $A = 1/2\ A_o$; after two half-lives $A = 0.75\ A_o$, etc. It is clear that if the starting activity is o, then in calculating the exposure to ^{222}Rn one cannot assume the activity to be the steady-state value (A_o) from the beginning: an allowance must be made for the initial period when the activity of the daughter (i.e. ^{222}Rn) was gradually rising to its steady-state value A_o.

A point of some interest, when radon is being accumulated from a nominally zero level to a steady-state level, e.g. the first time a steady-state Rn level is built up in a chamber using a ^{222}Rn source (e.g. pitchblende or a radium salt), is as follows. (It should be noted that the treatment does *not* apply when radon production has already reached a steady state - e.g. in a room or in a can buried in the soil. In such a case, the diffusion or transport of radon, from a reservoir at a steady-state activity level, into the dosimeter or the can is involved from the outset.) Until the steady-state is attained, the radon activity level slowly rises from 0 to an equilibrium value - when the decay and the build-up rates of ^{222}Rn balance each other. If, therefore, a detector is exposed to radon from the very beginning, it is not correct to assume that the detector has been collecting tracks (or receiving α-particles if it is a surface-barrier detector) corresponding to the decay-rate at the steady-state value throughout its exposure; a correction needs to be applied. For instance, if one takes a period of 4 $\tau_{1/2}$ of ^{222}Rn (i.e. 15.3 d) - where the final value will be 93.75% of the equilibrium value (i.e. 1 - $(1/2)^4 = 0.9375$) - the effective length of exposure is not 15.3 d at the steady-state value, but a shorter (effective) period. Thus, since the growth of the activity A of a radioactive daughter from a long-lived and abundant parent (in this case, ^{226}Ra with a half-life of $\sim$ 1600 yr) is given (see Fig. 2.1.7) by the equation ($A = A_o$ $(1 - e^{-\lambda t})$, where A_o is the equilibrium activity of the daughter, whose decay constant is λ (and the mean life is $\tau = \frac{1}{\lambda} = 5.5$ d), we have

$$\int_0^t (1 - e^{-\lambda t})dt = \left[t + \frac{e^{-\lambda t}}{\lambda}\right]_0^t = t - \left(\frac{1 - e^{-\lambda t}}{\lambda}\right) = t - \tau(1 - e^{-\lambda t}) \tag{2.1.4}$$

On substituting $t = 4\ \tau_{1/2}$, for instance, we get the total (effective) time of exposure at the full equilibrium value of A_o as

$$15.3 - 5.5(0.9375) = 15.30 - 5.16 = 10.14 \text{ days.} \tag{2.1.5}$$

Thus, the first ~15 days of growth is equivalent to only ~10 days at full or equilibrium value. This initial correction or deficit of ~5 days needs to be made for all finite exposures - e.g. 30 days of exposure, starting at zero level, is equivalent, roughly, to only 25 days of activity at equilibrium value. Where $t >> \tau$ (or $\frac{1}{\lambda}$), in the limit one gets $t \rightarrow t - \tau$ $(1 - e^{-\infty}) = t - 5.5$ d: i.e. the deficit tends to 5.5 d (the mean life of ^{222}Rn), and this correction should be applied to all exposures of over, say, 30 days.

It should be emphasized that, as stated above, when using a dosimeter in normal room air (or inside the soil), in which radon is assumed to be already at a steady-state level, the above correction is not necessary. It only applies when the equilibrium level of radon is being built up from nought in a 'closed system', and the detector is *in situ* from the beginning. Also, once the chamber has reached the steady-state equilibrium, i.e. after ~20 days, subsequent exposure does *not* require the (-ve) correction to be applied.

Acknowledgements. For critical reading of the manuscript the author is most grateful to R. L. Fleischer, R. Ilić and J. C. H. Miles.

References

Abdel-Naby A., Pálfalvi J. and Durrani S.A. (1986) Alpha-particle spectrometry with CR-39 using combined chemical and electrochemical etching. *Nucl. Tracks* **12**, 205-210.

Al-Najjar S.A.R., Bull R.K. and Durrani S.A. (1979) Electrochemical etching of CR-39 plastic: Applications to radiation dosimetry. *Nucl. Tracks* **3**, 169-183.

Cross W.G. and Tommasino L. (1970) A rapid technique for nuclear particle damage tracks in thin foils. *Radiat. Effects* **5**, 85-89.

Dreute J., Trakowski W., Schöfer B., Brechtman C., Drechsel H., Eversberg H., Fricke W., Beer J., Wiegel B. and Heinrich W. (1986) The Siegen automatic measuring system for nuclear track detectors: Status and new developments, *Nucl. Tracks* **12**, 1-6, 261-264.

Durrani S.A. and Bull R.K. (1987) *Solid State Nuclear Track Detection: Principles, Method and Applications.* Pergamon Press, Oxford.

Fews A. (1992) Flexible analysis of etched nuclear particle tracks. *Nucl. Instr. Meth.* **B 72**, 91-103.

Fleischer R.L. (1984) Theory of passive measurements of radon daughters and working levels by the nuclear track technique. *Health. Phys.* **47**, 263-270.

Fleischer R.L., Price P.B. and Walker R.M. (1975) *Nuclear Tracks in Solids: Principles and Applications.* University of California Press, Berkeley.

Harvey J.R. and Weeks A.A. (1982) A simple system for rapid assessment of etch pits. *Nucl. Tracks* **6**, 201-206.

ICRP (1993) Publication 65: *Protection Against Rn-222 at Home and at Work.* Annals of the ICRP **23**, No. 2.

Ilić R., Šutej T., Skvarč J. and Humar M. (1993) Current research on nuclear tracks in solids at J. Stefan Institute. *Nucl. Tracks Radiat. Meas.* **21**, 397-402.

Molnár J., Somogyi S., Szilágyi S. and Spesi K. (1984) Development of a CCD based system called DIGITRACK for automatic track counting and evaluation. *Nucl. Tracks* **8**, 243-246.

Skvarč J. (1993) Automatic image analysis system TRACOS. Informacije MIDEM (Ljubljana) **23**, 3, 201-205.

Tidjani A. (1990) Effect of UV light on the efficiency of alpha-particle detection of CR-39, LR-115 type II and CN-85. *Nucl. Tracks Radiat. Meas.* **17**, 491-495.

Wong C.F. and Tommasino L. (1982) Energy discrimination of alpha particles by electrochemical etching of track detectors. *Nucl. Tracks* **6**, 17-24.

Zamani M., Manopoulou M. and Charalambous S. (1986) Etching properties of CN-85 plastic track detectors. *Nucl. Tracks Radiat. Meas.* **11**, 39-43.

Additional References

Some further bibliography is given below.

Track formation theory

Fleischer R. L., Price P. B., Walker R. M. and Hubbard E. L. (1967) Criterion for registration in dielectric track detectors. *Phys. Rev.* **156**, 353-355.

Fleischer R. L. (1980) Nuclear track production in solids. In *Progress in Materials Science* (Chalmers Anniversary Volume), Pergamon Press, New York, pp. 97-123.

Etching methodology

Enge W. (1980) Introduction to plastic track detectors. *Nucl. Tracks* **4**, 283-308.

Somogyi G. (1977) Processing of plastic track detectors. *Nucl. Track Detection* **1**, 3-18.

Instrumentation

Afarideh H., Durrani S. A. and Al-Najjar S. A. R. (1987) Development of new techniques for thermal and fast neutron dosimetry based on scintillator-filled etched track counting. *Radiat. Prot. Dosim.* **19**, 23-27.

Al-Najjar S. A. R., Balcázar-Garcia M. and Durrani S. A. (1978) A multidetector electrochemical etching and automatic scanning system. *Nucl. Track Detection* **2**, 215-220.

Khayrat A. H. and Durrani S. A. (1997) Variation in alpha-particle track diameter in CR-39 as a function of residual energy and etching conditions. *Radiat. Meas.* (in press).

Malik S. R. and Durrani S. A. (1974) Spatial distribution of uranium in meteorites, tektites and other geological materials by spark counter. *Int. J. Appl. Radiat. Isotop.* **25**, 1-8.

Matiullah, Taylor C. and Durrani S. A. (1987) An integrated-circuits based variable power supply for electrochemical etching. *Nucl. Tracks Radiat. Meas.* **13**, 67-70.

Pálfalvi J., Eödögh I. and Verö B. (1980) Track density measurements using a VIDIMET-II A type image analyser. In *Proc. 10th Int. Conf. Solid State Nuclear Track Detectors* (eds. H. François *et al.*), Lyon, 1981, Pergamon Press, Oxford, pp. 503-507.

Wilson C. K. (1982) The jumping spark counter and its development as an aid to measuring alpha activity in biological and environmental samples. *Nucl. Tracks* **6**, 129-139.

Radon

Battencourt A. O. de, Galvao J. P., Lowder W., Olast M. and Sinnaeve J., eds. (1988) Natural Radioactivity. Proc. 4th Int. Symp. Natural Radiation Environment, Lisbon, 1987, *Radiat. Prot. Dosim.* **24**, 1-4, 1-559.

Campos-Venuti G., Janssens A., Olast M., Piermattei S., Sinnaeve J. and Tommasino L., eds. (1994) Indoor Radon Remedial Action. Proc. 1st Int. Workshop, Rimini, 1993, *Radiat. Prot. Dosim.* **56**, 1-4, 1-386.

Frank A. L. and Benton E. V. (1977) Radon dosimetry using plastic nuclear track detectors. *Nucl. Track Detection* **1**, 149-179.

O'Riordan M. C. and Miles J. C. H., eds. (1992) Radon 2000. Proc. Conf. London, 1992, *Radiat. Prot. Dosim.* **42**, 3, 145-262.

R. Ilić[1] and T. Šutej[2]

2.2 Radon Monitoring Devices Based on Etched Track Detectors

2.2.1 Introduction

Passive radon monitoring devices based on alpha-particle etched track detectors are very attractive for the assessment of long-term radon exposures. The potential of radon measurement with these detectors was recognised by Fleischer *et al.* (1965). The first devices were proposed in 1967 and patented by Alter and Price (1972) as Terradex TRACK-ETCH, and in 1968 in a patent disclosure by Becker (1970). Since then a variety of dosimeters have been developed or adapted for in-ground, in-air and under-water measurements by different laboratories. The most widely used are those of Urban and Piesch (1981) in Germany, and those described by Bartlett and Bird (1987) and Hardcastle *et al.* (1996) in the UK, Doi *et al.* (1992, 1994) in Japan and Tommasino (1988) in Italy for in-air measurements; and Fleischer and Mogro-Campero (1978) in the USA for in-ground measurements.

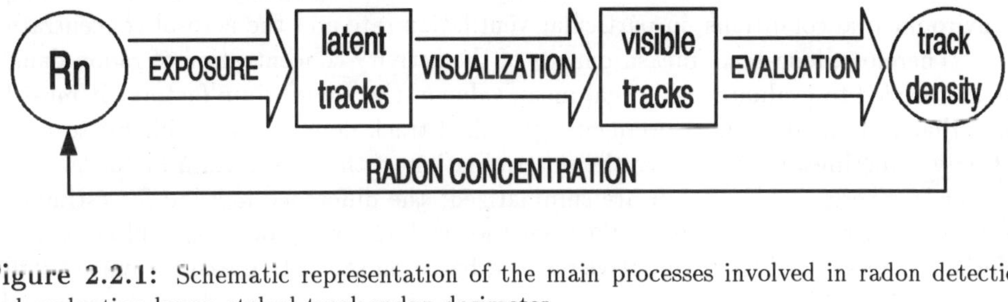

Figure 2.2.1: Schematic representation of the main processes involved in radon detection and evaluation by an etched track radon dosimeter.

The etched track radon dosimetry consists of different working phases (Fig. 2.2.1), and is based on complex nuclear, chemical, electrical and optical processes. The influence of the different parameters, such as the type of detector, etching conditions and track evaluation procedures on the dosimeter response are covered by Durrani and by Tommasino (sections 2.1 and 2.3 of the present book). The etched track detector used should be sensitive to the alpha-particles emitted by ^{222}Rn, ^{218}Po and ^{214}Po with initial energies of 5.5, 6.0 and 7.7 MeV, respectively. Because of its good sensitivity, stability against

[1]J. Stefan Institute, Ljubljana, and University of Maribor, Maribor, Slovenia
[2]Health Inspectorate, Ministry of Health, Ljubljana, Slovenia

various environmental factors and high degree of optical clarity, CR-39 has become the state-of-the-art etched track detector for environmental radon. The dosimeter is first exposed in air to radon and its progeny. Typical exposure time for diffusion and permeation samplers is from a few days to a year. Latent tracks are then revealed by chemical or electrochemical etching in alkaline solution and analysed by an optical microscope, electronic optical systems or non-microscopic track counting systems.

Radon monitoring devices can be classified into four broad categories: (i) envelope or bare detector samplers, (ii) diffusion samplers, (iii) permeation samplers, and (iv) radon collector samplers. The characteristics of these devices are well documented in the various Proceedings of International Conferences on Nuclear Tracks in Solids (e.g. Perelygin *et al.*, 1995); Proceedings of two workshops on Radon (Furlan and Tommasino, 1993; Tommasino *et al.*, 1990); and in several review papers (Fleischer, 1988; Khan *et al.*, 1993; Tommasino, 1988, 1990; Urban and Piesch, 1982).

In this section devices for in-ground, in-air and under-water measurements are described first. The section then presents physical processes involved in the detection process - i.e. radon diffusion in air, radon permeation through permeable membrane barriers, plate-out of radon decay products and adsorption of radon on the collector. A theoretical model, verified by experiments, which enables calculation of the sensitivity of such dosimeters, is also presented.

The performance of the dosimeters described in terms of the lowest detectable radon concentration in a given medium, and stability of the response to environmental conditions - u.v.radiation, oxygen, humidity, temperature - is then analysed. The equilibrium factor between radon and its short-lived daughters in air can strongly vary with changes of environmental conditions, in particular ventilation rate and the aerosol concentration in air. Therefore, long-term measurements of the activity concentration of radon daughters are needed to evaluate the actual mean value of the equilibrium factors. It must be noted that full daughter measurements by etched track detectors are difficult to make, and only sometimes reliable. For the determination of the equilibrium factor two concepts of long-term measurement are summarized: the difference method for estimating the total activity of ^{218}Po and ^{214}Po from the readings of an open and filtered radon diffusion chamber, and the alternative method based on separate measurements of ^{222}Rn and ^{214}Po. The characteristics of the commonly used dosimeters developed by some laboratories worldwide are surveyed. A discussion of the potential for future research and development concludes the section.

2.2.2 Dosimeters

2.2.2.1 Envelope type samplers

This type of radon dosimeter is used for the assesment of medium-term (a few days to one week) indoor radon measurements. Simply speaking, it comprises a bare detector foil enclosed in an envelope (Khan *et al.*, 1993; Miles *et al.*, 1984; Rannou *et al.*, 1986; Tommasino *et al.*, 1990) - see Fig. 2.2.2a. When the envelope is closed, the detector does not register alpha-particles, and the device is in the off-state. To turn it 'on', the envelope must be opened or - depending on design - one part of the detector foil must be transferred

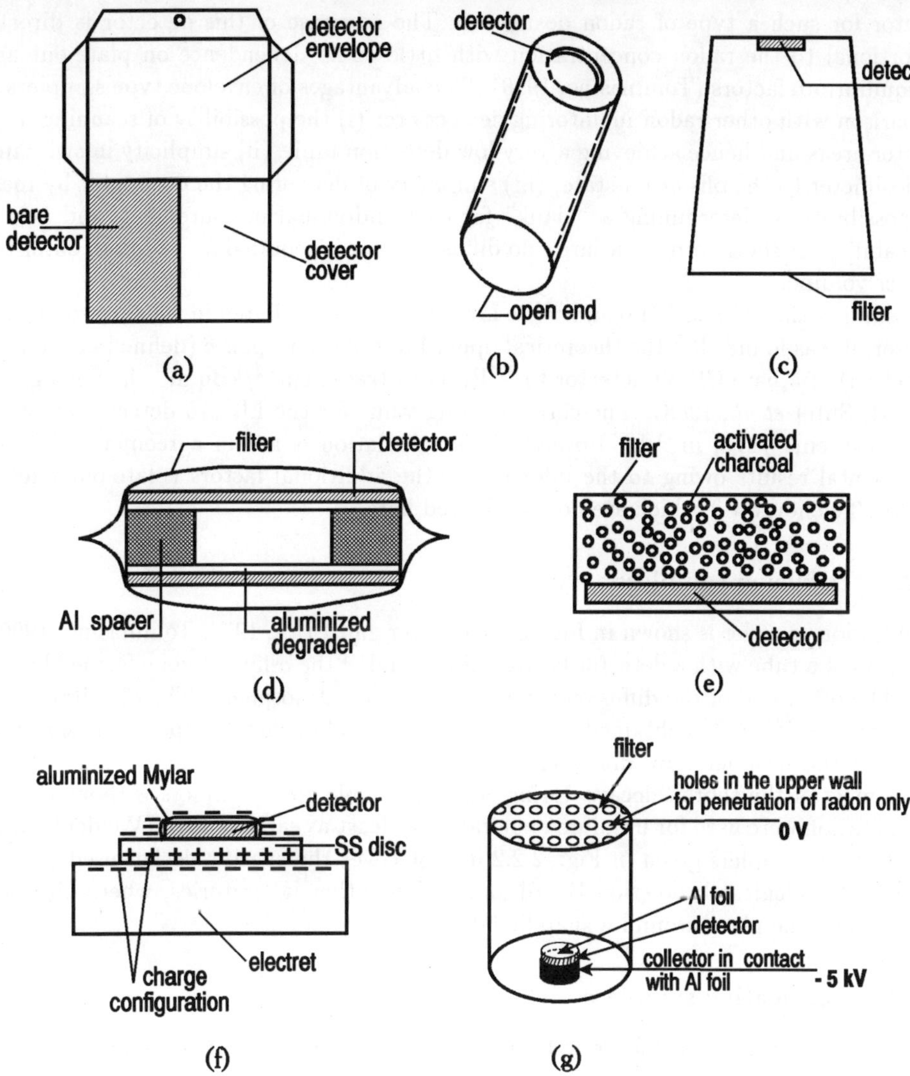

Figure 2.2.2: Radon monitoring devices based on etched track detectors: (a) the envelope type sampler, (b) diffusion sampler, (c) filter permeation sampler, (d) plastic bag permeation sampler, (e) charcoal collector sampler, (f) electret collector sampler, and (g) electrostatic collector sampler. Typical size of detector is 1 cm x 1 cm x 10-500 μm; and typical volume of dosimeter housing is 100 cm^3. See discussion in the text.

from the envelope to face the atmosphere to be monitored. The "transferred" part of the detector is thus used for the radon measurements, while the "remaining" part enclosed in the envelope is used for the "blank" assesment. LR 115 appears to be the most convenient detector for such a type of radon dosimeter. The response of this detector is directly proportional to the radon concentration with little or no dependence on plate-out and the equilibrium factors (Tommasino, 1990). The advantages of envelope type samplers in comparison with other radon monitoring devices are: (i) the possibility of scanning large detector areas and hence achieving a very low detection limit, (ii) simplicity in switching the dosimeter to the off- or on-state, (iii) simplicity of delivering the dosimeter by mail, (iv) possibility of determining a "blank" for each individual measurement, and (v) its applicability for short-term exposure - no diffusion time is required for radon to enter the sampler volume.

Assuming that ^{222}Rn, ^{218}Po and ^{214}Po are uniformly distributed in the air around the detector, it was found that the theoretical upper limit of the response (defined in the next subsection) of a bare CR-39 detector to ^{222}Rn is 14 tracks cm^{-2}/kBq m^{-3} h (Somogyi *et al.*, 1984; Šutej *et al.*, 1988). The corresponding value for the LR 115 detector amounts to 3 tracks cm^{-2}/kBq m^{-3} h. However, this calculation is not in agreement with the experimental results owing to the influence of the additional factors (plate-out effects, etching efficiency, etc.) which are not considered here.

2.2.2.2 Diffusion samplers

The diffusion sampler is shown in Fig. 2.2.2b (Alter and Price, 1972; Tommasino, 1990). It consists of a tube with a detector located at one end of the diffusion zone formed by the tube. The other end of the diffusion zone is open to the atmosphere. The discrimination against thoron (^{220}Rn) is obtained by delaying the arrival of that isotope in the sensitive volume of the dosimeter by diffusion, simply by the use of a sufficiently long tube - so that most of the thoron decays during transit. Simple open samplers without thoron discrimination were used for uranium exploration at least as early as 1970 (Weidenbaum).

Note that samplers given in Fig. 2.2.2 do not cover the specific design used by UK National Radiological Protection Board and various other laboratories, where diffusion of the radon take place around a snap-fit lid.

2.2.2.3 Permeation samplers

The devices commonly used for long-term (a few months to one year) monitoring consist of an enclosure that allows ^{222}Rn to enter through a permeable membrane. The two most popular designs for environmental monitoring of radon gas, cup-type and bag-type permeation samplers, are briefly surveyed below.

2.2.2.3.1 Membrane permeation samplers. The dosimeter is schematically shown in Fig. 2.2.2c. Here the permeable filter closes the open end of the cup (Ward *et al.*, 1977). The enclosure excludes condensed alpha emitters, and the filter area and thickness are designed to exclude unwanted ^{219}Rn (half-life, $\tau_{1/2}$ = 3.96 s) and ^{220}Rn ($\tau_{1/2}$ = 55.6

s) without unduly diminishing ^{222}Rn ($\tau_{1/2} = 3.82$ d) [illegible] daughters that are produced after ^{222}Rn enters the detection space. The [illegible] of a porous material such as fibreglass, microporous paper, or a plastic su[illegible]yethylene or PVC (≈ 10 μm thick). It was proved that this dosimeter gives a h[illegible] reproducible and unambiguous measure of ^{222}Rn. Several efforts have been mad[illegible]o further improve the response of the dosimeter and its stability against condensation of moisture, electrostatic effects, temperature fluctuations, etc. (Frank and Benton, 1982; Humar *et al.*, 1982; Likes *et al.*, 1979; Urban and Piesch, 1982).

2.2.2.3.2 Plastic bag permeation samplers. The concept of the permeation sampler is fully explored by the radon monitoring device developed by Tommasino *et al.* (1986) and shown in Fig. 2.2.2d. This permeation sampler is formed from a heat-sealed plastic bag (filter) made of polyethylene. There are two detector foils (usually LR 115), and aluminized polycarbonate degraders face the detector foils to optimize the detector responses and to make their surface conductive. The polyethylene bag protects the detector from humidity, dust, thoron and radon daughters. The advantages of such a type of dosimeter are: (i) simple heat sealing and hence low cost, (ii) small size and fast sampling time, (iii) high radon permeability, and (iv) elimination of water vapour. In addition, most of the radon dosimeters have a response which depends on the atmospheric pressure, since the sensitive volume (see subsection 2.2.3.2) depends on the range of alpha particles, which changes with pressure (Vasudevan and McLain, 1994). In contrast, the plastic bag sampler with polycarbonate degraders has little dependence on the atmospheric pressure, since the degrader is not the air but the plastic foil.

2.2.2.4 Radon collector samplers

Short-term exposures (from 1 hour to 1 day) are particularly useful for screening surveys to identify houses with high radon concentrations and to study time-dependent radon variations. Unfortunately, the types of radon dosimeters mentioned so far in this subsection are not sufficiently sensitive for short-term exposure measurements. To increase the sensitivity, three techniques based on the use of electrets, electrostatic collection, and activated charcoal have been proposed (Kotrappa *et al.*, 1982; Miyake *et al.*, 1987; Šutej *et al.*, 1991, 1993). The basic designs of dosimeters utilizing these techniques are briefly summarized below.

2.2.2.4.1 Charcoal collector sampler. The main parts of the dosimeter are shown in Fig. 2.2.2e (Šutej *et al.*, 1991). Here charcoal acts as a collector of radon from the air. An etched track detector (CR-39) records alpha particles emitted by radon and polonium isotopes adsorbed onto charcoal grains. It was found that for a short exposure time (1 h), the response is ≈ 500 tracks cm^{-2}/kBq m^{-3} h (if the detector foil is left inside the closed dosimeter for 2 days, after the exposure), and is nearly two orders of magnitude higher than that attainable with standard radon dosimeters utilizing only etched track detectors.

2.2.2.4.2 Electret collector sampler. An electret is an electrical analogue of a permanent magnet, and carries a permanent electric charge. Kotrappa *et al.* (1982) showed that teflon electrets with surface charge densities lying between 10^{-9} and 10^{-8} C cm^{-2} are suitable for the direct collection of radon decay products on an etched track detector - providing *in-situ* registration of the emitted alpha-particles - and are, therefore, useful in determining the short- and long-term average radon concentrations. The electret collector sampler is shown in Fig. 2.2.2f. Here a stainless-steel (SS) disc is fixed on top of the electret. Then the metal surface close to the electret has an induced positive charge, and a compensating negative charge appears on the top surface of the metal sheet. Any non-uniform charge distribution on the surface of the electret is not reflected in the charge density distribution on the surface of the metal sheet. The response of CR-39 detector in combination with an electret for ^{222}Rn was found to be 92 tracks cm^{-2}/pCi l^{-1} h (i.e. 2,486 tracks cm^{-2}/kBq m^{-3} h) (Kotrappa *et al.*, 1982). A sensitivity improvement compared to the permeation sampler (Fleischer *et al.*, 1980) - where there is no collection of decay products - by a factor of 728 was achieved by this dosimeter. This enhancement is useful when short-term measurements are to be made.

2.2.2.4.3 Electrostatic collector sampler. Determination of the radon concentration in air using an electrostatic method and etched-track detector has been considered by many authors such as Iida *et al.* (1988), Miyake *et al.* (1987) and Teterev *et al.* (1995). Here the Rn daughter products are collected by an electrostatic field on a thin (some μm) metal foil placed on the detector. A simplified design is shown in Fig. 2.2.2g (Teterev *et al.*, 1995). To explore the potential of this technique, further work related to collection of ions from the inner volume and elimination of decay products originating in the environmental air is required. A response as high as 5,000 tracks cm^{-2}/kBq m^{-3} h for a chamber with a sensitive volume of only 260 cm^3 was reported (Miyake *et al.*, 1987).

2.2.3 Physical Processes in Etched Track Dosimetry

2.2.3.1 Radon decay and plate-out of its products

To summarize the physical processes involved in etched track dosimetry, it is necessary to explain the behaviour of radon and its products in air. Various parameters influence the concentration of radon and its decay products in the air. The basic processes involved are attachment to atmospheric aerosols, deposition (plate-out), recoil of radon and radon daughters from aerosols and surfaces, and decay. The basic features of radon behaviour are described in the literature (e.g. Abu-Jarad *et al.*, 1981; Brager and Revzan, 1991; Brager *et al.*, 1991; Fleischer and Mogro-Campero, 1978; Fleischer *et al.*, 1980, 1984; Hadler and Paulo, 1994; Hadler *et al.*, 1992; Porstendörfer, 1984; Porstendörfer and Reineking, 1992; Pörstendorfer *et al.* 1987). Growth of the activity of the individual short-lived decay products, A_i, from a constant source of radon activity, A_o, is shown in Fig. 2.2.3. From this figure it is evident that ^{218}Po, owing to its short decay time ($\tau_{1/2}$ = 3 min), relatively quickly ($\approx$ 20 min) reaches equilibrium with ^{222}Rn activity. Activities of ^{214}Pb and ^{214}Bi (^{214}Po) are close to equilibrium with ^{222}Rn after approximately 3 h.

Other basic processes influencing the activity balance between radon and its progeny are illustrated in Fig. 2.2.4.

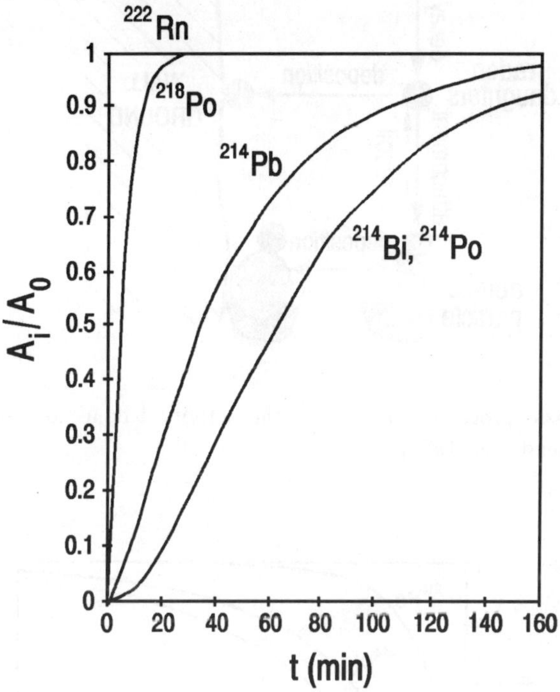

Figure 2.2.3: Growth of activity of individual short-lived radon decay products, A_i, from a constant source of radon activity, A_o. Time is denoted by t (adapted from Evans, 1969). The respective half lives ($\tau_{1/2}$) are: ^{222}Rn (3.82 d), ^{218}Po (3.11 min), ^{214}Pb (26.8 min), ^{214}Bi (19.9 min) and ^{214}Po (164 μs)

The mechanisms schematically shown in Fig. 2.2.4 are very simplified. Here ionisation, electrostatic forces, gravity, air flow and other important processes influencing the transport of radon and its decay products in air are neglected. If such processes are neglected and when only steady-state molecular diffusion is taken into account, with the diffusion coefficient equal for ^{222}Rn, ^{218}Po and ^{214}Po in air and being 0.06 cm^2 s^{-1}, we obtain a reduced activity concentration of radon decay products in the air near surfaces due to their progressive deposition on the surfaces. Activity concentration profiles for unattached radon progeny are shown in Fig. 2.2.5 (Brager and Revzan, 1991). It can be seen that their activity concentrations significantly decrease at distances less than about 10 cm from the walls.

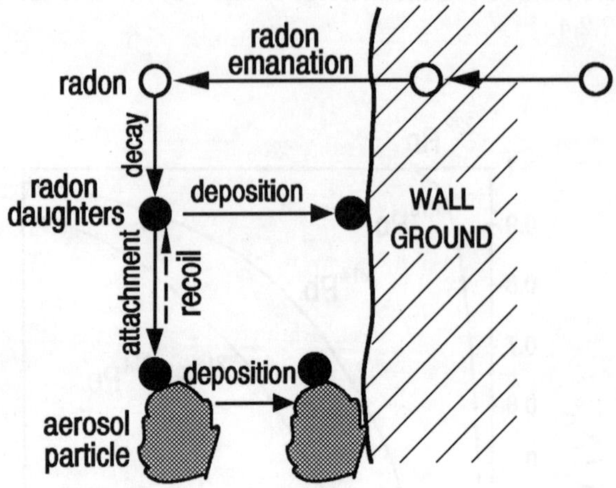

Figure 2.2.4: Basic processes influencing the activity balance of radon and its progeny (adapted from Porstendörfer, 1984).

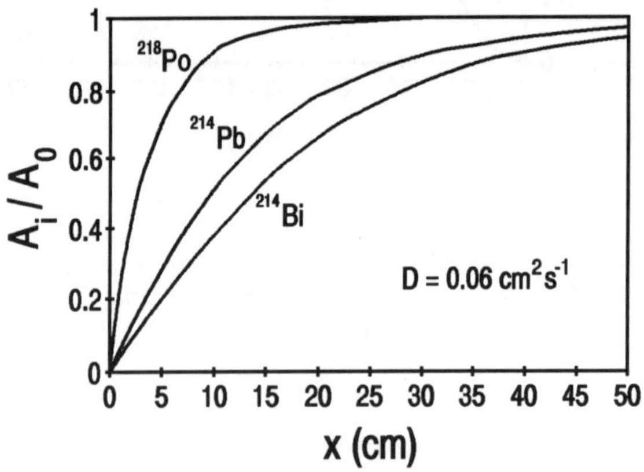

Figure 2.2.5: Activity concentration profiles for unattached radon decay products A_i. The distance from the surface is denoted by x. Radon activity and diffusion coefficient for radon in air are denoted by A_o and D respectively (adapted from Brager and Revzan, 1991).

It should be mentioned that previously described radon dosimeters cannot transform the measured signal (tracks cm^{-2}) to the equivalent lung dose (Sv). The lung dose is actually dependent on many other factors. Radon itself contributes a negligible dose because it is almost chemically inert, and is not deposited on surfaces in the lungs. A detailed description is given in this book by Muirhead (section 3.6). From the dosimetric point of view the attachment of these isotopes to aerosols, the breathing rate, age, sex, smoking habits, etc., of the subjects are very important. The ^{220}Rn daughters ^{216}Po, ^{212}Pb, ^{212}Bi can also significantly contribute to the lung dose (Doi *et al.*, 1994).

2.2.3.2 Detection efficiency of radon dosimeters

For dosimetric purposes it is important to measure the average radon and its progeny concentration over a time which should be long enough relative to the typical time scale of radon fluctuations caused by environmental conditions.

The signal measured by etched-track dosimeters is the integrated track density, ρ (tracks cm^{-2}). If the track density is uniform over the detector surface, the signal is defined by (Fleischer, *et al.*, 1980)

$$\rho = \frac{K}{\tau_r} \int_0^{t_e} C(t)dt \tag{2.2.1}$$

where τ_r is the radon decay time (5.5 days), *C(t)* is the radon concentration in the air around the detector (atoms m^{-3}) at time t, and t_e is the exposure time. K denotes the dosimeter response (calibration coefficient) and is defined by (Šutej *et al.*, 1988)

$$K = \frac{\tau_r}{C(t)} \frac{d\rho}{dt} = \frac{1}{A_o(t)} \frac{d\rho}{dt} \tag{2.2.2}$$

where $A_o(t)$ is the activity concentration of radon in air (Bq m^{-3}) at time t, and $d\rho/dt$ is the track density production rate (tracks cm^{-2} h^{-1}). The unit of K is tracks cm^{-2}/kBq m^{-3} h; it is usually expressed with the dimension of length (cm). Note that 1 cm times track per decay of geometrically determined K is equivalent to a value of 3.6 tracks cm^{-2}/kBq m^{-3} h.

For a diffusion chamber type dosimeter with a long exposure time ($t_e >> 3$ h), the response can be approximated by (Šutej *et. al.*, 1988)

$$K = \frac{\rho}{\overline{A_o} t_e} \tag{2.2.3}$$

where $\overline{A_o}$ denotes the average radon activity concentration.

In general, the total track density $\rho(\vec{r})$ at the point of observation in the detector defined by $\vec{r}$, is the result of an integration of all α-emissions which can produce observable tracks (Hadler and Paulo, 1994; Skvarč and Ilić, 1994)

$$\rho(\vec{r}) = \sum_i \rho_i^v(\vec{r}) + \sum_i \rho_i^p(\vec{r}) \tag{2.2.4}$$

where $\rho_i^v(\vec{r})$ is the track density contribution of Rn and Po isotopes present in the air, and $\rho_i^p(\vec{r})$ is the track density induced by the radon decay products plated on the detector itself and the inside wall of the dosimeter housing. For each alpha-emitting nuclide present in the air we can write (Fleischer and Mogro-Campero, 1978; Hadler and Paulo, 1994)

$$\rho_i^v(\vec{r}) = \int_o^{t_e} dt \int_{V_i} \frac{A_i(\vec{r}\,', t) cos\theta}{4\pi \mid \vec{r} - \vec{r}\,' \mid^2} dV \tag{2.2.5}$$

where $A_i(\vec{r}\,', t)$ is the activity concentration of i-th nuclide at time t at position $\vec{r}$; θ is the angle between the α particle's trajectory and the normal to the detector surface; $\mid \vec{r} - \vec{r}\,' \mid$ is the distance between the detector surface and the source volume dV; and V_i is the sensitive volume for the i-th nuclide. Sensitive volumes and surfaces relative to a point at the centre of the detector for two different conditions are shown in Fig. 2.2.6. The case shown in Fig. 2.2.6a could be applied to estimate the detection efficiency for Makrofol E and LR 115 detectors which are sensitive to the energy of alpha particles from 0 up to 2 MeV. The case shown in Fig. 2.2.6b is valid for the CR-39 detector, which is sensitive to all alpha particle energies emitted by ^{222}Rn, ^{218}Po and ^{214}Po.

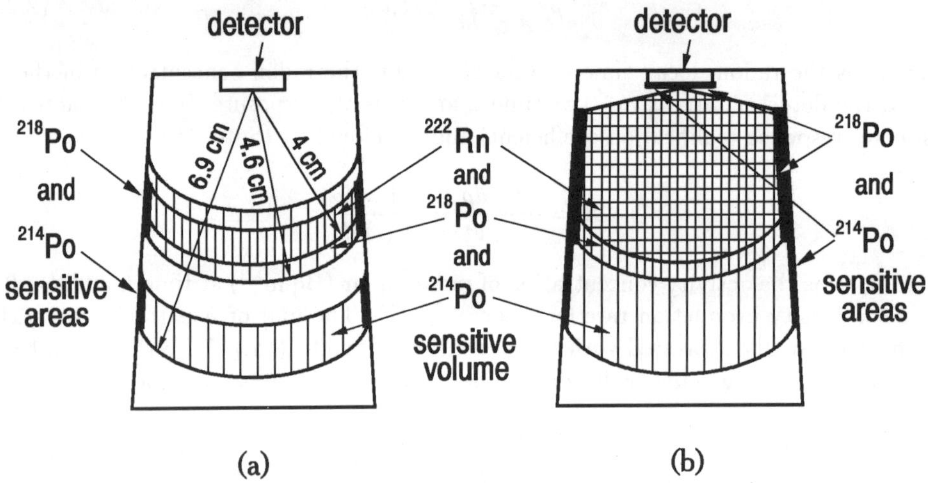

Figure 2.2.6: Sensitive volumes and surfaces relative to a point at the centre of the detector area. The detector is sensitive to alpha particles of energies of (a) 0-2 MeV, and (b) 0-7.7 MeV (adapted from Fleischer and Mogro-Campero, 1978).

When A_i is homogeneously distributed in the air and independent of time, and the detector is exposed in 2π geometry (bare detector), V_i is only limited by the effective range R_i of α-particles emitted by the i-th nuclide in the air, and the registration efficiency, which for constant etching rates is determined by the critical angle of etching, $\theta_{c,i}$. In this case, solution of the integral in equation (2.2.5) leads to

$$\rho_i^v = \frac{1}{4} A_i \, t_e \, R_i \, cos^2 \, \theta_{c,i} \tag{2.2.6}$$

which is in practice an often employed expression (Durrani and Bull, 1987; Fleischer and Mogro-Campero, 1978; Hadler and Paulo, 1994; Somogyi *et al.*, 1984), for the estimation of track density as a function of the radon activity and intrinsic detector efficiency. The range of alpha particles emitted by Rn isotopes and their decay products is given in Table 2.2.1.

Taking into account that θ in CR-39 detector under normal chemical etching conditions (see sections 2.1 and 2.3) is less than 20°, its influence on detection efficiency is a reduction of about 10 %. For LR 115 detector $\theta \approx 40^o$, and hence the detection efficiency is reduced by $\sim$ 40%. Besides this, LR 115 detector is sensitive to alpha particles only between 2 to 4 MeV. Considering these data from eqs. (2.2.3) and (2.2.6) we obtain a theoretical limit for the response of a bare detector, K_o, to ^{222}Rn

$$\begin{aligned} K_o &= \tfrac{1}{4} R_o cos^2 20^o = 0.88\,\mathrm{cm} = 3.2\ \mathrm{tracks\ cm^{-2}/kBq\ m^{-3}\ h}; \quad \text{for CR-39} \\ K_o &= \tfrac{1}{4} cos^2 40^o\,[R\,(4\ \mathrm{MeV}) - R\,(2\ \mathrm{MeV})] = 0.28\ \mathrm{cm} = \\ &= 1\ \mathrm{track\ cm^{-2}/kBq\ m^{-3}\,h}; \quad \text{for LR 115} \end{aligned} \tag{2.2.7}$$

Table 2.2.1: Range of alpha particles emitted by Rn and their decay products in air and in two commonly used detectors (Ilić, 1990).

Nuclide	α-particle energy (MeV)	Range		
		Air (cm)	CR-39 (μm)	Makrofol (μm)
^{222}Rn	5.48	4.0	35.5	38.7
^{218}Po	6.00	4.6	40.7	44.5
^{214}Po	7.70	6.9	59.8	65.6
^{220}Rn	6.28	4.9	43.6	47.7
^{216}Po	6.77	5.6	48.9	43.6
^{212}Po	8.78	8.5	75.8	83.3

The supposition that A_i is independent of distance may be valid only for ^{222}Rn. Due to plate-out effects, the radon-daughter activity is not homogeneous in air. The daughters are depleted near all material surfaces and deposited on interior dosimeter walls, including the detector surface itself. For the deposited nuclides (Fig. 2.2.6), we can write

$$\rho_i^p(\vec{r}) = \int_0^{t_e} dt \int_{S_i} \frac{G_i(\vec{r}\,'', t) cos\theta}{4\pi \mid \vec{r} - \vec{r}\,'' \mid^2} dS \tag{2.2.8}$$

where $G_i(\vec{r}\,'', t)$ is the surface activity concentration (Bq cm^{-2}) of the i-th nuclide at the detector and the wall of the dosimeter housing at the source area dS at $\vec{r}\,''$, and S_i is the effective area of the source of the i-th nuclide. Surface activity concentration can be expressed by (Šutej, 1995)

$$G_i(\vec{r}'', t) = \lambda_{i-1} n_{i-1} + j_i \tag{2.2.9}$$

where n_{i-1} and λ_{i-1} are the surface density (atoms cm^{-2}) and decay constant (s^{-1}) of deposited parents, and j_i is the flux (atoms cm^{-2} s^{-1}) of the i-th isotope on the surface of interest

$$j_i = -D\nabla C_i = -\frac{D}{\lambda_i}\nabla A_i \tag{2.2.10}$$

where D is the diffusion coefficient. The relations do not include recoil of ^{214}Po from the wall after decay of ^{218}Po.

Surface activity is strongly dependent on the dosimeter size and its shape, electrostatic forces, and other environmental parameters (Šutej, 1995). For a bare CR-39 detector the solution of eqs. (2.2.9) and (2.2.10) can be simplified by considering the plate-out of radon decay products. The surface activity concentration of deposited ^{218}Po in the steady state is (Šutej, 1995)

$$G_1 = j_1 = A_o\sqrt{\frac{D}{\lambda_1}} \tag{2.2.11}$$

Thus the detector response to ^{218}Po, K_1, considering 2π geometry and $\theta_1 = 20^o$, $D = 0.06$ cm^2 s^{-1} and $\lambda = 3.71{\cdot}10^{-3}$ s^{-1}, is

$$K_1 = \frac{1}{2}cos^2 20^o\sqrt{\frac{D}{\lambda_1}} = 1.78\,\text{cm} = 6.4 \text{ tracks cm}^{-2}/\text{kBq m}^{-3}\text{ h} \tag{2.2.12}$$

In a similar way the detector response to deposited ^{214}Bi (^{214}Po) can be estimated. Since the mathematical formulas are complicated, this is beyond the scope of the present book. Most recently (Nikezić *et al.*, 1996), the Monte Carlo method was applied for the determination of the response of diffusion sampler type radon dosimeters. It should be mentioned that plate-out considerations are invalid in practice, because accurate values of the diffusion coefficients are unknown without elaborate measurements of the size and activity distributions of the aerosols (Fleischer *et al.*, 1984). Plate-out determination of radon decay products on solid surfaces by autoradiography is described by Šutej and Ilić (1991).

The radon concentration inside the housing of the dosimeter also depends on its permeation through the filter (for a permeation sampler). This process is surveyed in the next subsection.

2.2.3.3 Radon permeation through membranes

In general, membranes reduce the radon concentration inside the dosimeter housing because they delay its entry by the mean diffusion time through the membrane.

The radon concentration inside the dosimeter housing, C, can be described by the differential equation

$$\frac{dC}{dt} = -\frac{C}{\tau_r} - \frac{C_o - C}{\tau_m} \tag{2.2.13}$$

where C_o is the radon concentration outside the dosimeter housing, τ_r is the mean decay time of ^{222}Rn (5.5 d), and τ_m is the mean permeation time. The build-up of radon inside the dosimeter housing (Fig. 2.2.7) is (Fleischer *et. al.*, 1980)

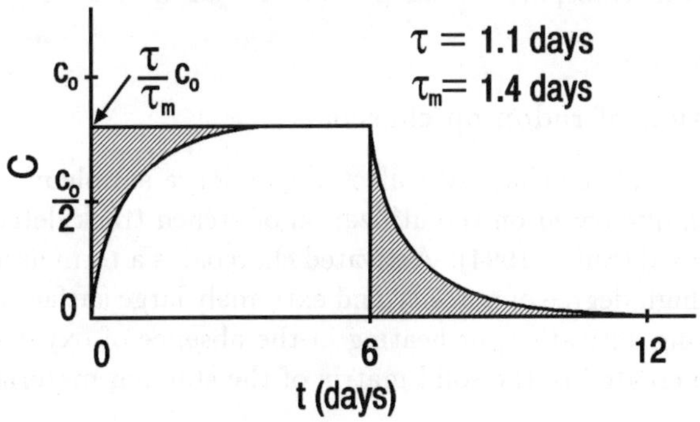

Figure 2.2.7: Variation of ^{222}Rn concentration, C, with time, *t*, for a filter permeation sampler (cup). Filter ^{222}Rn delay (or permeation) time and mean time (see definition below) are denoted by τ_m and τ respectively (adapted from Fleischer *et al.*, 1980).

$$C = C_o \frac{\tau}{\tau_m}(1 - e^{-t/\tau}) \tag{2.2.14}$$

where the mean time τ is given by

$$\frac{1}{\tau} = \frac{1}{\tau_r} + \frac{1}{\tau_m} \tag{2.2.15}$$

The loss of radon into a radon-free atmosphere is thus *($t > t_o$; t_o is exposure time)*

$$C = C_o \frac{\tau}{\tau_m}(1 - e^{-t_o/\tau})e^{-(t-t_o)/\tau} \tag{2.2.16}$$

The mean permeation time is given by

$$\tau_m = \frac{V_d \delta}{P S_f} \tag{2.2.17}$$

where P is the radon permeability of the filter, δ and S_f its thickness and area, and V_d the interior volume of the dosimeter. Values of permeabilities for various materials can be found in Fleischer (1988) and Hafez and Somogyi (1986). From Fig. 2.2.7 it can be seen that in the case when the detector is removed immediately after the end of the exposure, t_o, the track density (area under the curve in Fig. 2.2.7) is not directly proportional to $(\tau/\tau_m)C_o t_o$, but is reduced by a factor of $(\tau/t_o)(1\text{-}e^{t_o/\tau})$, which is proportional to the shaded area (on the left) above the curve. However, if the detector is left in the dosimeter housing for a time much longer than τ (i.e. $t-t_o >> \tau_m$), additional decays in the housing contribute just the same number of tracks that were lost because of the delay in radon entering the housing at the beginning of exposure.

If a layer of charcoal is placed in front of the detector foil (charcoal collector sampler), the adsorption of radon inside this layer should be considered. This is done in the next subsection, where the transport of radon from the air into a charcoal layer is briefly described.

2.2.3.4 Adsorption of radon on charcoal

To study the response of the charcoal collector sampler, a simple model of the detection process in the technique based on the utilization of etched track detectors and activated charcoal was developed (Šutej, 1994). Activated charcoal is a term used to describe materials that exhibit a high degree of porosity and extremely large surface area (Scarpitta and Harley, 1990). During activation, or heating in the absence of oxygen, both macropores and micropores are created in the solid matrix of the starting material.

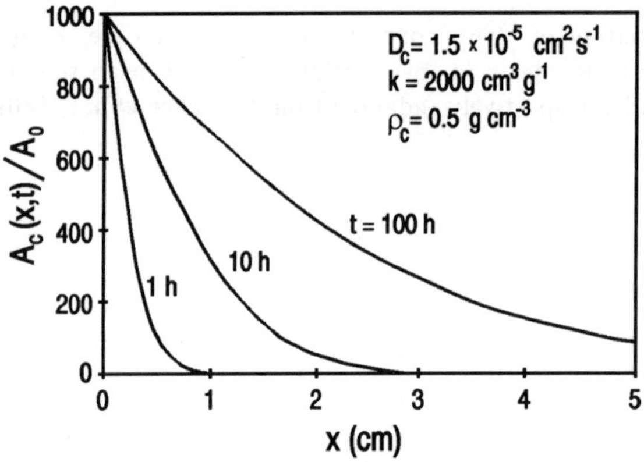

Figure 2.2.8: Radon activity, A_c, plotted against depth, x, in an infinite medium for different values of time, t. Radon activity in air, diffusion coefficient for radon in charcoal, adsorption coefficient, and density of the charcoal are denoted by A_o, D_c, k, and ρ_c respectively (adapted from Šutej, 1994).

Equilibrium adsorption of radon by charcoal can be described by (Šutej, 1994)

$$A_c = k\rho_c A_o \tag{2.2.18}$$

where A_c and A_o are the activity concentrations of radon in the charcoal and in the air (Bq m^{-3}), k is the adsorption coefficient (cm^3 g^{-1}) and ρ_c is the density of the charcoal (g cm^{-3}).

A one-dimensional approximation to the diffusion process governed by Fick's law and the decay law is (Cohen and Cohen, 1983; Šutej, 1994)

$$\frac{\partial A_c}{\partial t} = D_c \frac{\partial^2 A_c}{\partial x^2} - \lambda_o A_c \tag{2.2.19}$$

where A_c is the activity of radon in the charcoal as a function of time, t, and depth, x; D_c is the diffusion coefficient for radon in the charcoal (cm^2 s^{-1}); and λ_o is the decay constant of ^{222}Rn ($2.1 \cdot 10^{-6}$ s^{-1}).
The approximate solution for $t << \lambda_o{}^{-1}$ and $x << \sqrt{D_c/\lambda_o}$ is (Šutej, 1994)

$$A_c(x, t \to 0) = k\rho_c A_o \left[1 - erf\left(\sqrt{\frac{x^2}{4D_c t}}\right)\right] \tag{2.2.20}$$

The solution of eq. (2.2.19) for a steady-state and for an infinite thickness of charcoal layer is

$$A_c(x, t \to \infty) = k\rho_c A_o e^{-x\sqrt{\lambda_o/D_c}} \tag{2.2.21}$$

If the layer of charcoal has a finite thickness, d, the solution for a steady state is

$$A_c(x) = A_o k \rho_c \frac{ch[\sqrt{\lambda_o/D_c}(d-x)]}{ch(\sqrt{\lambda_o/D_c}d)} \tag{2.2.22}$$

The radon concentration A_c vs depth x in an infinite medium for different values of time t is shown in Fig. 2.2.8. Typical data for D_c, k and ρ_c are listed in Fig. 2.2.8, from which it can be seen that the radon concentration on the surface of the charcoal layer, A_c (x=0), is 1000 times higher in comparison with the radon concentration in the air, A_o. Additionally, the depth of radon penetration amounts to ~ 5 mm and 5 cm for exposure times of 1 h and 100 h, respectively.

2.2.4 Performance of Radon Dosimeters

Evaluation of the performance of a variety of radon monitoring devices based on etched track detectors was undertoken under laboratory and field conditions. Details of the performance studies are given in this book by Miles (section 3.4). It is emphasized that calibration, intercomparison and quality assurance are prerequisites for reliable data collection (Culp, 1993; Miles and Olast, 1990; Miles *et al.*, 1984, 1996). The main task of future research work should be the comprehensive investigation of all sources of error for the dosimeters described. This is briefly outlined in this subsection. Subsequently, an overview of the potential applications of dosimeters in radon studies, and studies related to the behaviour of the detectors in different environmental conditions, are summarized.

2.2.4.1 Response of radon dosimeters and detection limits

The sensitivity of a dosimeter utilizing etched track detectors, defined as the smallest detectable radon activity concentration, A_{min} (Bq m^{-3}), can be calculated (Humar *et al.*, 1992; Ilić *et al.*, 1990) by

$$A_{min} = \frac{L_d}{KS_s t_e} \qquad (2.2.23)$$

where L_d is the lower limit of counted tracks for qualitative detection (tracks), and S_s is the scanned area of the detector (cm^2). The response K and exposure time t_e are already defined. In general, L_d is defined (Currie, 1968) as

$$L_d = \kappa^2 + 2\kappa\sigma_{bl} \qquad (2.2.24)$$

where κ is the confidence factor and σ_{bl} is the standard deviation of the blank defined (Ilić and Rant, 1991; Ilić *et al.*, 1990) as

$$\sigma_{bl} = \sqrt{(\rho_b + \mu_b t_e)S_s + (sS_s)^2} \qquad (2.2.25)$$

where ρ_b is the background track density (tracks cm^{-2}), μ_b is the mean rate of background radiation tracks (tracks cm^{-2} h^{-1}) and s (tracks cm^{-2}) is a constant that describes the variation of the background track density due to the variation of detector sensitivity, etching and counting technique. Finally, we get (Šutej *et al.*, 1993)

$$A_{min} = \frac{\kappa^2 + 2\kappa\sqrt{(\rho_b + \mu_b t_e)S_s + (sS_s)^2}}{KS_s t_e} \qquad (2.2.26)$$

For a well known blank, defined as the signal resulting from an experiment in which conditions are indentical to the experiment of interest, except that no radon gas is present, the value of κ amounts to 1.65 for the 95% confidence level, and normal distribution.

As discussed in subsection 2.2.3.2 the response of an etched track radon dosimeter is dependent on its design, the type of detector and the detector processing technique. Typical values of parameters in eq. (2.2.26), are ρ_b = 10 tracks cm^{-2}, S_s = 1 cm^2, s = 5 cm^{-2}, $\mu_b = 10^{-5}$ tracks cm^{-2} s^{-1}. The response of some radon dosimeters is given in Table 2.2.2. The response of 50 dosimeters from 48 laboratories can be found in a recent publication prepared by Miles *et al.* (1996). Since $K = 10^{-3}$ tracks cm^{-2}/Bq m^{-3} h for diffusion and permeation samplers, we can see that the lowest detectable radon concentration in air for an exposure time of 1 day is 1000 Bq m^{-3}. The response of the charcoal collector sampler for a short exposure time (1 h) is two orders of magnitude higher than that attainable with a standard dosimeter using etched track detectors alone. This leads to a high sensitivity. Radon activity concentrations as low as $\sim$ 30 Bq m^{-3} can be measured in a short exposure time (3 h).

2.2.4.2 Effects of atmospheric parameters

It was noted earlier that environmental conditions strongly influence the behaviour of radon daughters in air and their deposition on surfaces (walls, samplers, and detector

Table 2.2.2: Response of some radon dosimeters utilizing etched track detectors.

Type of dosimeter	Institution[1] (Trade name)	Detector	Response $\left[\frac{\text{tracks cm}^{-2}}{\text{kBq m}^{-3}\text{ h}}\right]$	Reference
Bare detector (theoretical limit)			14	Šutej *et al.* (1988)
Diffusion	NRPB	CR-39	2.5	Miles *et al.* (1996)
Membrane Permeation	TRACK ETCH	CR-39	7.1	Nelson (1987)
	IJS	CR-39	6.2	Whysall *et al.* (1991)
	KfK	Makrofol E	0.67	Miles and Olast (1990)
	SUMS	Makrofol E	1.64	Doi *et al.* (1994)
Bag Permeation	ENEA	LR 115	0.49	Miles and Olast (1990)
Charcoal Collection	IJS	CR-39	545	Šutej *et al.* (1993)
Electret Collection	BARC	CR-39	2,486	Kotrappa *et al.* (1982)
Electrostatic Collection	KUMM	CR-39	5,000	Miyake *et al.* (1987)

[1] NRPB - National Radiological Protection Board (UK),
TRACK ETCH - Trade name of a dosimeter developed by Terradex Co. (USA),
IJS - "Jožef Stefan" Institute (Slovenia),
KfK - Kernforschungszentrum Karlsruhe (Germany),
SUMS - Shiga University of Medical Science (Japan),
ENEA - Ente Per le Nuove Technologie, L'Energia e L'Ambiente (Italy),
BARC - Bhabha Atomic Research Centre (India),
KUMM - Kobe University of Mercantile Marine (Japan).

foils). The most important parameters are (Brager and Revzan, 1991; Nazaroff and Nero, 1988): (i) electric field and ionization in air, (ii) concentration, size and content of airborne particles in the air, (iii) concentration of trace gases (CO_2, NO_x, SO_x, etc.) in the air, and (iv) thermodynamic parameters (temperature, humidity, precipitation, pressure, air flow). In addition, the material properties (shape of surface, composition, etc.) also influence the radon daughters' interaction with surfaces.

The stability of the dosimeter response is also dependent on the environmental conditions. The dependence of the stability of dosimeter response on the manufacturing process, ultraviolet radiation, oxygen, humidity and temperature have been investigated (Csige *et al.*, 1988, 1991; Frank *et al.*, 1986; Heins and Enge, 1986; Šutej *et al.*, 1988; Tidjani, 1991). However, the physics of these processes is still not fully understood. Large and uncontrolled variations in the efficiency for alpha particles emitted by radon and its decay products were observed in some detectors exposed to solar light (Šutej *et al.*, 1988). Fading effects which change the properties of polymeric track detectors such as etching rates during heating have been observed. O_2 treatment deteriorates the surface quality of nondoped CR-39, creates bad transparency and therefore results in decreased accuracy of measurements when using transmitted illumination (Heins and Enge, 1986). The beneficial effect of antioxidant doping of CR-39 on the stability of the material and of the latent tracks was confirmed (Frank *et al.*, 1986). It was observed that the etch induction time increases and the etch rate ratio, V_T/V_B (where V_T is the track etch rate and V_B is the bulk etch rate), decreases if the detector is irradiated in a vacuum, because of the outgassing of oxygen from the detector itself (Csige *et al.*, 1988). UV exposure can drastically change the properties of polymeric track detection (Tidjani, 1991). It can sensitize detectors owing to an increase in etch rate ratio. If the UV irradiation is too strong, V_B can drastically increase and hence the detector sensitivity is decreased. Various chemical and mechanical treatments were shown to have negligible effects in filter-permeation samplers (Hess *et al.*, 1985).

Since the fading of latent tracks is enhanced at higher temperatures (Durrani and Bull, 1987), the response of a given dosimeter may be somewhat different in different geographical regions.

2.2.4.3 Separate measurement of radon isotopes

Radon isotopes can be separated most effectively by the so-called barrier technique (Ward *et al.*, 1977). While passing from one volume to a second across an appropriate barrier, the shorter-lived radon isotopes largely decay before reaching the detector placed in the second volume. Precise ^{220}Rn plus ^{222}Rn measurements can be made by the same device as used for a ^{222}Rn only measurement, with the single substitution of a permeable filter that allows free entry of all radon isotopes, but prevents radioactive aerosols and unattached daughters from entering the detector's sensitive volume. From eqs. (2.2.14) and (2.2.15) we can estimate that a barier with a mean delay time, for instance, of 1.7 days, reduces the ^{222}Rn concentration to a value of 76% of that in air. Recently Doi and Kobayashi (1994) improved the characteristics of the passive radon dosimeter developed by Urban and Piesch (1981) and produced a passive radon/thoron discriminative dosimeter which can differentially and accurately measure low levels of both radon and thoron in the outdoor environment.

2.2.4.4 Radon daughter measurements and determination of the equilibrium factor

Measurement of the radon concentration in air can be performed using the dosimeters described in subsection 2.2.2. However, the radiation-induced damage of the human respiratory tract is mainly the result of short-lived radon daughters, characterized by the potential alpha energy concentration. Thus estimation of the radiation exposure of the public requires long-term measurements of the potential alpha energy concentration of radon daughters during a period of some months. Since such measurements are difficult, in most cases the potential alpha energy is calculated from the measured radon concentration assuming an appropriate equilibrium factor. In general, the equilibrium factor F is defined by

$$F = \frac{W}{W_{eq}} \tag{2.2.27}$$

where W is the total potential alpha energy of the actual daughter mixture and W_{eq} is the total potential alpha energy concentration of the radon daughters in equilibrium with radon. The above equation can be transformed to (Dörschel and Piesch, 1993)

$$F = \frac{0.105\,A_1 + 0.516\,A_2 + 0.379\,A_3 + 5.254 \cdot 10^{-8}\,A_4}{A_0} \tag{2.2.28}$$

where A_0, A_1, A_2, A_3 and A_4 are the activity concentrations of ^{222}Rn, ^{218}Po, ^{214}Pb, ^{214}Bi and ^{214}Po, respectively.

The equilibrium factor can be estimated by means of short-term measurements of the radon daughter concentrations of ^{218}Po and ^{214}Po. Short-term measurements of radon and radon decay product concentrations in air can be performed by a variety of active techniques, described by Monnin and Seidel (section 1.2). However, such a value of F is not representative for longer periods of time.

The theoretical basis of radon daughter measurements using a purely passive system of etched track detectors was developed by Fleischer (1984) and Somogyi *et al.* (1984). In contrast to short-term measurements, many authors (e.g. Planinić and Faj, 1991; Šarenio and Guhr, 1991) have employed this concept in a so-called conventional difference method which estimates the total activity concentration of both alpha active radon daughters, ^{218}Po and ^{214}Po. Here the detector system usually consists of a diffusion sampler and a bare detector to determine A_o and $A_o + A_1 + A_4$, respectively, from the measured track densities. This method is characterized by an unsatisfactory uncertainity in the measurement of the derived $F = f((A_1 + A_4)/A_0)$. Recently a new method based on separate measurements of ^{222}Rn and ^{214}Po was proposed (Dörschel and Piesch, 1993). Taking into account the uncertainties of estimating the equilibrium factor, and also by considering the effect of the varying unattached fraction of radon daughters, conclusions were drawn on the suitability of both methods for practical use (Dörschel and Piesch, 1994). It was found that the uncertainty in F as estimated with the difference method is about a factor of 3 higher compared to the new method. The results are summarized in Fig. 2.2.9. In the latest intercomparison of passive radon detectors organized by the

Commission of the European Communities (Miles *et al.*, 1996), only 5 passive detector systems for radon daughter determination were tested. Out of 30 measurements only one result was in satisfactory agreement with those obtained by active measuring systems. This indicates the need for further research and development of radon-daughter dosimeters utilizing etched track detectors. There is an additional problem with radon daughter measurements using passive detectors, as illustrated by Fig. 2.2.5. Radon daughter levels are depleted by plate-out effects close to surfaces; so the measuring device alters the atmosphere it is measuring. Human beings alter the atmosphere in a different way: they are active sampling devices, removing the depleted layer as they breathe. Passive devices can never reproduce these conditions.

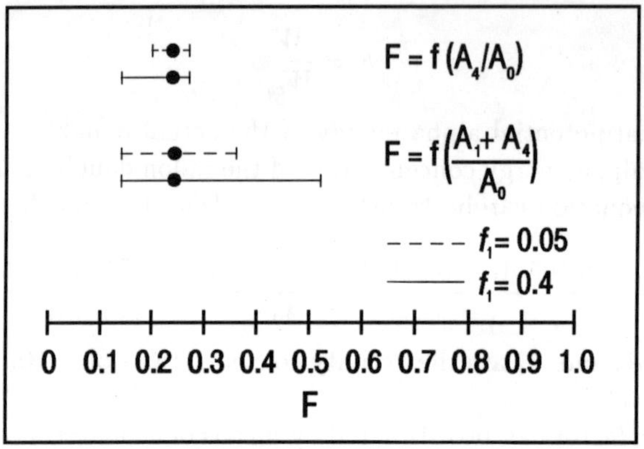

Figure 2.2.9: Measurement uncertainty for the estimation of the equilibrium factor, $F = 0.25$, by two methods, taking into account two different fractions, f_1, of unattached ^{218}Po (adapted from Dörschel and Piesch, 1994). The equilibrium factor is described by the function (f) of the activity concentrations of ^{222}Rn (A_o), ^{218}Po (A_1) and ^{214}Po (A_4) respectively. With a fixed value of $f_1 = 0.05$ the uncertainty for the equilibrium factor results in + 48% and - 40% for the 1st method ($F = f(A_1 + A_4)/A_o$); and +12% and -16% for the second method ($F = f(A_4)/A_o$) respectively.

The separate determination of radon and radon daughters is possible by a new promising technique involving charged particle spectroscopy based on simultaneous measurements of surface track size parameters (area, minor axis) and grey level value averaged over a single track using a microcomputer control system (Ilić *et al.*, 1993). The energy resolution was found to be 0.35 MeV, which is sufficient to discriminate between alpha-particle tracks induced by radon and radon daughters (Šutej, 1995).

2.2.4.5 Applicability of radon dosimeters

2.2.4.5.1 In-air measurements. Recent results of the performance of a variety of radon monitoring devices based on etched track detectors showed that a precision of 20%

appears to be a reasonable standard for indoor air (Culp *et al.*, 1993). The reliability of outdoor radon concentration measurements may, however be significantly affected by the environmental conditions (Šutej *et al.*, 1988). It has been pointed out that strong and uncontrolled variations in efficiency for the detection of alpha-particles emitted by radon and its decay products were observed in detectors exposed to solar light. Therefore the detectors must be protected from sunlight for quantitative and reproducible outdoor radon determinations. A holder for this purpose is shown in Alter *et al.* (1981). Moisture is also a potential problem for outdoor measurements. Namely, if the humidity is high a fall in temperature could cause condensation of moisture droplets on the detector foil, which would act as absorbers for alpha-particles, and thus lower the track densities in an uncontrolled way (Fleischer, 1988). By the use of a desiccant (Likes *et al.*, 1979), thermally insensitive material and/or appropriate design it is possible to prevent the condensation of moisture.

2.2.4.5.2 In-ground measurements. Subsurface radon monitoring by track-etching techniques is described in Chapter 4 of the present book by Balcázar; Fleischer; Hakl *et al.*; Khan; and Monnin and Seidel. Briefly, for geoscientific or geoengineering purposes the dosimeter is buried in the soil. The most useful and practical depth of the hole is in the range 60-70 cm. It should be mentioned that moisture can cause severe problems for in-ground measurements. With new improved dosimeters, however, month-to-month reproducibility of measurement is typically ± 13% (Fleischer, 1988).

2.2.4.5.3 Under-water measurements. Using most of the dosimeters described above, ^{222}Rn can be measured under water. For instance, for a membrane type permeation sampler it is necessary only to weight the cup so as to assure that the filter-closed end is submerged (Fleischer, 1988). For precise radon measurements the water temperature should simultaneously be measured, because the ^{222}Rn (water)/^{222}Rn (air) ratio varies with temperature.

2.2.5 Conclusion

Radon monitoring devices based on alpha-particle etched track detectors, on account of their excellent performance and unique characteristics, can be considered to be the main radon dosimeters nowadays. Their application in the following types of radon surveys is recommended: Screening Surveys, Confirmatory Measurements, Epidemiological Studies, Mine-Screening, Mine-Personnel, Mine-Area, Screening of Other Workplaces, Schools, etc.

Among the possible areas of further enquiry we cite: (i) development of high-sensitivity radon dosimeters based on utilization of a radon collector and etched track detector, (ii) research and development of radon-daughter dosimeters, and (iii) study of the interaction of radon with materials which can lead to a better understanding of some intrinsic properties of radon: e.g. with a "metalloid" - an element which lies on the diagonal of the Periodic Table between the true metals and nonmetals and which exhibits some of the characteristics of both. Finally, laboratories not previously engaged in this field/subject area are encouraged to consider including it in their service and research capabilities.

Acknowledgement. For critical reading of the manuscript the authors are most grateful to S. A. Durrani, R. L. Fleischer, J. C. H. Miles and L. Tommasino.

References

Abu-Jarad F., Wilson C.K. and Fremlin J.H. (1981) The registration of the alpha-particles from polonium isotopes plated-out on the surface of the plastic detectors LR 115 and CR-39. *Nucl. Tracks* **5**, 3, 285-290.

Alter H. W., Fleischer R. L., Gingrich J. E. and Murdock S. (1981) Passive integrating measurements of radon and thoron: Methods and field observations. In *Radiation Hazards in Mining* (ed. M. Gomez), AIME, New York, pp. 581-585.

Alter H.W. and Price P.B. (1972) Radon detection using track registration material. *U.S. Patent 3,665,194*.

Bartlett D. T. and Bird T. V. (1987) Technical specification of the NRPB radon personal dosemeter. NRPB-R208, London, HMSO.

Becker K.H. (1970) Personnel radon dosimeter. *U.S. Patent 3,505,523.*

Brager G.S. and Revzan K.L. (1991) Transport and deposition of indoor radon decay products - II. Influence of environmental conditions. *Atmosph. Environment* **25B**, 3, 359-368.

Brager G.S., Nero A.V. and Tien C.L. (1991) Transport and deposition of indoor radon decay products - I. Model development and validation. *Atmosph. Environment* **25B**, 3, 343-358.

Cohen B.L. and Cohen E.S. (1983) Theory and practice of radon monitoring with charcoal adsorption. *Health Phys.* **45**, 2, 501-508.

Csige I., Hunyadi I., Somogyi G. and Fujii M. (1988) Vacuum effect on the etch induction time and registration sensitivity of polymer track detectors. *Nucl. Tracks Radiat. Meas.* **15**, 1-4, 179-182.

Csige I., Hunyadi I. and Charvát J. (1991) Environmental effects on induction time and sensitivity of different types of CR-39. *Nucl. Tracks Radiat. Meas.* **19**, 1-4, 151-154.

Culp T., Conway R., Smith K. and Miller M. (1993) A performance study of four passive environmental radon detectors. *Radioac. Radiochem.* **4**, 3, 50-59.

Currie L.A. (1968) Limits for qualitative detection and quantitative determination. Application to radiochemistry. *Anal. Chem.* **40**, 3, 586-593.

Doi M. and Kobayashi S. (1994) The passive radon - thoron discriminative dosimeter for practical use. *Hoken Butsuri* **29**, 155-166.

Doi M., Kobayashi S. and Fujimoto K. (1992) A passive measurement technique for characterization of high-risk houses in Japan due to enhanced levels of indoor radon and thoron concentrations. *Radiat. Prot. Dosim.* **45**, 1-4, 425-430.

Doi M., Fujimoto K., Kobayashi S. and Yonehara H. (1994) Spatial distribution of thoron and radon concentrations in the indoor air of a traditional Japanese wooden house.

Health Phys. **66**, 1, 43-49.

Dörschel B. and Piesch E. (1993) A new approach to estimating the equilibrium factor between radon and its daughters. *Radiat. Prot. Dosim.* **48**, 2, 145-151.

Dörschel B. and Piesch E. (1994) Effect of varying unattached fraction of radon daughters on the measurement of the equilibrium factor using nuclear etched track detectors. *Radiat. Prot. Dosim.* **54**, 1, 41-45.

Durrani S.A. and Bull R.K. (1987) *Solid State Nuclear Track Detection: Principle, Methods and Applications.* Pergamon Press, Oxford.

Evans R.D. (1969) Engineers' guide to the elementary behaviour of radon daughters. *Health Phys.* **17**, 229-252.

Fleischer R.L. (1984) Theory of passive measurement of radon daughters and working levels by the nuclear track technique. *Health Phys.* **47**, 2, 263-270.

Fleischer R.L. (1988) Radon in the environment - opportunities and hazards. *Nucl. Tracks Radiat. Meas.* **14**, 4, 421-435.

Fleischer R.L. and Mogro-Campero, A. (1978) Mapping of integrated radon emanation for detection of long-distance migration of gases within the earth: Techniques and principles. *J. Geophys. Res.* **83**, B7, 3539-3549.

Fleischer R.L., Price P.B. and Walker R.M. (1965) Solid state track detectors: applications to nuclear science and geophysics. *Ann. Rev. Nucl. Sci.* **15**, 1-28.

Fleischer R.L., Turner L.G. and George A.C. (1984) Passive measurement of working levels and effective diffusion constants of radon daughters by the nuclear track technique. *Health Phys.* **47**, 1, 9-19.

Fleischer R.L., Giard W.R., Mogro-Campero A., Turner L.G., Alter H.W. and Gingrich J.E. (1980) Dosimetry of environmental radon: Methods and theory for low-dose integrated measurements. *Health Phys.* **39**, 957-967.

Frank C. and Benton E.V. (1982) Properties of a small radon diffusion chamber with plastic track detectors. In *Proc. 11th Int.Conf. on Solid State Nuclear Track Detectors* (eds. P.H. Fowler and V.M. Clapham), Bristol, 1981, Pergamon Press, Oxford, pp. 531-538.

Frank A.L., Ogura K. and Benton E.V. (1986) Fading of latent tracks in CR-39 due to environmental conditions. *Nucl. Tracks* **12**, 1-6, 83-86.

Furlan G. and Tommasino L., eds. (1993) *Proc. 2nd Workshop on Radon Monitoring in Radioprotection, Environmental and/or Earth Sciences.* Trieste, 1991, World Scientific, Singapore.

Hadler N.J.C. and Paulo S.R. (1994) Indoor radon daughter contamination monitoring: The absolute efficiency of CR-39 taking into acount the plate-out effect and environmental conditions. *Radiat. Prot. Dosim.* **51**, 4, 283-296.

Hadler N.J.C., Innes P.J., Osorio A.M. and Paulo S.R. (1992) The influence of the size of materials surfaces on radon daughters plate-out. *Nucl. Instr. Meth.* **A313**, 290-294.

Hafez A-F. and Somogyi G. (1986) Determination of radon and thoron permeability through some plastics by track technique. *Nucl. Tracks Radiat. Meas.* **12**, 1-6, 697-700.

Hardcastle G. D., Howarth C. B., Naismith S. P., Algar R. A. and Miles J. C. H. (1996) NRPB etched-track detectors for area monitoring of radon. NRPB-R283 HMSO, London.

Heins T. and Enge W. (1986) Oxygen effect on the etch rate in CR-39 plastic detector. *Nucl. Tracks* **12**, 1-6, 87-89.

Hess C. T., Fleischer R. L. and Turner L. G. (1985) Field and laboratory tests of etched track detectors for ^{222}Rn: Summer-vs-winter variations and tightness effects in Maine houses. *Health Phys.* **49**, 65-79.

Humar M., Najžer M. and Ilić R. (1982) Moisture-insensitive probes for separate measurements of radon isotopes. In *Proc. 11th Int. Conf. on Solid State Nuclear Track Detectors* (eds. P.H. Fowler and V.M. Clapham), Bristol, 1981, Pergamon Press, Oxford, pp. 549-552.

Humar M., Šutej T., Skvarč J., Mljač L., Radež M. and Ilić R. (1992) Indoor and outdoor radon survey in Slovenia by etched track detectors. *Radiat. Prot. Dosim.* **45**, 1-4, 549-552.

Iida T., Ikebe Y., Hattori T., Yamanishi H., Abe S., Ochifuji K. and Yokoyama S. (1988) An electrostatic integrating ^{222}Rn monitor with cellulose nitrate film. *Health Phys.* **54**, 2, 139-148.

Ilić R. (1990) Damage track detectors for alpha particle registration-track formation and detector processing. In *Proc. Int. Workshop on Radon Monitoring in Radioprotection, Environmental Radioactivity and Earth Sciences* (eds. L. Tommasino *et al.*), Trieste, 1989, World Scientific, Singapore, pp. 133-144.

Ilić R. and Rant J. (1991) The search for cold nuclear fusion with track-etch and bubble damage detectors. *Nucl. Tracks Radiat. Meas.* **19**, 1-4, 619-625.

Ilić R., Rant J., Šutej T., Doberšek M., Krištof E., Skvarč J. and Koželj M. (1990) Investigation of deuterium-deuterium fusion reaction in cast, annealed and cold rolled paladium, *Fusion Techn.* **18**, 3, 505-511.

Ilić R., Šutej T., Skvarč J. and Humar M. (1993) Current research on nuclear tracks at J. Stefan Institute. *Nucl. Tracks Radiat. Meas.* **21**, 3, 397-403.

Khan H.A., Qureshi I.E. and Tufail M. (1993) Passive dosimetry of radon and its daughters using solid state nuclear track detectors (SSNTDs). *Radiat. Prot. Dosim.* **46**, 3, 149-170.

Kotrappa P., Dua S.K., Pimpale N.S., Gupta P.C., Namb K.S.V., Bhagwat A.M. and Samon S.D. (1982) Passive measurement of radon and thoron using TLD or SSNTD on electrets. *Health Phys.* **43**, 3, 399-404.

Likes R. S., Mogro-Campero A. and Fleischer R. L. (1979) Moisture - insensitive monitoring of radon. *Nucl. Instrum. Meth.* **159**, 395-400.

Miles J.C.H. and Olast M. (1990) Results of the 1989 CEC intercomparison of passive radon detectors. *UK National Radiation Protection Board*, Chilton, NRPB, M-251.

Miles J.C.H., Stares E.J., Cliff K.D. and Sinnaeve J. (1984) Results from an international intercomparison of etching techniques for measuring radon and radon decay products. *Radiat. Prot. Dosim.* **7**, 169-173.

Miles J. C. H., Algar R. A., Howarth C. B., Hubbard L., Risica S., Kies A. and Poffijn A. (1996) *Results of the 1995 European Commission intercomparison of passive radon detectors.* European Commission, Directorate - General XII, Brussels, EUR 16949 EN.

Miyake H., Oda K. and Michijima M. (1987) Portable and high-sensitive apparatus for measurement of environmental radon using CR-39 track detector. *J. Appl. Phys.* **4**, 607-610.

Nazaroff W.W. and Nero A.V., eds. (1988) *Radon and its Decay Products in Indoor Air.* John Wiley and Sons, New York.

Nelson R.A. (1987) Measurement uncertainties of long term ^{222}Rn averages at environmental levels using α-track detectors. *Health Phys.* **53**, 447-453.

Nikezić D., Baixeras C. and Kostić D. (1996) Sensitivity determination and optimisation of a cylindrical diffusion chamber, for radon measurements, with a CR-39 detector. *Nucl. Instr. Meth.* **A373**, 290-298.

Perelygin V.P., Vater P., Ilić R. and Durrani S.A., eds. (1995) *Proc. 17th Int. Conf. on Nuclear Tracks in Solids.* Dubna, 1994, *Radiat. Meas.* **25**.

Planinić J. and Faj Z. (1990) Equilibrium factor and dosimetry of Rn by a nuclear track detector. *Health Phys.* **59**, 3, 349-351.

Porstendörfer J. (1984) Behaviour of radon daughters products in indoor air. *Radiat. Prot. Dosim.* **7**, 1-4, 107-113.

Porstendörfer J., Reineking A. and Becker K.H. (1987) Free fractions, attachment rates, and plate-out rates of radon daughters in houses. *Radon and its decay products* (ed. P.K. Hopke), American Chemical Society, Washington DC, Series 33, pp. 285-300.

Porstendörfer J. and Reineking A. (1992) Indoor behaviour and characteristics of radon progeny. *Radiat. Prot. Dosim.* **45**, 1-4, 303-311.

Rannou A., Jeanmarie L., Tymen C., Mouden A., Naour E., Parmentier N. and Renouard H. (1986) Use of cellulose nitrate as radon and radon daughter detectors for indoor measurements. *Nucl. Tracks Radiat. Meas.* **12**, 1-6, 747-750.

Sarenio O. and Guhr A.A. (1991) A passive individual dosimeter for integrating measurements of the radon daughter products contents in air. *Nucl. Tracks Radiat. Meas.* **19**, 1-4, 395-400.

Scarpitta S. C. and Harley N. H. (1990) Adsorption and desorption of noble gases on activated charcoal: I. ^{133}Xe studies in monolayer and packed bed. *Health Phys.* **59**, 4, 383-392.

Skvarč J. and Ilić R. (1994) Radonography. *Nucl. Instr. Meth.* **B88**, 430-434.

Somogyi G., Paripás B. and Varga Zs. (1984) Measurement of radon, radon daughters and thoron concentrations by multi-detector devices. *Nucl. Tracks Radiat. Meas.* **8**, 1-4, 423-427.

Šutej T. (1994) Sensitivity of a combined charcoal and etched track technique for ^{222}Rn. *Radiat. Prot. Dosim.* **56**, 1-4, 255-258.

Šutej T. (1995) *Interaction of Radon and its Progenies with Solid State Surfaces.* Ph.D. Thesis, University of Maribor, Maribor (in Slovene).

Šutej T. and Ilić R. (1991) Plateout determination of radon decay products by autoradiography. *Nucl. Tracks Radiat. Meas.* **19**, 1-4, 425-426.

Šutej T., Ilić R. and Najžer M. (1988) Response of track-etch dosimeters to environmental radon. *Nucl. Tracks Radiat. Meas.* **15**, 1-4, 547-550.

Šutej T., Križman, M. and Ilić R. (1991) A new technique for radon measurements based on combination of track-etch detector and activated charcoal. *Nucl. Tracks Radiat. Meas.* **19**, 1-4, 423-424.

Šutej T., Ilić R. and Križman M. (1993) Response of a new charcoal-etched track radon dosimeter. *Nucl. Tracks Radiat. Meas.* **21**, 3, 441-443.

Teterev Yu., Tretyakova S.P., Golovchenko A.N., Ilić R. and Skvarč J. (1995) Determination of the absolute radon concentration in air using an electrostatic method and CR-39 detector. *Radiat. Meas.* **25**, 1-4, 645-646.

Tidjani A. (1991) Property modifications in UV irradiated polymeric track detectors. *Nucl. Instr. Meth.* **B58**, 43-48.

Tommasino L. (1988) Assessment of natural and man-made alpha emitting radionuclides. *Nucl. Tracks Radiat. Meas.* **15**, 1-4, 555-565.

Tommasino L. (1990) Radon monitoring by alpha-track detection. *Nucl. Tracks Radiat. Meas.* **17**, 1, 43-48.

Tommasino L., Cherouati D.E., Seidel J.L. and Monnin M. (1986) A plastic-bag sampler for passive radon monitoring. *Nucl. Tracks Radiat. Meas.* **12**, 1-6, 681-684.

Tommasino L., Furlan G., Khan H.A. and Monnin M., eds. (1990) *Proc. Int. Workshop on Radon Monitoring in Radioprotection, Environmental Radioactivity and Earth Sciences.* Trieste, 1989, World Scientific, Singapore.

Urban M. and Piesch E.A. (1981) Low level environmental radon dosimetry with a passive track etch detector device. *Radiat. Prot. Dosim.* **1**, 2, 97-109.

Urban M. and Piesch E.A. (1982) A passive radon dosimeter for survey programs in dwellings. In *Proc. 11th Int. Conf. on Solid State Nuclear Track Detectors* (eds. P.H. Fowler and V.M. Clapham), Bristol, 1981, Pergamon Press, Oxford, pp. 577-580.

Vasudevan L. and McLain M. E. (1994) Atmospheric pressure effects on the calibration constant of alpha-track radon detectors. *Health Phys.* **66**, 3, 318-326.

Ward W.J. III, Fleischer R.L. and Mogro-Campero A. (1977) Barrier technique for separate measurement of radon isotopes. *Rev. Sci. Instrum.* **48**, 11, 1440-1441.

Weidenbaum B. (1970) *Radon Emanometry Test with Track-Etch Film.* Rob Claims Cities Service Minerals, General Electric Vallecitos Nuclear Laboratory Report, October, Pleasanton, CA.

Whysall K., Miles J.C.H. and Olast M. (1996) *Results of the 1991 CEC Intercomparison of Passive Radon Detectors.* European Commission, Directorate - General XII, Brussels, EUR 15156 EN.

L. Tommasino[1]

2.3 Track Registration: Etching and Counting Methods for Nuclear Tracks

2.3.1 Introduction

The first intentional etching of particle tracks in insulators was made by Young (1958), but his work escaped general notice until later. It was Price and Walker (1962), who, apparently unaware of Young's work, showed that etching solutions preferentially attack the damaged regions, producing enlarged pits that can be easily observed under an optical microscope, appearing similar to tracks in nuclear emulsions. Since then, etchable tracks have been observed in many crystals, in glasses and in a wide variety of plastics (Fleischer *et al.*, 1975). Different materials vary considerably in sensitivity, with minerals and glasses the least sensitive and organic materials being the most sensitive. The organic materials, typically used for the detection of radon and its daughters (Frank and Benton, 1977), are made of cellulose nitrate (LR 115, CN 85), bisphenol-A polycarbonate (Lexan, Makrofol), and polyallyl di-glycol carbonate (PADC). The PADC, also known as CR-39 (Columbia Resin-1939 from Pittsburg Glass Co.), is by far the most sensitive material, which is capable of recording alpha particles with a wide range of energies (Cartwright *et al.*, 1978; Henshaw *et al.*, 1982; Durrani and Bull, 1987).

2.3.2 Chemical Etching

2.3.2.1 Track formation and etchability

When a charged particle traverses matter, the distribution of ionisation or deposited energy as a function of distance from the particle trajectory is largely a function of the particle velocity and its charge. A low-velocity particle loses its energy through many low-energy collisions; this energy is confined to a small cylindrical volume around the particle trajectory. This contrasts sharply with fast particle interactions, where a substancial fraction of the energy loss is by production of high-energy electrons. Because of their considerable range, these electrons extend the ionisation or deposited energy over a much larger volume. The quantity which determines the chemical track reactivity is the energy deposited in the close vicinity of the particle trajectory. In fact, the most important characteristics of damage track detectors is that there is a minimum density of damage along the core of the track that will permit the track to be enlarged by etching. The whole of the energy loss does not contribute to this damage. That part delivered by

[1]ANPA - Agenzia Nazionale per la Protezione dell'Ambiente, Rome, Italy

high-energy delta rays creates damage that is too far from the track core to be effective, so far as etching is concerned.

In Fig. 2.3.1, the primary ionisation, which is proportional to the energy deposited close to the track, is shown plotted against the particle's velocity for different ions (Fleischer *et al.*, 1967).

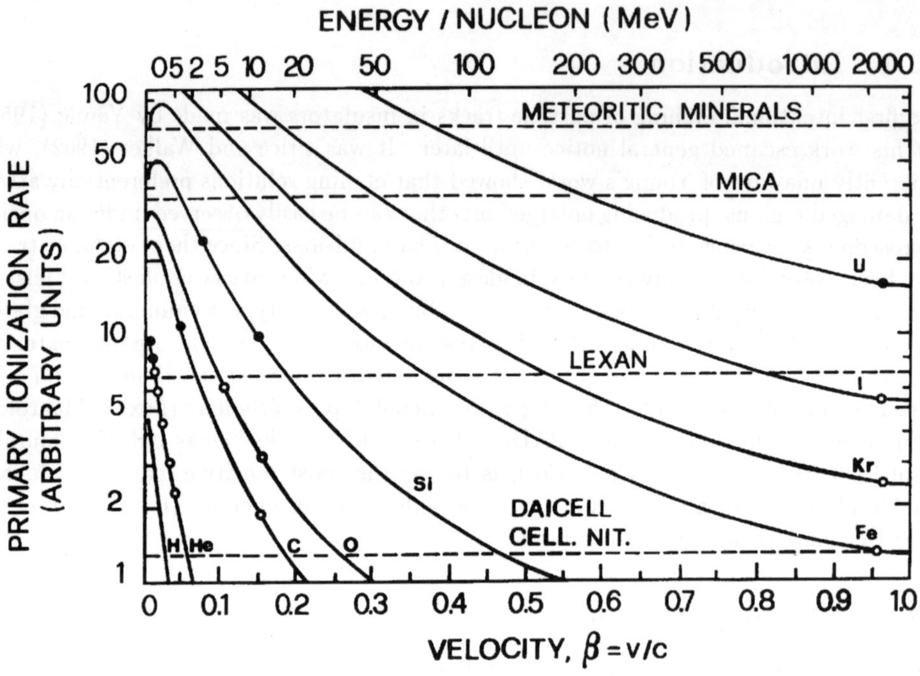

Figure 2.3.1: Primary ionization rate for various heavy ions (solid curves) versus particle velocity and track etch threshold for various materials (dashed lines).

The solid points indicate that tracks were found in Lexan polycarbonate, while the open circles indicate that these particles failed to register. The horizontal dashed lines show that there is a threshold for the primary ionisation rate for different materials, whose value (especially for organic polymers) depends on such details as manufacturing processes of the plastic, the etching conditions and thermal treatments pre- and post-irradiation (Griffith and Tommasino, 1990). The existence of this threshold is one of the most valuable characteristics of damage track detectors, since particles whose primary ionisation rate is below the critical value of a given material will not be registered. This explains their ability to discriminate against large fluxes of lighter energetic particles, electrons, gamma rays, etc.

2.3.2.2 Etching mechanisms

When a polymer is subjected to an etchant (that is, a polymer degrader rather than a solvent), the surface of the polymer is removed with a bulk etch rate V_B. As a result of the passage of a highly ionizing particle in a polymer, a chemically reactive damage trail may be produced, characterized by a different etching rate known as the track etch rate V_T.

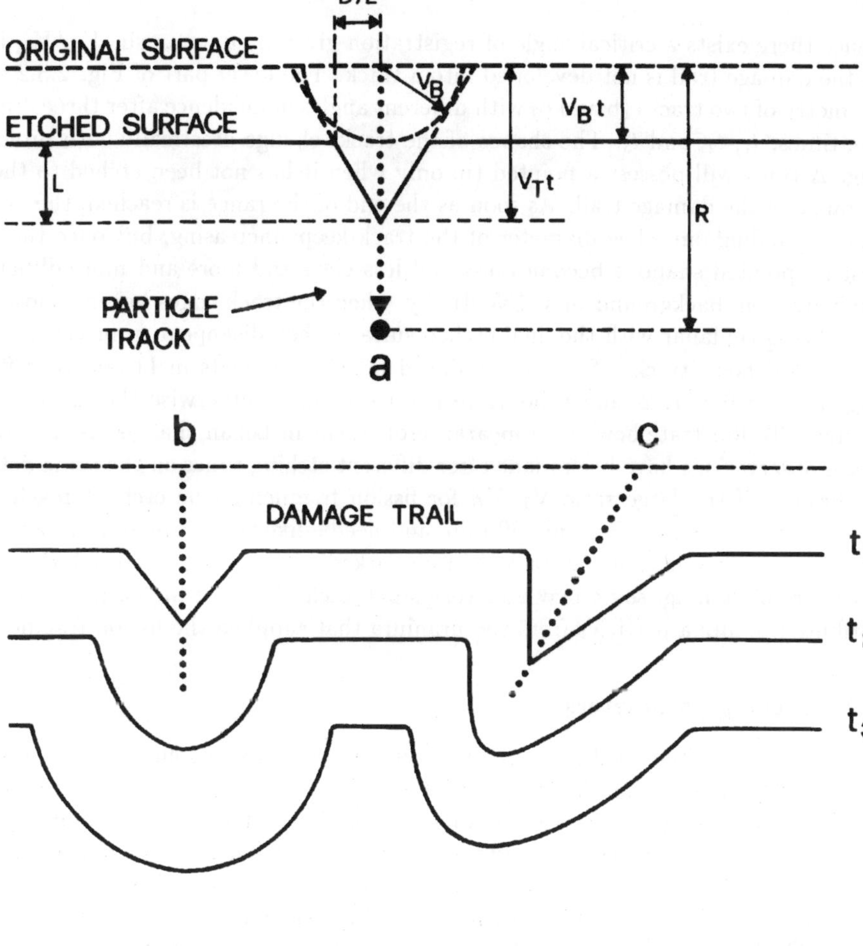

Figure 2.3.2: Track etching geometry in isotropic detector materials, where: R - range of the particle, D - diameter of the etched track, L - length of the etched track, t - etching time.

A track will be enlarged by etching only if the rate of etching along the track, V_T, exceeds the rate at which the surface is etched, V_B. If the track is not perpendicular to the surface then the component of V_T perpendicular to the surface must exceed V_B for registration to occur.

The upper part of Fig. 2.3.2 shows the track- etching geometry for a particle, normally incident on the detector surface. In this simple case the track-etching rate, V_T, is assumed constant and the bulk etching rate, V_B, isotropic.

For tracks incident on the surface at a dip angle (which is the angle between the track and the surface), the component of V_T perpendicular to the surface, $V_T sin\theta$, must exceed V_B, in order to produce a track

$$V_T sin\theta > V_B; \quad sin\theta > V_B/V_T \tag{2.3.1}$$

Hence there exists a critical angle of registration given by $\theta_c = \arcsin V_B/V_T$, below which the damage trail is not developed into a track. The lower part of Fig. 2.3.2 shows the geometry of two tracks (b and c) with different angles of incidence after three different etching times: t_1, t_2, and t_3. The shapes of the tracks change drastically with prolonged etching. A track will possess a pointed tip only when it has not been etched to the end of the range of the damage trail. As soon as the end of the range is reached, the pointed tip starts rounding out. The diameter of the track keeps increasing, but once the track has lost its pointed shape it becomes less and less clear and more and more difficult to discriminate from background pits. Eventually, when the track edges become closer and closer to being coplanar with the final etched surface, they disappear from view.

Fig. 2.3.3 shows tracks of neutron-induced fission fragments in Lexan for different etching times of 0.5, 1, 2, and 6 hours respectively, under otherwise the same etching parameters. To illustrate how the appearance of tracks in Lexan changes as the etching proceeds, observations have been made after different etching times on the same detector area. Because of the large ratio V_T/V_B for fission fragments, the etchant reaches the end of the damage region after only 30 min, and needle-like tracks appear. An extensive etching of 6 hours rounds out the track tip and markedly increases its width. Incidentally, after 6 hours of etching, some new and very small etch pits start appearing. They can be attributed to alpha particles from the uranium that supplied the fission fragments.

2.3.2.3 Etching parameters

The etchants most often used for organic polymers are aqueous solutions of potassium hydroxide (KOH) and sodium hydroxide (NaOH). The reactivity of alkaline etchants can be increased by adding alcohol to the aqueous solutions. For example, Somogyi and Gulyas (1972) have first shown that adding alcohol to KOH solution increases both the etching speed and the sensitivity of the polycarbonate detectors. However, when the same etchant is used for CR-39, the etching speed increases but the detector sensitivity decreases (Griffith and Tommasino, 1990; Somogyi and Hunyadi, 1980).

A large variety of data available in the literature deal with the studies of experimental etching parameters for the more common polymeric detectors (Enge *et al.*, 1974; Fleischer *et al.*, 1975; Hilderbrand and Benton, 1980). In practice, the parameters most important for control of the etching speed of the detector are the temperature and concentration of the etchant. While the V_T and V_B can be easily altered, the effects of temperature and

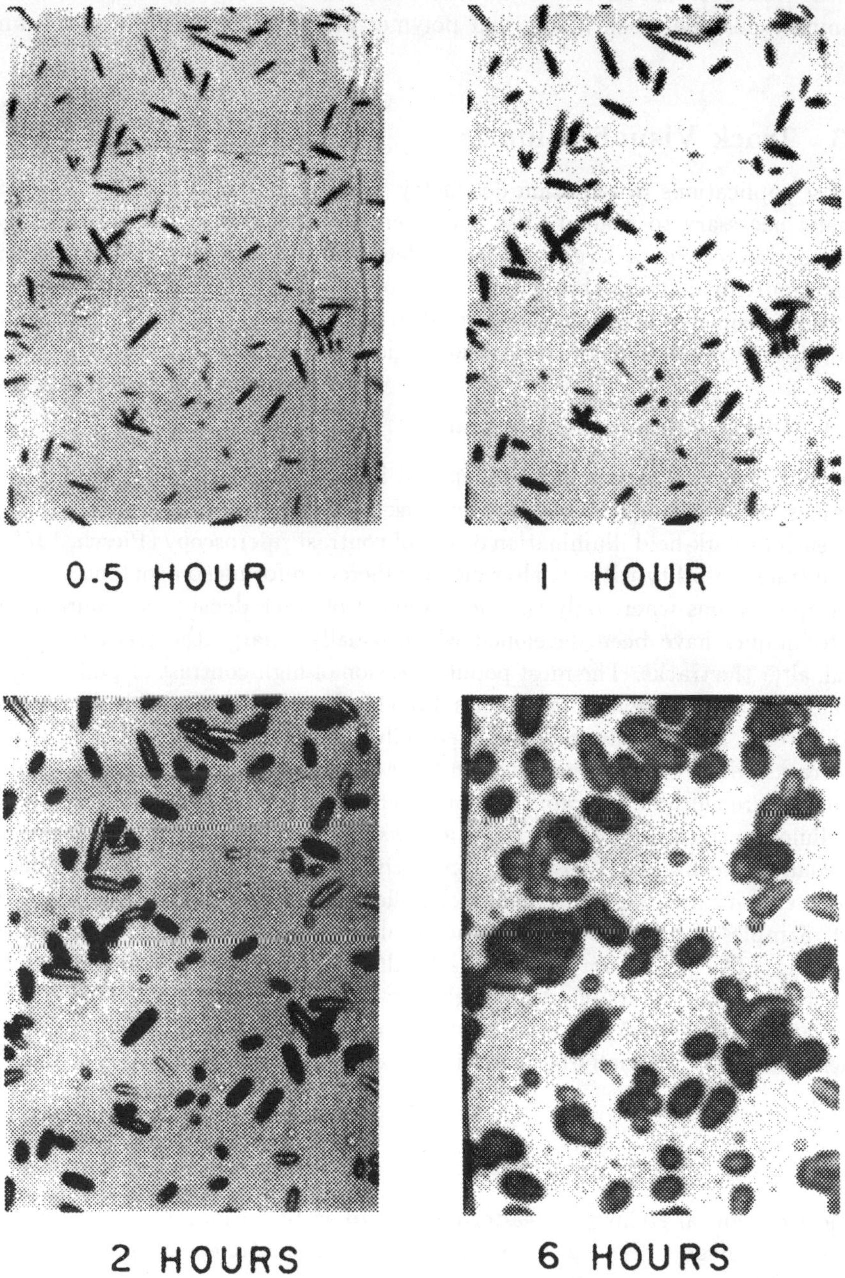

Figure 2.3.3: Tracks of neutron-induced fission fragments in Lexan for different etching times.

concentration of the etchant on sensitivity of the detector (V_T/V_B) are much less pronounced. However, care should be taken to avoid excessive etching temperatures in order to minimise possible track fading and polymer degradation (Griffith and Tommasino, 1990).

2.3.3 Track Visualization

For most applications in radiation dosimetry, and in particular in radon monitoring, it is usually necessary to evaluate low track densities (tens to a few hundred tracks per square centimeter) for a large number of detector foils. In this case, the readout by counting individual tracks through a microscope is a tedious process. Because of these difficulties, dosimetry applications have taken on new impetus with the development of simple, automatic methods of track visualization and counting.

2.3.3.1 Image enhancement of tracks

Images of tracks can be enhanced by special illumination techniques, by high contrast techniques, or by adding materials to the tracks themselves (Somogyi *et al.*, 1979). Techniques such as dark-field illumination or use of contrast microscopy (Piesch, 1971) simply increase track visibility without altering the inherent information content.

For applications where only the measurement of track density is required, high contrast techniques have been developed which usually enlarge the track images but, in general, alter the tracks. The most popular version of high-contrast or pinhole imaging is the commercial cellulose nitrate detector LR 115 (Barbier, 1970). This detector consists of a thin (6 or 12 μm) intensely red-dyed cellulose nitrate coating on a clear polyester base. Tracks which etch through the red layer appear as clear spots against a red background and are suitable for rapid readout methods. In one version of the LR 115, the thin cellulose nitrate film can be conveniently stripped after etching and counted by the spark counting technique (Cross and Tommasino, 1970), which will be described below.

An interesting image enhancement technique was first proposed by Dixon and Williams (1976). Samples of polycarbonate plastic are illuminated from the edge following chemical etching. The light from the edge is totally reflected internally, except at the etch pits or tracks. This results in a tiny bright spot that enhances the visibility significantly. Harvey and Weeks (1985) have successfully applied this technique for the registration of tracks on CR-39, specially because of the excellent optical properties of this detector material.

2.3.3.2 Track enlargement by electrochemical etching

The electrochemical etching (hereafter referred to as ECE) makes it possible to enlarge all types of damage tracks up to any size so that automatic counting becomes a simple procedure (Tommasino, 1970).

Electrochemical etching is obtained whenever damage track detectors are stressed by AC electric fields during the chemical etching (Sohrabi, 1974; Tommasino *et al.*, 1981).

For this type of etching, the irradiated plastic foil is sandwiched between two separate volumes of electrolyte, of which at least one is an etchant. An alternating high voltage is applied across the plastic foil by two electrodes which are immersed in the electrolyte solutions. The electrochemical etching conditions most often used are electric fields of the order of a few tens of kV cm^{-1} rms at frequencies of 1 to 10 kHz (Griffith and Tommasino, 1990).

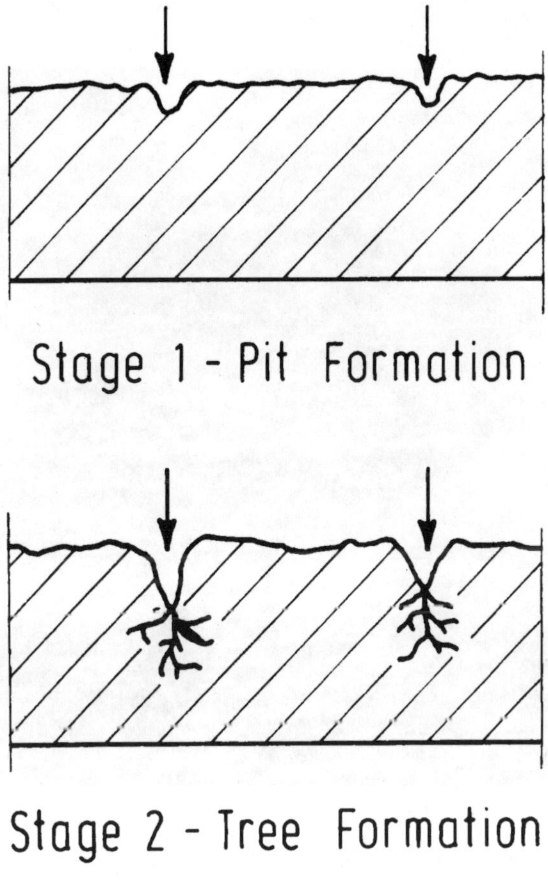

Figure 2.3.4: Electrochemical etching mechanisms.

While a complete understanding of the ECE phenomenon is very difficult, a simplified model has been proposed, which may be very useful in practice (Tommasino *et al.*, 1981). The electrochemical etching processes occur in two stages, as shown in Fig. 2.3.4. At the beginning of the electrochemical etching, the most important process is the formation of a track or an etch pit. This first step occurs under similar conditions as those required to etch tracks only chemically, as described earlier. Once the track has been formed, the electric field at the tip of the conductive needle can be many times greater than the average electric field applied to the insulator. The very high localised stress gradients can initiate electrical phenomena, which, when combined with the chemical action of the electrolyte, produce tree-shaped damage at the track tip. Fig. 2.3.5 shows one of these tree-types of damage, obtained at the tip of a plutonium alpha track in CR-39.

Figure 2.3.5: "Tree" induced at the tip of an alpha-particle track.

The set of ECE parameters to be chosen for any detector material can only be determined empirically. Just as a rule of thumb, the parameters can be separated roughly into two categories according to the two different stages of the ECE processes. For ECE track formation, the most important parameters are the same as those for chemical etching, while for tree initiation and propagation the electric field and frequency become predominant.

2.3.3.3 Alpha particle registration versus etching processes and detector materials

Among the three most common plastic detectors, LR 115 and PADC can easily register alpha particle tracks by chemical etching processes. By contrast, the third, i.e. polycarbonate detectors can be conveniently used for the registration of alpha particles only if electrochemically etched.

Because of its high sensitivity, CR-39 registers full-energy alpha-particles from radon-daughters plated-out on its surface, while the cellulose nitrate and the polycarbonate detectors are both insensitive to full-energy α's from plate-out. These latter characteristics are very important for the measurement of radon daughters or, in general, when using track detectors in the bare configuration(Domanski *et al.*, 1982). However, CR- 39 could be made insensitive to plate-out, if processed in such a way that the tracks of alpha particles (with energies of 6 and 7.7 MeV) from radon daughters are not revealed.

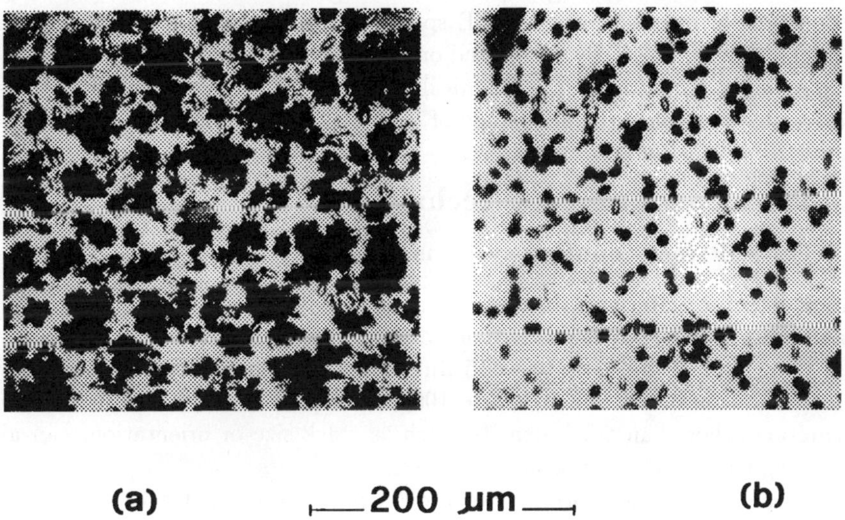

Figure 2.3.6: (a) Treeing components of electrochemically etched tracks; (b) Track components of electrochemically etched spots. For viewing conditions, see text.

This can be achieved by a two-steps etching consisting respectively of a chemical etching step for the revelation of only those tracks with energy below 6 MeV, followed by an electrochemical etching step to enlarge only those tracks which have been pre-etched (Espinosa *et al.*, 1981; Dajko and Turek, 1989).

The most attractive characteristic of ECE processes involves the possibility of using a variety of combinations of etching parameters and/or chemical and electrochemical etching processes for the purpose of enhancing the desired detector response.

For example, it is possible to achieve energy discrimination of alpha particles using different chemical etchings prior to the electrochemical etching of polycarbonate (Dörschel *et al.*, 1993; Hassib *et al.*, 1980).

Wong and Tommasino (1982) suggested the measurement of the diameters of tracks in polycarbonate etched only electrochemically for alpha-particle spectrometry.

The ECE-processed detectors can be used successfully when the track density is less than a few thousand tracks per square centimetre. For higher track densities both the spot diameter and ECE efficiency decrease as the track density increases(Al-Najjar *et al.*, 1979).

However, the combination of a chemical etching step for track formation prior to the electrochemical etching may be useful to increase the response linearity (Espinosa *et al.*, 1981).

Fig. 2.3.6 shows electrochemically etched tracks of alpha particles on CR-39, where the track formation has been obtained by a chemical pre-etching. When the CR-39 foil is observed under an optical transmission microscope, tracks appear as dark spots, as shown on the left-hand side of Fig. 2.3.6. By focusing close to the surface (specially with a reflection type of microscope) the ECE spots will disappear and only the chemically etched tracks appear, as clearly illustrated on the right-hand side of Fig. 2.3.6.

The possibility to separate the chemically etched tracks from the ECE tree-spots makes it feasible to exploit the advantage of both types of etching.

2.3.4 Track Measurement Techniques

2.3.4.1 Optical track counting systems

Since track detector materials are transparent, optical transmission microscopes are the easiest and most widely used counting tool. Most chemically etched tracks range in length from less than 1 μm to perhaps 10 or 15 μm. Therefore, the microscope must have an overall magnification in the range 100x to 1000x.

Information other than track density, such as track size or orientation, can also be obtained using an optical microscope.

In the case of radon monitoring, it is often required to count tracks in many microscopic fields for a large number of samples, so that the effort involved in the manual counting is worth avoiding.

Commercial pattern recognition systems have been developed for particle characterization problems in biology, metallurgy, aerosol physics, and other applications that require determination of the size and shape factors of objects in an image. The applications of these systems to problems of track counting and sizing is straightforward.

The least expensive class of commercial pattern recognition systems is the relatively simple bacterial colony counter, which is suitable for the counting of the large electrochemically etched spots (Griffith *et al.*, 1984).

The other end of the commercial image analysis spectrum includes some very sophisticated and powerful systems, which make it possible to obtain detailed information from a large number of samples (Benton *et al.*, 1982).

The ability of any system to count high track densities reproducibly depends on the resolving power, or ability to separate adjacent objects. In practice, as the track density increases, the relationship between instrumental and real track densities becomes less linear. The linear region can be extended by using higher magnifications, thus improving the resolution. However most systems should be able to count up to 1,000-2,000 tracks per cm^2 (typical range of tracks for radon measurements) without becoming nonlinear.

Additional attractive features of automatic systems include the ability to differentiate track size rapidly and automatically. This has value in distinguishing tracks from different types of particles, or particles with different energies (Henshaw, 1982).

2.3.4.2 Spark counting system

The most commonly used non-optical system for alpha particle registration is the spark counting (Cross and Tommasino, 1970) using the thin cellulose nitrate film $\approx$ 6 to 13 μm from strippable LR 115 detectors.

In this case, the film with etched-through holes is placed between two electrodes forming a simple capacitor. As a dc voltage is applied to the capacitor, the etched tracks begin to break-down sequentially. If both electrodes were solid metal, a given hole could continue to spark or break down. Therefore, a metallized film is used as the top electrode (typically aluminized Mylar). When breakdown occurs, the spark vaporizes a hole in the aluminium, effectively removing the conductor on one side (Cross and Tommasino, 1970). In addition, an RC circuit is needed to remove the applied voltage at the time of sparking and to avoid spark propagation in the neighboring holes (Tommasino, 1980). The linearity of spark counters depends on a number of variables, including thickness of the metallized layer on the electrode, RC circuit parameters, etc. A few thousand tracks per square centimetre is the typical upper limit.

References

Al-Najjar S.A.R., Bull R.K. and Durrani S.A. (1979) Electrochemical etching of CR-39 plastic: Applications to radiation dosimetry. *Nucl. Tracks* **3**, 169-172.

Barbier, J. (1970) Contrast improvement of images obtained in cellulose nitrate films by track etching methods. *Trans. Am. Nucl. Soc.* **13**, 530-539.

Benton E.V., Henke R.P., Hildebrand D.J., Casson R.M., Dreschel H., Heinrich W. and Rudot R. (1982) Computerized track analysis of high-energy accelerated argon-projectile fragments. In *Proc. 11th Int. Conf. on Solid State Nucl. Track Detectors* (eds. P. H. Fowler and V. M. Clapham), Bristol, 1981, Pergamon Press, Oxford, pp. 881-884.

Cartwright B.G., Shirk E.K. and Price P.B. (1978) A nuclear-track-recording polymer of unique sensitivity and resolution. *Nucl. Instr. Meth.* **153**, 457-460.

Cross W. G. and Tommasino L. (1970) Rapid reading technique for nuclear particle tracks in thin foils. *Radiat. Eff.* **5**, 85-89.

Dajko G. and Turek K. (1989). Electrochemical etching of CR-39 detectors irradiated with alpha particles. *Nucl. Tracks Radiat. Meas.* **16**, 241-246.

Dixon G.F. and Williams J.C. (1976) Track revelation in solid state track recorders and utilizing the light guide principle. *Nucl. Instr. Meth.* **135**, 293-298.

Domanski T., Chruscielewski W. and Swiatnicki G. (1982) The performance of passive differentiating track detectors containing a diffusion barrier. *Radiat. Prot. Dosim.* **2**, 27-33.

Dörschel B., Burgkhardt B., Lewitz J., Piesch E. and Streubel G. (1993) Experimental determination of the alpha energy window in makrofol etched track detectors for the separate measurement of the radon decay product Po-214 in air. *Radiat. Prot. Dosim.* **50**, 5-12.

Durrani S. and Bull R.K. (1987) *Solid State Nuclear Track Detection: Principle, Method and Application.* Pergamon Press, Oxford.

Enge W., Grabisch K., Beaujean R. and Bartholoma K. P. (1974) Etching behaviour of a cellulose nitrate plastic detector under various etching conditions. *Nucl. Instr. Meth.* **115**, 263-268.

Espinosa G., Garcia G., Golzarri J. L., Tommasino L. and Raponi F. (1981) Electrochemical etching registration and track-formation time. In *Proc. 11th Conf. on Solid State Nuclear Track Detectors* (eds. P. H. Fowler and V. M. Clapham), Bristol, 1981, Pergamon Press, Oxford, pp. 241-245.

Fleischer R.L., Price P.B., Walker R.M. and Hubbard E.L. (1967) Criterion for registration in dielectric track detection. *Phys. Rev.* **156**, 353-357.

Fleischer R.L., Price P.B. and Walker R.M. (1975) *Nuclear Tracks in Solids: Principles and Applications.* University of California Press, Berkeley.

Griffith R.V., McMahon T.A. and Espinosa G. (1984) A commercial bacterial colony counter for semiautomatic track counting. *Nucl. Tracks Radiat. Meas.* **8**, 215-220.

Griffith R.V. and Tommasino L. (1990) Etch track detectors in radiation dosimetry. In *Dosimetry of Ionizing Radiation* (eds. K. Kase *et al.*), Academic Press, Inc., vol. 3, pp. 323-426.

Frank A.L. and Benton E.V. (1977) Radon dosimetry using plastic nuclear track detectors. *Nucl. Track Detection* **1**, 149-154.

Harvey J.R. and Weeks A.L. (1985) A Neutron Dosimetry System Based on the Chemical Etch of CR-39. Comm. Eur. Communities Rep. EUR-9762, vol. 1, 519-524.

Hassib G.M., Piesch E. and Messera G.E. (1980) Electrochemical etching of alpha particles in polycarbonate and applications. In *Proc. 10th Int. Conf. on Solid State Nucl. Track Detectors* (eds. H. Francois *et al.*), Lyon, 1979, Pergamon Press, Oxford, pp. 239-332.

Henshaw D.L. (1982) Applications of CR-39 nuclear track detector in medicine and technology. *Phys. Technol.* **13**, 266-277.

Henshaw D.L., Amin S., Clapham V.M., Fowler P.H., Webster D.J., Thompson A. and O'Sullivan D. (1982) The form of the CR-39 response curve and its effects on particle resolution and areas of applications. In *Proc. 11th Int. Conf. on Solid*

State Nuclear Track Detectors (eds. P. H. Fowler and V. H. Clapham), 1981, Pergamon Press, Oxford, pp. 115-118.

Hilderbrand D. and Benton E.V. (1980) The chemical etching behaviour of cellulose nitrate. *Nucl. Tracks* **4**, 77-85.

Piesch E. (1971). The use of interference-contrast microscopy for nuclear particle recording in solids. *Zeiss Inf.* **76**, 58-62.

Price P. B. and Walker R.M. (1962) Electron microscope observation of etched tracks from spallation recoils in mica. *Phys. Rev. Lett.* **8**, 217-222.

Sohrabi M. (1974) The amplification of recoil particle tracks in polymers and its applications in fast neutron personnel dosimetry. *Health Phys.* **27**, 598-604.

Somogyi G. and Gulyas J. (1972) Methods for improving radiograms in plastic detectors. *Radioisotopy* **13**, 549-555.

Somogyi G. and Hunyadi I. (1980) Etching properties of the CR-39 polymeric nuclear track detector. In *Proc. 10th Int. Conf. on Solid State Nuclear Track Detectors* (eds. H. Francois *et al.*), Lyon, 1979, Pergamon Press, Oxford, pp. 443-452.

Somogyi G., Toth-Szilagyi M., Monnin M. and Gourcy J. (1979) Non-etching nuclear track visualization in polymers: Fluorescent and dyed tracks. *Nucl. Tracks* **3**, 151-156.

Tommasino L. (1970) Electrochemical etching of damage track detectors by pulse and sinusoidal waveforms. In *Proc. 7th Int. Colloquium on Corpuscular Photographs and Solid State Nuclear Track Detectors*, Barcellona, vol. 1, pp. 213-217.

Tommasino L. (1980) Solid dielectric detectors with breakdown phenomena and their applications in radioprotection. *Nucl. Instr. Meth.* **173**, 73-80.

Tommasino L., Zapparoli G. and Griffith R.V. (1981) Electrochemical etching I. Mechanisms. *Nucl. Tracks* **4**, 191-196.

Wong C.F. and Tommasino L. (1982) Energy discrimination of alpha particles by electrochemical etching of track detectors. *Nucl. Tracks* **6**, 17-24.

Young D. A. (1958) Etching of radiation damage in Lithium Fluoride. *Nature* **182**, 375-379.

J. C. H. Miles[1]

2.4 Calibration and Standardisation of Etched Track Detectors

2.4.1 Introduction

The principle of operation of passive radon detectors is very simple, and it might be thought that there is little to go wrong with them. However, international intercomparisons have shown that even experienced laboratories can make errors. The quality measures required to maintain satisfactory standards of performance can be considered within two areas: quality control within the laboratory and external quality assurance such as testing schemes and intercomparisons. Both of these are essential if accurate results are to be obtained.

2.4.2 Sources of Errors

2.4.2.1 Detector materials

A study of the performance of different laboratories in international intercomparisons of passive radon detectors (Miles and Sinnaeve, 1986) showed that LR-115, polycarbonate, and poly allyl diglycol carbonate (PADC or CR-39) could all be used to make accurate measurements. Differences in accuracy between laboratories were caused by differences in quality control rather than the materials used.

It has been the experiencc of many laboratories that different batches of detector material, and even different sheets within a batch, can vary in background and sensitivity. Variations in sensitivity are usually due to the manufacturing process, so are outside the control of the laboratory using the material. Background tracks may arise from flaws in the material or from accidental exposures to radon and its decay products. Exposure of bare detector materials to air even for a few hours can greatly increase background, and they should be kept covered as much as possible. Storage in radon-free air (aged in the cylinder for several weeks) or in nitrogen is preferable. It has also been found that conditions during cutting of the sheets can produce high backgrounds.

2.4.2.2 Design of detector

There are two basic designs of passive radon detectors, open and closed. Closed detectors admit radon gas but exclude radon decay products and dust. Radon gas inside the enclosure decays through its chain of decay products, producing a track density which is

[1]National Radiological Protection Board, Chilton, UK

proportional to the radon gas exposure of the detector. It is simpler to maintain quality control for this type of detector, as the sensitive element is protected from most environmental contaminants and light, and it is clear what quantity the device is measuring: integrated exposure to radon. There are various designs of closed detector. Providing they meet the requirement of admitting gases while excluding particulates, there seems little to choose between them for accuracy of radon measurement. It has been found that electrostatic charges on the inside surfaces of a closed detector can affect its performance (Urban, 1986). Radon decay products are electrically charged when formed, so electrostatic charges affect where they plate out. Different charges on the holders can cause different track densities to be found on detectors which were exposed together. To prevent this, holders made from metal or conducting plastic may be used, or holders may be dipped in a solution of a material which allows charge to leak away, such as detergent. The surface of the detector must also be made conducting either by covering it with a thin layer of conducting material or dipping it in an anti-static solution. A solution of 1 part of detergent (such as that used for washing dishes) in 500 parts of water has this effect.

Open detectors are more sensitive than closed detectors to radon and its decay products in the air. Their response depends on the detector material used. Open CR-39 detectors record tracks due to all alpha emitters suspended in the air in front of them and plated out on their surfaces. Their response varies widely, depending on atmospheric conditions (Miles *et al.*, 1984) and they will not be considered further here. It is not advisable to use bare CR-39 detectors. Kodak LR-115 is often used for open detectors. This material has the property of detecting only those alpha particles whose energies lie within a range of about 1.7-4.1 MeV (Jönsson, 1981). This means that radon decay products which plate out on the surface of open detectors made from LR 115 will not be detected, as their alpha particles have too high an energy. Only alpha particle emitters suspended in the air at some distance from the detector surface will be detected, as their alpha particles will be degraded in energy by their passage through the air. In practice, detectors of this type have a response which is quite closely proportional to their exposure to radon gas independent of atmospheric conditions and equilibrium factor, and indeed is similar to that of the closed detectors. Being open, however, they are more vulnerable to damage and the effects of environmental contaminants. They must not be exposed to sunlight for long periods, as this affects their properties.

2.4.2.3 Transit exposures

When the time spent by the detector in transit is very short compared to the exposure period, transit exposures are usually ignored as they will be insignificant. If the exposure in transit may be significant, the detectors can be sent out in heat-sealed radon-proof bags. Radon can diffuse easily through most materials, including polyethylene and PVC, so bags made of these materials should not be used. Aluminium foil is impermeable to radon but is easily torn and cannot be sealed effectively, so is also unsuitable. Heat-sealable plastic sheets or bags used commercially for preserving perishable foods are often made from polyamide or polyethylene terephthalate and are usually radon-proof, but should be tested before use. Transparent or aluminised foils and bags are available. They need to be heat sealed to exclude radon. Dudeny *et al.* (1995) showed that sealed bags leak in some cases, and recommended testing by placing the bags in a chamber with slightly reduced pressure. Bags which do not swell when the pressure is reduced are leaky.

2.4.2.4 Environmental effects

It has been shown (Homer and Miles, 1986) that heat and humidity in the presence of oxygen in the air can affect the sensitivity of etched-track detectors while they are being used to measure radon. The magnitude of the effect depends on the detector material and the etching and counting techniques used. In one study (Wrixon *et al.*, 1988) it was shown that when tracks in CR-39 were counted using a semi-automatic system, the sensitivity of the detector declined by 18% during exposure in homes for 6 months, resulting in an average sensitivity 9% lower than that initially measured. Corrections for such effects must be included in calculations of radon concentrations.

It is important that detectors are placed so as to ensure that the location of the measurement is representative, and to minimise any environmental effects. In particular, locations close to sources of heat and draughts are best avoided. Open LR 115 detectors should be placed so that direct sunlight does not reach them at any time of day.

2.4.2.5 Processing and counting of detectors

The choice of processing technique will affect the accuracy of the measurement. Track detectors are inherently linear in response - each unetched track is independent and remote from other tracks except at extremely high track densities. Etching and counting the tracks may however introduce some non-linearity, depending on the techniques used. In a study of the results from two Commission of the European Communities (CEC) intercomparisons, Miles and Sinnaeve (1986) found that non-linearity of response could be a problem when electrochemical etching is used. This is a well-recognised problem (Al-Najjar *et al.*, 1979), and linearity corrections are normally applied to results from electrochemical etching. The fact that some results from electrochemical etching were non-linear after application of a linearity correction implies that the non-linearity may be variable, and possibly sheet-dependent.

Whether chemical or electrochemical etching is used, close control of the etching process is required. Bulk chemical etching rates typically change by about 7% for each 1 K change in temperature. Since the surface area of a track is proportional to the square of the horizontal distance etched away, track areas typically change by about 15% for each 1 K change in etching temperature. The situation is more complex for electrochemical etching, but again the visible track size is strongly affected by temperature. If track counting techniques are employed which use visible track area to identify genuine tracks, small changes in etching temperature may result in significant changes in sensitivity. Such temperature changes may be caused by differences in the numbers of detectors etched together, or the material of the etching vessel. It is essential that the conditions of etching are closely monitored. The consistency of the etching can be checked by including a standard sample in each etching batch and checking that a parameter such as mean track area remains within predetermined limits.

When tracks are counted it is important that ones at the edge of a counting field are treated consistently: for instance a track may be counted only if the majority of its area is within the field. If this restriction is not observed, and for instance all tracks within or crossing the field are counted, the effective area counted will be larger than the field. It has generally been found that track counting by a trained microscopist is the most accurate technique, but it is slow and tedious. A microscopist can tell the difference between genuine tracks and other features better than an automatic system. The main requirement for accurate counting is experience: the microscopist should know the range of track sizes and shapes that are produced by alpha particles to ensure that spurious tracks are rejected. There is usually a range of track sizes going down to very small, faint tracks. The microscopist needs to be consistent in deciding what is and what is not a track, and needs consistent conditions (the same microscope, magnification and lighting) to achieve this. Different microscopists may have different criteria, and produce a different sensitivity for the same detectors.

Automatic systems for track counting vary widely, but generally use the darkness, area and often the shape of tracks to decide whether they are genuine or not. Such systems are very sensitive to the lighting conditions and focus of the microscope: it is important that all conditions are kept constant and regular checks are made on the conditions and the accuracy of counting using standard detectors.

Chemically etched track detectors that are automatically counted will give sub-linear results at high track densities, where the tracks overlap. This normally applies to a very small proportion of detectors in radon work, and either a linearity correction is applied or detectors are recounted by eye. As in the case of electrochemically etched detectors, the non-linearity may vary depending on the etching conditions and from sheet to sheet.

2.4.3 Quality Control Procedures

2.4.3.1 Acceptance criteria

The sensitivity and background track density of detector materials should be measured before detectors are issued and materials rejected if necessary. Elements taken systematically from across each sheet are used. One of the most important tests that laboratories apply before issuing passive radon detectors is a check on the mean background of each sheet of detector material. A commonly used criterion for judging sheets is to take the minimum detectable track density due to radon exposure as being twice the standard deviation on the background track density. Sheets which have too high a minimum detectable track density are rejected. Eg., if the background is 40 ± 25 cm^{-2}, then minimum detectable track density would be ≈ 50 cm^{-2}.

Sensitivity is normally less variable than background, but a check on this is desirable before issue of detectors. Sensitivity may be checked using a radon exposure or an exposure from an alpha-particle source such as ^{241}Am. However, there are potential dangers in using anything other than radon as the only means of calibration. The distribution of alpha energies and angles from a source will be different from that in a radon exposure, and the variation in sensitivity from sheet to sheet found with an alpha-particle source may be less than that found with radon exposures. Similarly, the reproducibility may be better when tested using an alpha particle source than when using radon. However, an alpha particle source can be useful for a quick check on the properties of a sheet of detector material and for tests of the approximate range of track densities over which the response of the system is linear.

Routine measurement of the percentage standard deviations on sets of passive detectors exposed together to radon is important to allow the estimation of the uncertainty on measurements made in the field. An increase in the standard deviation for the same number of counted tracks is a useful pointer to the existence of problems in the processing of detectors, and investigation of the source of the increase will be indicated.

2.4.3.2 Etching and counting checks

Because of the sensitivity of the etched track technique to etching conditions, these must be monitored carefully. This is particularly important if the tracks are to be counted automatically. The etching solution should be obtained from a reliable, consistent source; or if it is made up at the time of etching it should be tested. The temperature of the etching solution should be measured at the start and end of etching. It is helpful to include a test element exposed to alpha particles in each etching batch to monitor any variation in the etching conditions. The test elements for successive etching batches should all be taken from one sheet.

The best check of counting efficiency is to keep a standard set of detectors which are regularly recounted (at least daily for automatic systems, less often for microscopists). The mean size of tracks on exposed detectors should be compared with that of the test set to ensure that the track acceptance criteria remain appropriate. Mellander and Enflo (1992) describe methods used to optimise an automatic readout system for accurate radon measurements.

2.4.3.3 Blind tests

In order to check that the measurements are performed equally well for detectors exposed in the field and for those where the expected result is known, blind tests may be carried out. Some detectors issued for field exposure are diverted from normal, exposed without the knowledge of the issuing laboratories, and reintroduced into the system. One way of using these detectors is to expose them in pairs in the same location in a building as an additional test of reproducibility and field performance. Alternatively, they may be given a known radon exposure in the laboratory.

2.4.3.4 Tests on linearity, ageing, fading

There are various tests which need to be performed when setting up a radon measurement service, and which should be repeated from time to time to ensure that the service is performing as it should. These include tests of linearity and the effects of the environment on detectors.

Linearity checks may be carried out using detectors exposed to alpha-particle sources such as ^{241}Am, as it is difficult to achieve the range of exposures required using radon. Because of differences in track sizes between detectors exposed to radon and those exposed to alpha particle sources, this type of check is not suitable for calculating linearity corrections for non-linear systems, but only for establishing the approximate range of track densities over which a system responds linearly.

Measurements of the effects of the environment on the sensitivity of detectors can also be carried out. In one study (Wrixon *et al.*, 1988), pairs of detector elements were exposed within the same holder in homes over six months. One of each pair of elements was given an extra radon calibration exposure on return to the laboratory. After etching, the latter comparison with the results from the other element of the pair and with results from elements retained at the laboratory under nitrogen allowed the decline in sensitivity to be estimated. A simpler study can be carried out in the laboratory, though the conditions may not be quite the same as in homes. In this type of investigation, detectors are sealed in air in radon-proof bags. After storage, their performance is compared with detectors stored under nitrogen.

2.4.4 Routine Calibration

Attempts have been made to calculate the theoretical efficiency of passive radon detectors, but because the number of tracks counted depends on the etching and counting conditions these have not been useful for practical calibration. They can, however, give some insight into how the efficiency varies as the shape and size of the holder is changed.

Calibration of passive radon detectors needs to be traceable to a recognised standard such as those now provided by the National Institute of Standards and Technology in the USA and the National Physical Laboratory in the UK. This is normally accomplished by monitoring calibration exposures with active measurement equipment which has itself been compared with a recognised standard.

Such calibrations are usually carried out in specially constructed calibration chambers. International Atomic Energy Agency (IAEA, 1995) has identified three basic types of radon calibration chamber: research chambers used mainly by reference laboratories; large-volume chambers with the necessary technical support for large intercomparisons; and low-cost chambers that are readily accessible to organisations and users in different regions of the world. Such chambers provide users with access to instrument calibrations and evaluations based on primary standards.

A range of radon exposures may be obtained by the use of a constant radon concentration and different exposure times. To calibrate open detectors, ventilation and aerosol conditions should be adjustable to permit several different equilibrium factors to be obtained. A high aerosol concentration allows decay products to become attached to aerosols, retaining them in the air and resulting in a high equilibrium factor. Low aerosol concentrations leave the unattached decay products free to diffuse to surfaces, producing a low equilibrium factor. In the case of many low-cost chambers it will not be necessary to control temperature, humidity and ambient aerosols, but they should be monitored. Simple particle generators such as a wax candle may be used to generate high particle concentrations if required and inlet filters will reduce the aerosol concentration.

There should be enough mixing within the chamber to obtain homogeneity of the conditions. If only radon is of interest, natural convection may be sufficient. Care must be taken to ensure that radon decay products concentrations are uniform within the test chamber if open detectors are being calibrated. Equilibrium factor should be regulated by controlling the aerosol concentration and aerosol residence time.

2.4.5 Intercomparisons

Intercomparisons of passive radon detectors are important to ensure that radon measurements made by different laboratories, often in different countries, are carried out to a high and comparable standard. The laboratory tests in an intercomparison may include tests of linearity and of response to different equilibrium factors between radon and its decay products. The analysis of these results allows conclusions to be drawn about detector materials and processing and counting techniques, as well as the information about the performance of individual laboratories.

In order to assist in assuring reliabilty and comparability of radon measurements obtained by different institutions worldwide, the International Radon Metrology Programme (IRMP), a system of Reference, Technical Support and Regional Coordinating Laboratories has been established (IAEA, 1996). Under the coordination of the IAEA and the European Communities (EC), and with the assistance of the University of Salzburg (Scientific Secretariat), the IRMP has implemented the structure shown in Fig. 2.4.1.

The structure includes reference laboratories to provide guidance on the scientific issues concerning the metrology of radon, thoron and their decay products, in particular in the areas of laboratory and field calibration of measurement devices, field sampling and survey methods and analytical procedures. Technical support laboratories provide technical support to the Regional Coordinating Laboratories in the form of inter-calibrated exposure chambers, used to conduct intercomparison exercises for radon, thoron and their decay products under defined laboratory conditions. Regional coordinating laboratories provide logistical and, on a limited scale, also technical assistance in the coordination and conduct of regional activities related to Quality Assurance Programmes concerning radon, thoron and their decay products.

The IRMP is organized in accordance with a charter which outlines the objectives, scope and requirements for participation in IRMP activities. In particular the following procedures are established:

> End users requiring calibration of passive devices will pass them on to their national laboratories. The national laboratories may calibrate the detectors or pass them on to the Regional Coordinating Laboratory for calibration by the Technical Support Laboratory.

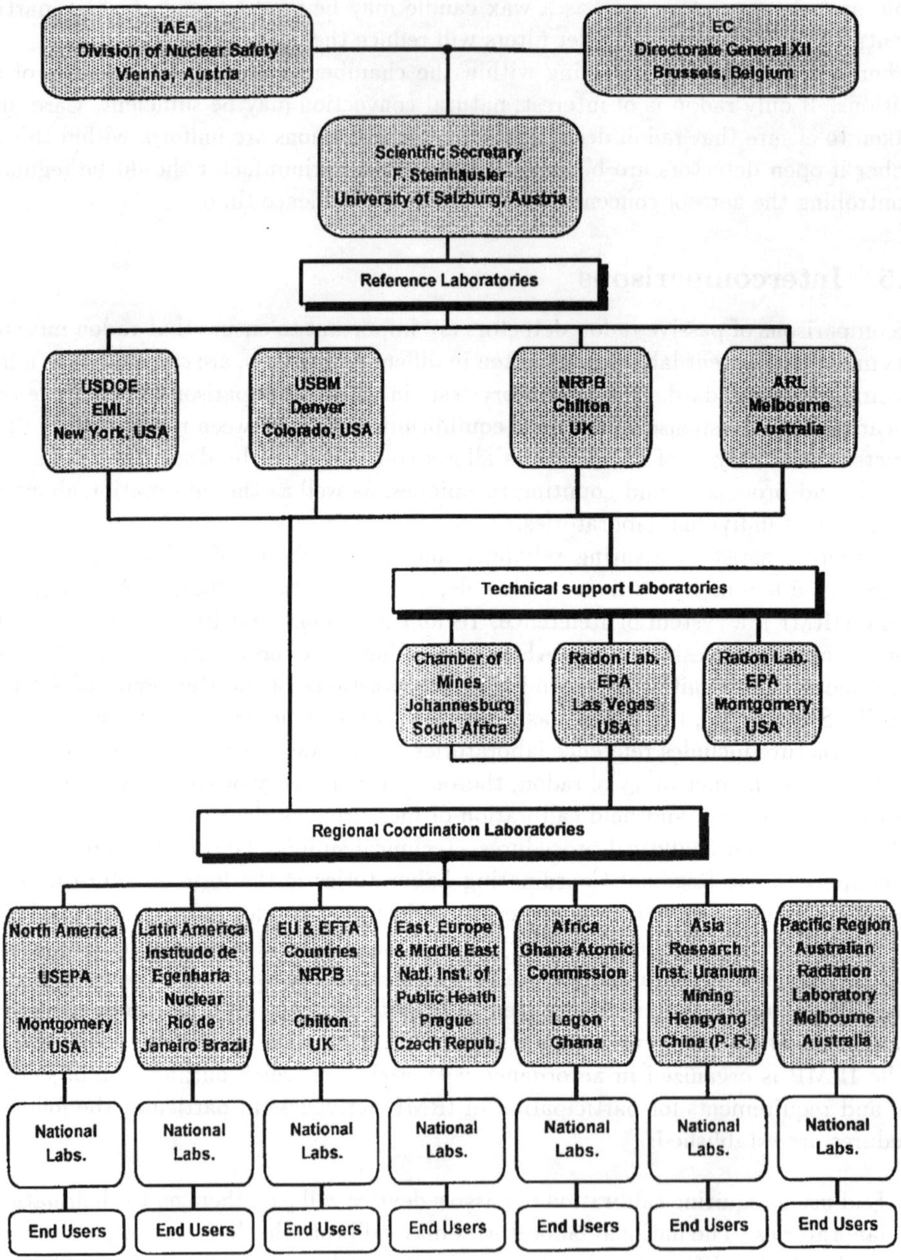

Figure 2.4.1: International Radon Metrology Programme.

Technical Support Laboratories will run calibration exercises at their facilities. They will also intercompare their measurement techniques with the Reference Laboratories.

Intercomparisons of passive detectors in Europe, America and Australia are reported by Miles *et al.* (1984), Knutson (1988), Solomon *et al.* (1988), Miles and Olast (1990), Hopper *et al.* (1993), Whysall *et al.* (1996), and Sensintaffar and Steinhäusler (1996). These intercomparisons have been particularly useful in identifying problems with radon measurements in individual laboratories and in providing a common standard of calibration.

The tests in such an intercomparison may be carried out in the laboratory and in homes. Exposing large numbers of detectors in a home over several months and monitoring their radon exposure accurately poses problems. In the CEC series of intercomparisons the tests in homes did not reveal any problems with closed detectors which were not found in the laboratory tests, so this type of test is one which may be omitted if only closed detectors are used.

2.4.6 Testing and Approval of Processing Laboratories

More formal testing schemes are operated to demonstrate compliance with minimum standards in some countries. For measurements of radon in homes in the USA, the Environmental Protection Agency has produced protocols for radon measurement procedures and quality control (US EPA, 1993). Compliance with these protocols is not compulsory, but does allow a laboratory to be included on an approved list. The EPA requirements place considerable emphasis on appropriate measurement procedures and record keeping. To check the accuracy of the measurements, EPA has established a Radon/Radon Progeny Measurement Proficiency Evaluation and Quality Assurance Program.

In the UK, there is one scheme for radon detectors issued to workers, and one for detectors issued to homes. Exposure of workers to radiation is controlled by the Health and Safety Executive (HSE, 1985). HSE requires a laboratory to demonstrate its competence to make the measurements by submitting detailed information on staff, facilities, methods and record keeping, and by taking part in regular performance tests. A laboratory which does not meet the HSE requirements cannot be used to assess the exposure of workers to radon.

NRPB operates a scheme for laboratories offering radon detectors for use in homes in the UK which is similar to the EPA one (Cliff *et al.*, 1991). Participation in this scheme is not compulsory, but local authorities in the UK have been requested by the government to give favourable consideration only to grant applications supported by monitoring results from laboratories successfully participating in the scheme. In Sweden there is a system of detailed requirements for different types of detectors and measurements (Falk *et al.*, 1994).

Various requirements must be met before a laboratory is regarded as certified. The requirements are laid down by the certifying authority, usually a national body. Certification schemes typically require:

A quality control manual from the laboratory containing details of:
- Measurement protocol,
- Reporting procedures,
- Equipment used,
- Processing procedures,
- Quality control procedures,
- Personnel (experience, qualifications).

Testing by the validating authority of:
- Accuracy,
- Precision,
- Linearity,
- Dynamic range,

(as appropriate for the purposes of the measurement).

Table 2.4.1: Radon reference levels given by different countries and organisations. The meaning of a reference level varies from country to country (see text).

Country, organisation	Radon reference level (Bq m^{-3})
Netherlands	20
Luxembourg	150
United States of America	150
Australia	200
Ireland	200
Norway	200
Sweden	200
United Kingdom	200
Germany	250
Austria	400
Belgium	400
Finland	400
Switzerland	400
Canada	800
European Community	400
Basic Safety Standards for Radiation (International Atomic Energy Agency, World Health Organisation and others)	200-600
International Commission on Radiological Protection	200-600

Before approval, the validating authority ensures that the laboratory's quality control manual and its performance in tests meet the requirements of the measurements proposed. Regular retesting and reexamination of the quality control manual are required at intervals of not more than 1 year, preferably less.

2.4.7 Radon Reference Levels, Action Levels and Regulations

Radon exposures in mines are regulated in many countries, usually following the guidelines given by the International Commission for Radiological Protection (ICRP 1986, 1993). Limits are usually quoted in the archaic units of Working Level Months, where 1 WLM = 630,000 Bq m^{-3} h Equilibrium Equivalent Radon.

Attitudes of governments to radon in homes vary enormously: some countries have no guidance, some have advisory levels, and two (Sweden and Switzerland) have both an advisory level and a compulsory upper limit. Table 1 shows reference (or advisory or action or guidance) levels for radon in existing houses. The data for this table were taken from a survey carried out recently (Colgan and Gutierrez, 1995).

The status of the reference level for the Netherlands is different from that in other countries. The Netherlands has generally very low radon concentrations in houses, and has chosen of 20 Bq m^{-3} as a target for the average level in homes to provide a level of risk comparable with other environmental and everyday risks.

References

Al-Najjar S.A.R, Bull R.K. and Durrani S.A. (1979) Electrochemical etching of CR-39 plastic: Applications to radiation dosimetry. *Nucl. Tracks* **3**, 169-183.

Cliff K.D., Miles J.C.H. and O'Riordan M.C. (1991) Validation scheme for laboratories making measurements of radon in dwellings. NRPB-M276, NRPB, Chilton, UK.

Colgan P.A. and Gutierrez J. (1995) National approaches to controlling exposure to radon. Presented at the 6th Symposium on the Natural Radiation Environment, Montreal, 1995.

Dudeney C.S., Wilson D.L., Bertini H., Gammage R.B. and Otten, J.A. (1995) On the reduction of error in alpha track detector measurements of indoor ^{222}Rn. *Health Phys.* **69**, 501-507.

Falk R., Hagberg N., Mjönes L., Möre H., Nyblom L. and Swedjemark G.A. (1994) Standards, calibration and quality assurance of radon-222 measurements in Sweden. *Nucl. Instr. Meth.* **A339**, 254-263.

Homer J.B. and Miles J.C.H. (1986) The effects of heat and humidity before, during and after exposure on the response of PADC (CR-39) to alpha particles. *Nucl. Tracks* **12**, 1-6, 133-136.

Hopper R.D., Levy R.A. and Steinhäusler F. (1993) IAEA-EPA International climatic test program for integrating radon detectors, US-Environmental Protection Agency (EPA) Report, Las Vegas. N.Y.

HSE (1985) Guidance on approval of dosimetry services under the ionising radiations regulations 1985. *Health and Safety Executive*, HMSO, London.

IAEA (1996) Instrumentation and methods for assessment of radon in the indoor environment, to be published.

ICRP, Radiation protection of workers in mines (1986) *ICRP Publication 47, Annals of the ICRP* **16**, No. 1.

ICRP. Protection against ^{222}Rn at home and at work. (1993) *ICRP Publication 65, Annals of the ICRP* **23**, No. 2.

Jönsson G. (1981) The angular sensitivity of Kodak LR-film to alpha particles. *Nucl. Instr. Meth.* **190**, 407-414.

Knutson E.O. (1988) The OECD/CEC radon measurements intercomparisons: a further analysis of the results. *Radiat. Prot. Dosim.* **24**, 277.

Mellander H. and Enflo A. (1992) The alpha track method used in the Swedish radon epidemiological study. *Radiat. Prot. Dosim.* **45**, 65-71.

Miles J.C.H and Olast M. (1990) Results of the 1989 CEC intercomparison of passive radon detectors. Chilton, NRPB M-251.

Miles J.C.H., Stares E.J., Cliff K.D. and Sinnaeve J. (1984) Results from an international intercomparison of techniques for measuring radon and radon decay products. *Radiat. Prot. Dosim.* **7**, 169-173.

Miles J.C.H. and Sinnaeve J. (1986) The performance of different types of etched-track radon dosemeters in two international intercomparisons. *Nucl. Tracks* **12**, 1-6, 735-738.

Sensintaffar E.L. and Steinhäusler F. (1995) First IAEA-EPA International Intercomparison Exercise. US-Environmental Protection Agency (EPA) Report, Montgomery, Al. (in press).

Solomon S. B., Peggie J. R. and Gan T. H. (1988) Pacific region radon intercomparison. *Proc. 7th Int. Congress of the International Radiation Protection Association*, Sydney, 1988.

Urban M. (1986) Passive one-element track etch detector for simultaneous measurement of radon, thoron and decay products in air. *Nucl. Tracks* **12**, 685-688.

US EPA (1993) *Protocols for Radon and Radon Decay Product Measurements in Homes.* Environmental Protection Agency, EPA 402-R-93-003, Washington D.C.

Whysall K. Miles J.C.H. and Olast M. (1995) Results of the 1991 CEC Intercomparison of Passive Radon Detectors, Commission of the European Communities/DG XII Report, Brussels, in press.

Wrixon A.D., Green B.M.R., Lomas P.R., Miles J.C.H., Cliff K.D., Francis E.A., Driscoll C.M.H., James A.C. and O'Riordan M.C. (1988) Natural radiation exposure in UK dwellings. NRPB R-190, HMSO, London.

3.

APPLICATIONS IN RADIATION PROTECTION AND THE ENVIRONMENT

G. Jönsson[1]

3.1 Indoor Radon Surveys

3.1.1 Radon or Radon Daughter Levels

Radon (^{222}Rn) belongs to a naturally-occurring disintegration series (in the decay chain of ^{238}U) where the radon atom in turn is transformed into the various radon daughters. The gas comes into equilibrium with its daughters if the condition is fulfilled that the air exchange is zero.

Indoors, some ventilation normally exists, which means a certain degree of air exchange; so that individual atoms of the disintegration series have some probability of leaving the indoor air. The radon atom is the longest-lived in the series immediately following ^{226}Ra. The further down among the daughters the shorter is the time of existence of an atom. Atoms further down in the series have lower activity concentration than those close to the radon atom because the equilibrium state between radon and daughters is not normally reached. The methods of measurement can be separated into three main classes:

1. Measurements of radon gas,
2. Measurements of radon daughters,
3. Measurements of both radon gas and radon daughters.

Although most methods of measurement determine only the radon concentration, models for the evaluation of the risk of lung cancer and other cancers to humans are based on knowledge about the deposition of the radon decay products in the body (ICRP, 1987; WHO, 1988). In large-scale measurements, the classes 1 and 3 using plastic films as time-integrating detectors are most convenient.

It is possible to calculate the radon daughter level in Equilibrium Equivalent Radon (EER) Concentration from the radon gas level if the equilibrium factor F is known (Swedjemark, 1983). The EER value in Bq m^{-3} is given by the formula

$$EER = 0.104A_1 + 0.514A_2 + 0.382A_3 \tag{3.1.1}$$

where A_1, A_2 and A_3 are the activity concentrations in Bq m^{-3} of the individual radon daughters ^{218}Po, ^{214}Pb and ^{214}Bi respectively. The contribution from ^{214}Po is negligible. The factor F is defined as

$$F = EER/A_{Rn} \tag{3.1.2}$$

where A_{Rn} is the ^{222}Rn concentration in Bq m^{-3} (Swedjemark, 1983).

The F-factor indoors is expected to be 0.3-0.7 depending on air exchange rate, the aerosol concentration in the air and the degree of plate-out on solid surfaces in the room.

[1]University of Lund, Lund, Sweden

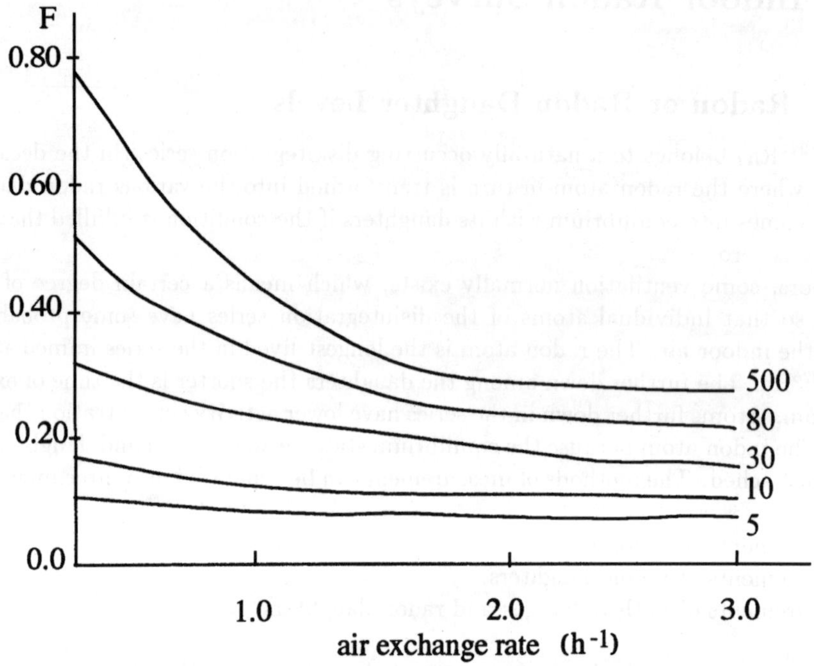

Figure 3.1.1: The F-factor for different air exchange rates (Swedjemark, 1983). The figures 5, 10, 30, 80 and 500 are the values of the attachment rate h^{-1} of the free disintegration products to the aerosols in the indoor air. The calculations are based on a certain radon exhalation rate into the indoor air.

Table 3.1.1: Equilibrium factors F as measured in dwellings in some countries (Ramachandran and Subba Ramu, 1994).

Country	F	
	Average	Interval
Canada	0.5	0.2-0.7
Finland	0.5	0.3-0.6
Sweden	0.4	0.1-0.8
Norway	0.5	0.3-0.8
USA	0.5	0.3-0.8
Germany	0.3	0.2-0.7
India	0.5	0.1-1.0

If the radon daughters level is 200 Bq m^{-3} the radon gas level A_{Rn} = 400 Bq m^{-3} when $F = 0.5$; and A_{Rn} = 500 Bq m^{-3} when $F = 0.4$. Fig. 3.1.1 shows how the F-factor is dependent on the air exchange rate. The F-factor varies from one country to another because of typical ventilation rates, size and structure of dwellings, etc. Table 3.1.1 shows intervals of F-factors in some countries and some median values (Ramachandran and Subba Ramu, 1994). There may be a seasonal variation of F as shown in Table 3.1.2, which is mainly dependent on the ventilation patterns in houses.

Table 3.1.2: Seasonal variation of F in India (Ramachandran and Subba Ramu, 1994).

Season (months)	Air conditioned room	Chawl type
Summer (March -May)	0.14	0.26
Rainy (June-August)	0.20	0.22
Autumn (September-November)	0.21	0.44
Winter (December-February)	0.25	0.57

3.1.2 Indoor Radon from Building Materials

One source of indoor radon is the building material. All materials contain ^{238}U and are consequently potential radon emitters. However, some materials have higher concentrations of ^{238}U and ^{226}Ra. This is valid both for natural and artificial materials. Typical uranium-bearing natural materials are certain granites; but it is always possible to find uranium-rich bedrocks of different types locally used as building material. Certain soils, for example some clays, used as building material may have enhanced uranium content. Typical artificial materials are different types of concrete or light-weight concrete in which ashes, alumshale or uranium-rich material from mining activities are constituents.

Gamma-ray measurement on walls is an easy way to localize or identify the walls as important potential radon emitters in a house. The energy window of the gamma ray detector should be between 50 keV and 3 MeV. If the gamma ray intensity exceeds say 0.25 μSv h^{-1} for a wall, it is an important radon source. However, this fact does not mean that the radon concentration of the indoor air is significant, because the air exchange rate has a great influence on the concentration. As some materials may be present only as thin plates, the total mass is too small to constitute an effective radon source. Some types of gypsum have enhanced amounts of radium, which can be detected by a gamma detector.

If a uranium-rich material is crushed and used as insulation in walls or floors, the gamma-ray intensity from the material may be too low for it to be located by a gamma-ray detector. Insulation layers made from crushed material of uranium-rich bedrock or uranium-rich light-weight concrete may contribute considerably to the indoor radon levels. The radon exhalation from the grains depends on the size of the grains. It can be expected that

$$E = cD^{-x} \tag{3.1.3}$$

where E is the exhalation rate, c is a constant and D is the mean diameter of the grain; and x is between 0.5 and 1 depending on the grain density (Tanner, 1964). In theoretical calculations D should be understood as the statistical distribution of the grain size.

Tables 3.1.3 and 3.1.4 give the radium (^{226}Ra) and thorium (^{232}Th) contents of some building materials, in terms of their specific activities.

Table 3.1.3: ^{226}Ra and ^{232}Th contents of some Swedish building materials (Möre, 1985). These isotopes reflect the possible emission of radon and thoron from materials.

Material	^{226}Ra (Bq kg^{-1})	^{232}Th (Bq kg^{-1})
Brick	40-164	71-180
Limestone	7-15	4-10
Concrete	31-63	46-127
Light-weight concrete, sand based	3-132	4-157
Light-weight concrete, alumshale based	788-2,627	19-96
Gypsum, natural	2-9	0-12
Clinker of clay	137-189	161-184

Table 3.1.4: ^{226}Ra and ^{232}Th contents in some German building materials and wastes (Winter and Wicke, 1993).

Material	^{226}Ra (Bq kg^{-1})	^{232}Th (Bq kg^{-1})
Granite	30-500	17-311
Brick	10-200	12-200
Limestone	4-41	2-20
Gypsum	2-70	2-100
Concrete	7-92	4-71
Light-weight concrete	6-80	1-60
Old copper slag	861-2100	18-78
Tin slag	1,000-1,200	230-340
Brown coal ash	4-200	6-150

Fig. 3.1.2 shows the typical distribution of radon levels indoors in Swedish houses where the building material is the main radon source. The building materials do not normally give rise to very high radon concentrations.

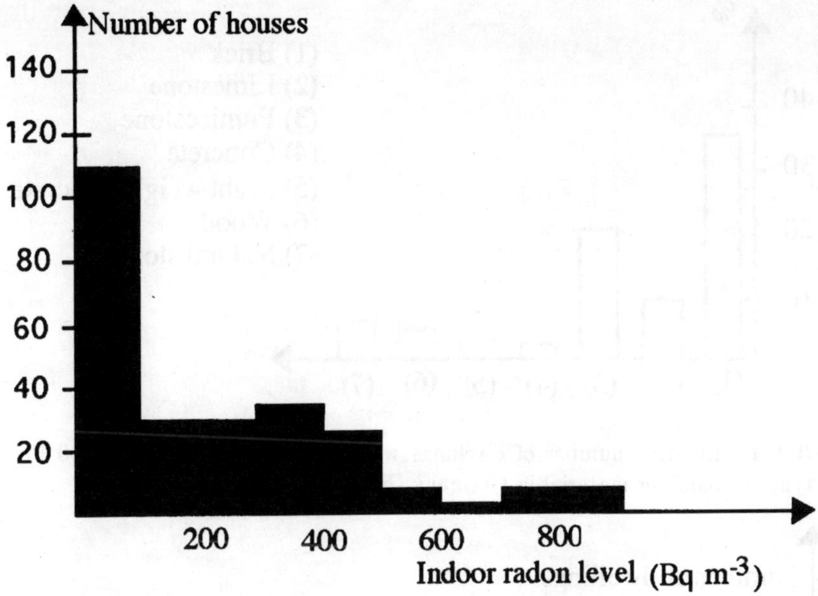

Figure 3.1.2: Radon levels in 251 selected houses in a south Swedish community where the building material (light-weight concrete) is the main radon source (Jönsson, 1992).

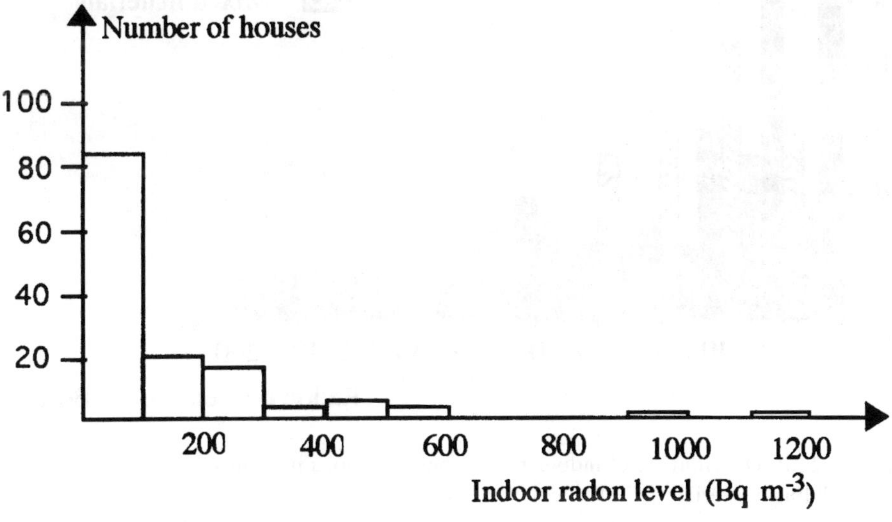

Figure 3.1.3: Radon levels in 137 selected houses in a south Swedish community. The radon source is a type of light-weight concrete which is only a faint radon source in these houses. Two high-radon-level houses probably get additional radon from the soil.

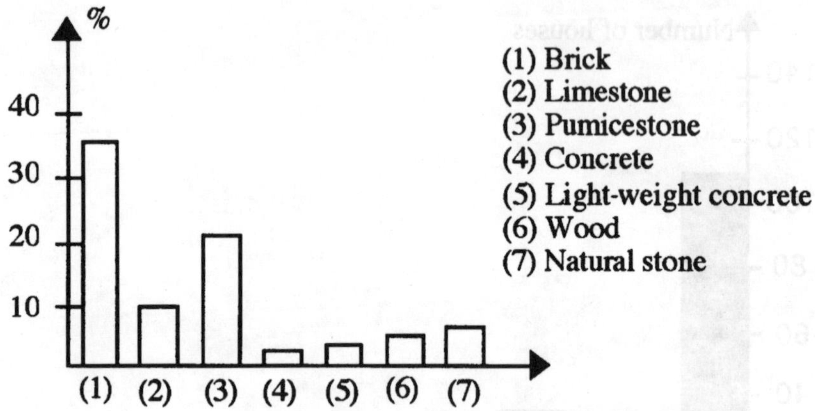

Figure 3.1.4: Relative number of dwellings with radon levels exceeding 100 Bq m^{-3} for different types of building material in Germany (Heller and Poulheim, 1992).

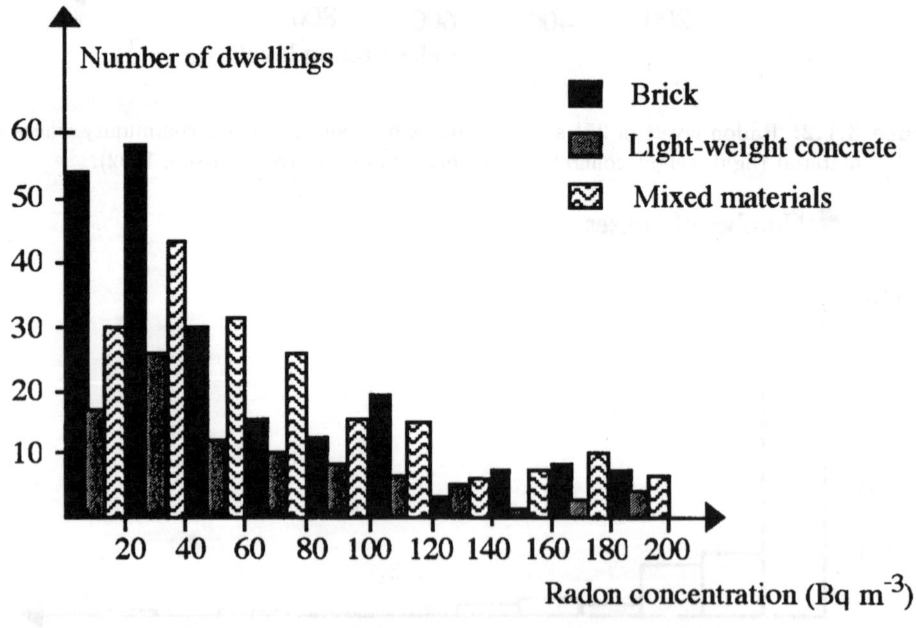

Figure 3.1.5: Distribution of indoor radon concentrations in family houses in Slovakia using various building materials (Winter and Wicke, 1993).

Fig. 3.1.3 shows a typical distribution of radon levels indoors in Swedish houses where the building material (light-weight concrete) is a faint radon source. The distribution shows two houses with particularly high radon levels, where the ground is probably the additional radon source. Fig. 3.1.4 shows the relative number of German dwellings with radon levels exceeding 100 Bq m^{-3} for different building materials. All types of dwellings are represented. The distribution in Fig. 3.1.4 may be different in other countries. Fig. 3.1.5 shows a similar example from Slovakia, focusing on the importance of different building materials.

3.1.3 Indoor Radon from Soil and Bedrock

The soil and the bedrock beneath a building are in general the main source of indoor radon. The basis for this radon source is the presence of uranium-rich bedrock or soil. There are many types of bedrock containing uranium; but, for it to be a significant source of radon it is essential that the uranium concentration is over, say 50 ppm, especially if the bedrock is deep down (deeper than 100 m). The radon gas follows the structure of the bedrock along the different layers of the bedrock and along the cracks and the zones of cracks. The soil is a product of the bedrocks in the vicinity, because of erosion caused by the ice movements or deposits from the water flow in big rivers. If the material is rich in uranium, the soil itself is a radon source. Table 3.1.5 shows the uranium concentration in some Swedish rock types.

Table 3.1.5: Uranium and thorium concentrations in some Swedish bedrocks (Swedish radon commission, 1983).

Type of bedrock	Uranium concentration ($g/10^3$ kg)	Thorium concentration ($g/10^3$ kg)
Granite, normal	2-10	5-20
Granitc, "radioactive"	8-40	10-90
Sandstone	0.5-5	1-10
Limestone	0.5-2	1-2
Shale	1-10	2-15
Alumshale	10-350	2-10

If the soil is permeable, like gravel, the radon gas produced in underlying bedrock is easily transported upwards through the soil. Presence of layers of ground water, wet clay or other non-permeable soils means that the radon gas is more or less prevented from flowing upwards. A special type of radon source in the ground is the crack zones and faults, which may offer a very effective passage for the transport of radon gas upwards. Some of the big faults are also earthquake zones. One such huge fault is the San Andreas fault in California. It is, however, very rare to find houses situated in fracture zones that have enhanced indoor radon concentrations. An example of this type of radon source on the small scale is when a certain house is the only one in a row of houses which has a radon problem. This house may be situated on such a fracture zone. Conditions which must be fulfilled in order to get enhanced indoor radon levels in this case are that there be no effective sealing of the gas against the ground and that there be a negative pressure gradient in the indoor air.

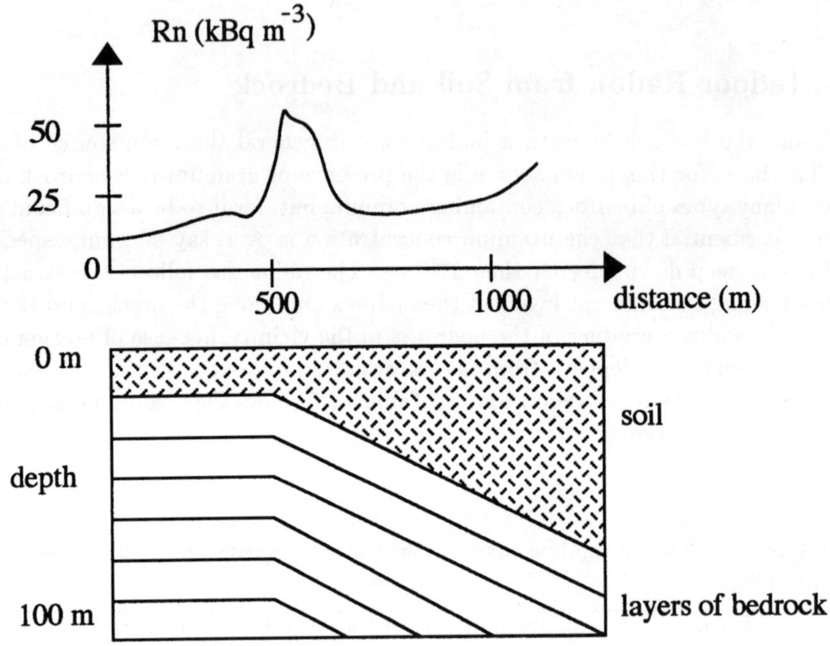

Figure 3.1.6: There is an upflow of radon gas where the structure of the bedrock is broken. The radon gas coming along the layers of bedrock is streaming through the overlying soil.

Fig. 3.1.6 shows an example of the soil radon levels along a line across a crack zone. Houses lying on the ground at the distance 500-700 m in Fig. 3.1.6 have a higher potential risk of high indoor radon levels.

Fig. 3.1.7 shows an example of the importance of the presence of uranium in the soil (Jönsson, 1991). Measurements are made in the soil in low and in high risk areas. Measurements are also made indoors in houses lying on these two types of ground.

Table 3.1.6 shows the result of indoor radon measurements in 11 single-unit houses in south Sweden. The indoor radon levels have been measured twice. The soil radon levels and the mean gamma ray intensity of the building material are also given. House no 5 has the highest soil radon concentration. House no 7 has the highest gamma ray intensity. Both houses have the highest indoor radon levels.

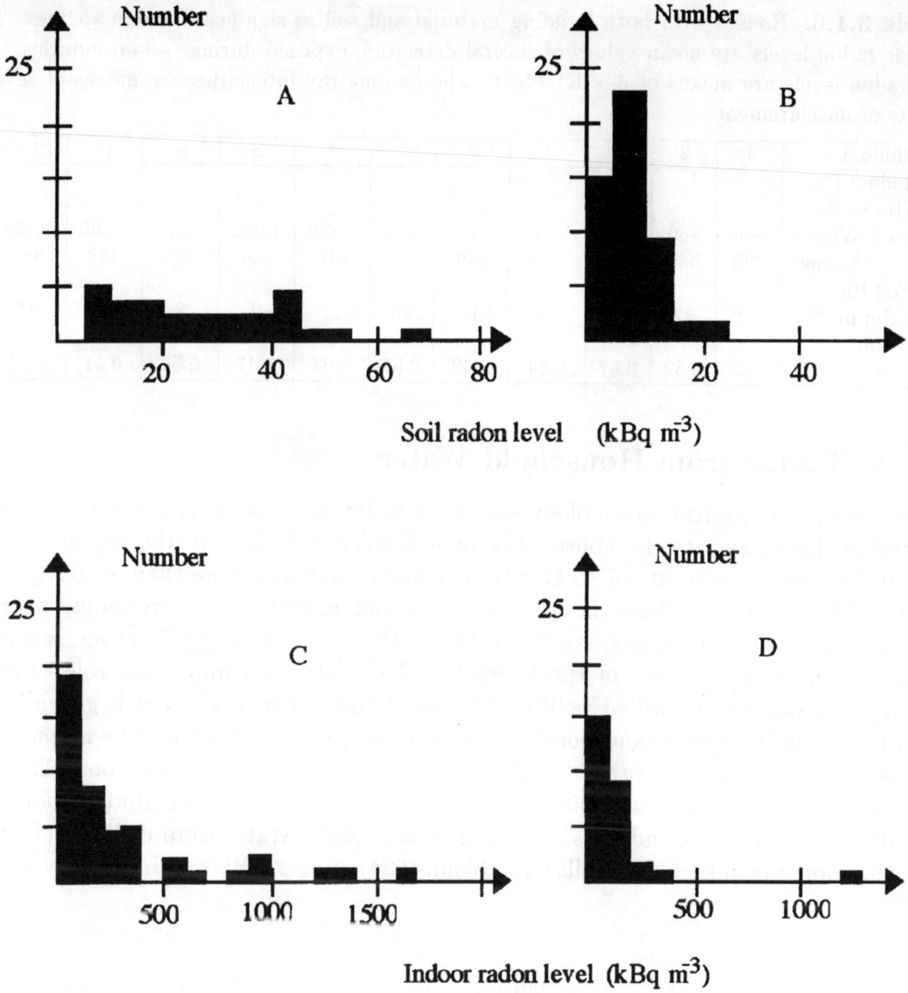

Figure 3.1.7: Measurements of radon indoors and in soil in a south Swedish community (Jönsson, 1991).
A: Soil radon levels in uranium-rich soil. (U concentration 20-200 ppm), 26 measurements.
B: Soil radon levels in uranium-poor soil. (U concentration < 50 ppm), 49 measurements.
C: Indoor radon levels in houses on ground corresponding to A. 47 houses.
D: Indoor radon levels in houses on ground corresponding to B. 27 houses.

Table 3.1.6: Houses with both building material and soil as significant radon sources. The indoor radon levels are mean values of several detectors, exposed during ~ two months. The soil radon levels are means of 3-5 detectors. The gamma ray intensities are means of several points of measurement.

House no.	1	2	3	4	5	6	7	8	9	10	11
Indoor Rn (Bq m^{-3})											
Winter	840	860	820	1,140	1,220	340	1,380	1,050	1,070	1,160	1,160
Spring	280	540	600	310	630	-	610	460	520	430	340
Soil Rn (kBq m^{-3})	10	22	18	24	44	26	12	9	8	22	34
Gamma (μSv h^{-1})	0.33	0.32	0.57	0.44	0.42	0.57	0.63	0.47	0.50	0.41	0.57

3.1.4 Radon from Household Water

Radon gas is, in physical terminology, soluble in water in the sense that the gas is transported in the water into the house. The mole fraction solubility of the gas in water is $2.3 \cdot 10^{-4}$ at the temperature of 15°C, which is about ten times more than for oxygen gas (Lide, 1993). The gas follows the water from its source, where the surrounding bedrock and soil are producing the gas. As ^{226}Ra, the mother atom of the ^{222}Rn atom, is soluble in water, the ^{226}Ra content in the bedrock and soil plays an important role. Consequently, the water from drilled wells in ^{238}U and ^{226}Ra rich ground, that is ground with granite, permatite, syenite and porphyry, is expected to contain radon. The radon gas is released where the water is tapped in the dwelling, that is kitchen, shower room, laundry, etc. A rule of thumb is that 1,000 Bq l^{-1} of radon in the water contributes about 100 Bq m^{-3} of radon to the indoor air (Kulich *et al.*, 1988). Water from drilled wells contains far more radon than dug wells (Åkerblom, 1994). A dug well contains mainly surface

Table 3.1.7: ^{222}Rn and ^{226}Ra concentrations in drinking water in Finland (Asikainen and Kahlos, 1980).

Water supply	Number	^{222}Rn (Bq l^{-1})	^{226}Ra (Bq l^{-1})
Dug wells	690	4*-1,600	0.004*-0.3
Drilled wells	570	4-44,000	0.004-7.5

*Detection limit

water and is not so deep. Water from public water supplies has very small radon levels normally, which is partly due to the treatment of the water in the supply plants. There, because of other reasons, the water is treated in open air and the radon gas is released from the water. The private household water supply is more or less a closed system from the drilled hole to the tap point indoors. Infiltrated surface water or lake water is often the main or additional water supply for the big public water plants, which means low or no radon content in the water. Table 3.1.7 shows an example of the difference of radon concentration in water between dug wells and drilled wells.

The treatment of water in public water service plants means a release of radon gas from the water into the indoor air of the plants. Radon levels in this air of around 20,000 Bq m^{-3} are found in Sweden in areas with high radon concentration in the water.

3.1.5 General Surveys

General surveys of indoor radon levels in houses of all types give distributions like those of Fig. 3.1.8. The radon level is seldom zero, and there is always a tail towards higher concentrations. Table 3.1.8 shows the median radon levels in some countries. This median does not tell us the range or spread of the measured radon levels in a region but it is helpful for the calculations of the lung cancer risk for the population on the average.

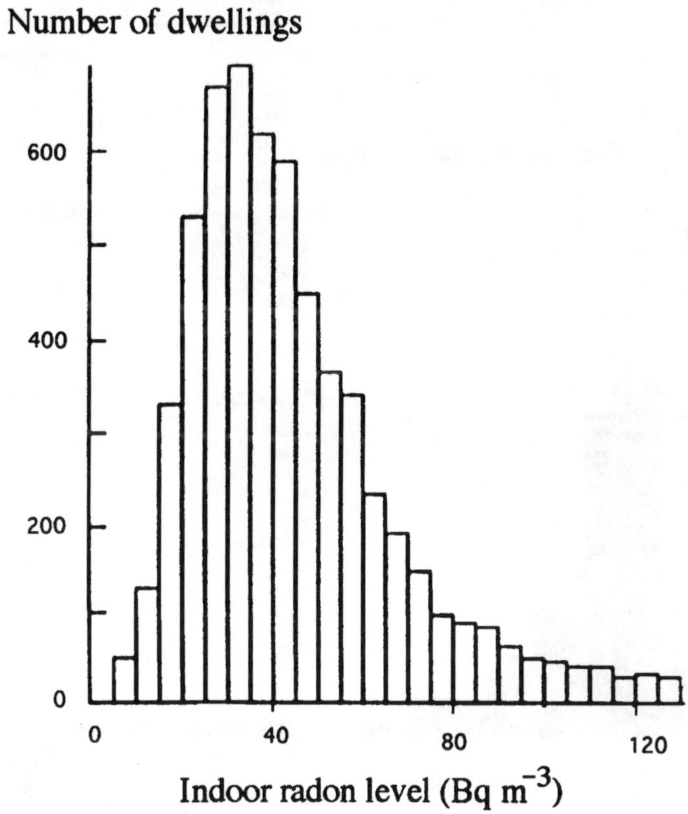

Figure 3.1.8: Typical distribution of indoor radon levels in 5,970 dwellings in Germany. The median value is 40 Bq m^{-3} and the arithmetic value is 50 Bq m^{-3}, showing that the distribution is skew (Reiners *et al.*, 1992).

Table 3.1.8: Median radon levels in dwellings in some countries (WHO, 1988). (The number of measurements being the basis for the median value, is very different).

Country	Median radon level ($Bq\ m^{-3}$)
Belgium	41
Finland	63
France	44
Germany	40
Iceland	61
Italy	25
Netherlands	24
Switzerland	60
Sweden	75 (multistoreys) 141 (detached houses)
UK	22*

* arithmetic mean

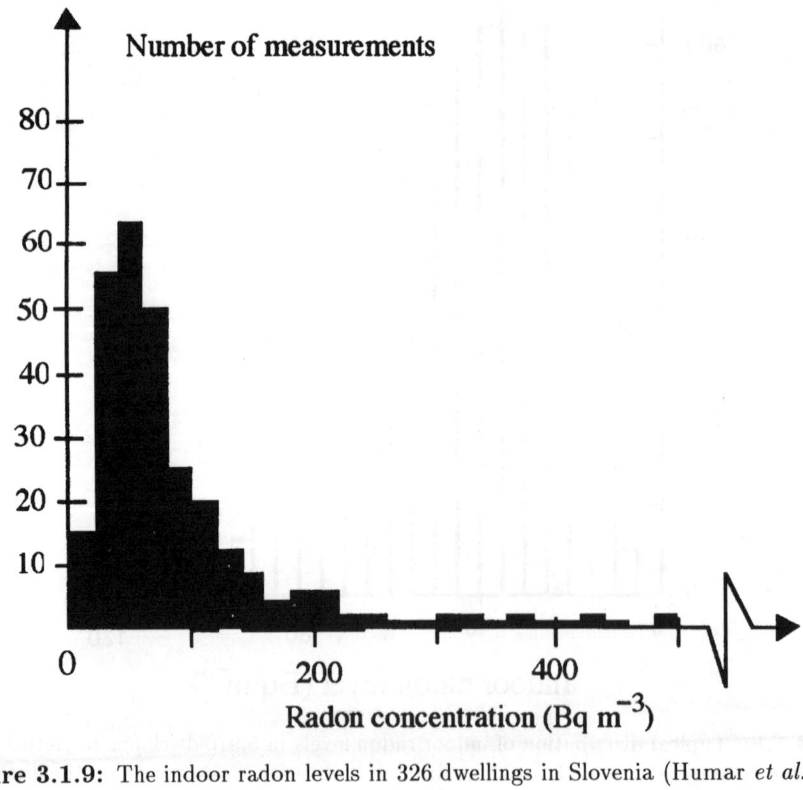

Figure 3.1.9: The indoor radon levels in 326 dwellings in Slovenia (Humar *et al.*, 1992).

Another typical distribution from a general survey of indoor radon in dwellings is given in Fig. 3.1.9 for Slovenia. The radon source is both the soil and the building material.

3.1.6 Radon Levels in Apartment Houses and in Single-Unit Houses

It is evident from measurements that the high indoor radon levels are found in single unit houses. An inferior ventilation system and a closer contact with the ground are two reasons for higher indoor radon levels, on average in single-unit houses compared with apartment houses. The statement is supported by Table 3.1.9, which shows that nearly 37% of the detectors register radon levels exceeding 200 Bq m^{-3} in single-unit houses compared to only about 11% in apartment houses in Sweden. The houses are selected for measurement on the basis of the building material.

A German investigation (Heller and Poulheim, 1992) gives the median values of indoor radon in single unit houses and in apartment houses. The median values are 43 Bq m^{-3} and 34 Bq m^{-3} respectively.

Table 3.1.9: A comparison between indoor radon levels in apartment houses (1,006 detectors) and in single-unit houses (3,268 detectors) (Jönsson, 1988). The investigation is made in Sweden.

Radon concentration (Bq m^{-3})	0-100	100-200	200-300	300-400	> 400
Single unit houses (%)	27.5	35.8	20.1	9.2	7.3
Apartment houses (%)	48.7	38.0	10.6	0.02	0.01

3.1.7 Radon Distributions Within a House

Although the radon source is often concentrated in one room or in one part of the house, the gas streams into the other parts of the house, mainly because of the ventilation or in spite of the ventilation. A single door kept open or closed continuously is enough to affect the radon transport within the house. The gas coming into a space with no or very small air exchange will stay in that space. In self-ventilated houses the gas follows the air streaming by convection upwards from the lower part to the upper part of the house. The convection does not work, however, if the temperature inside the house is about the same as or lower than outside the house.

Fig. 3.1.10 shows the distribution of the ratio of Rn (upper floor)/Rn (ground floor) in 488 single-unit houses in Sweden. The median value of the distribution is 0.8, which confirms the statement that the lower part of a house has higher radon levels than the upper part of the house. If the ground beneath the house is the radon source, this statement is obvious unless the gas is stored up in the upper floor because of missing outlets. The result of Fig. 3.1.10 is also confirmed by the data in Table 3.1.10 taken from a national investigation on indoor radon in Germany.

Table 3.1.10: Median radon levels at different floors in German houses from a nationwide investigation (Heller and Poulheim, 1992).

Floor →	Cellar	Ground floor	1st floor	2nd floor	Outside
Median radon level ($Bq\ m^{-3}$)	52	43	38	33	14

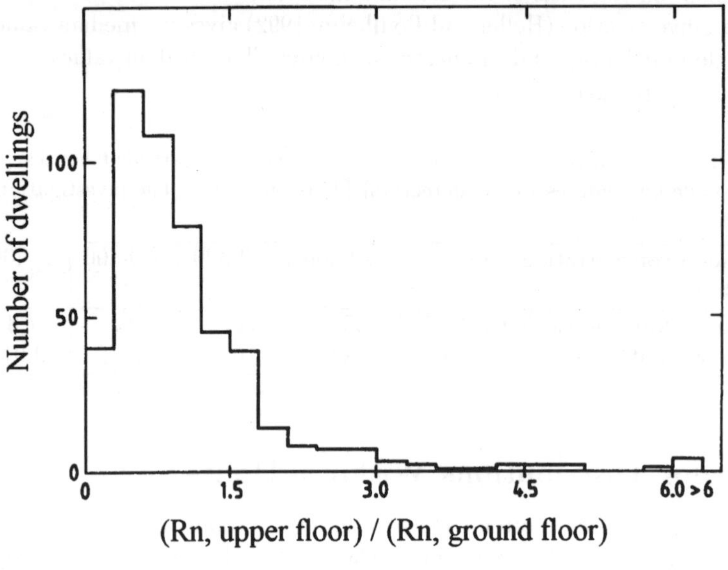

Figure 3.1.10: A comparison between radon levels in the upper floor and ground floor in 488 single unit houses in Sweden (Jönsson, 1987).

Among the different rooms in a house the cellar normally has the lowest air exchange rate and is also closest to the ground. The ground is the most powerful radon source. On the other hand, the kitchen has the best ventilation, normally, and should have the lowest radon levels unless bottled natural gas is used for cooking. Table 3.1.11 shows that in general the cellar has the highest radon level. The bedroom, normally on the 1st floor in Germany, has the same radon level as the other rooms (except the cellar).

In Fig. 3.1.11 the distribution of the ratio Rn [illegible] has a median value of 1.0, which means that the bedroo[illegible] rooms in a house except for the cellar.

Table 3.1.11: Radon levels in different rooms of a house in the mean in a nationwide German investigation (Heller and Poulheim, 1992).

Room	Bedroom	Livingroom	Kitchen	Childrens room	Cellar
Radon level ($Bq\ m^{-3}$)	37	41	37	37	52

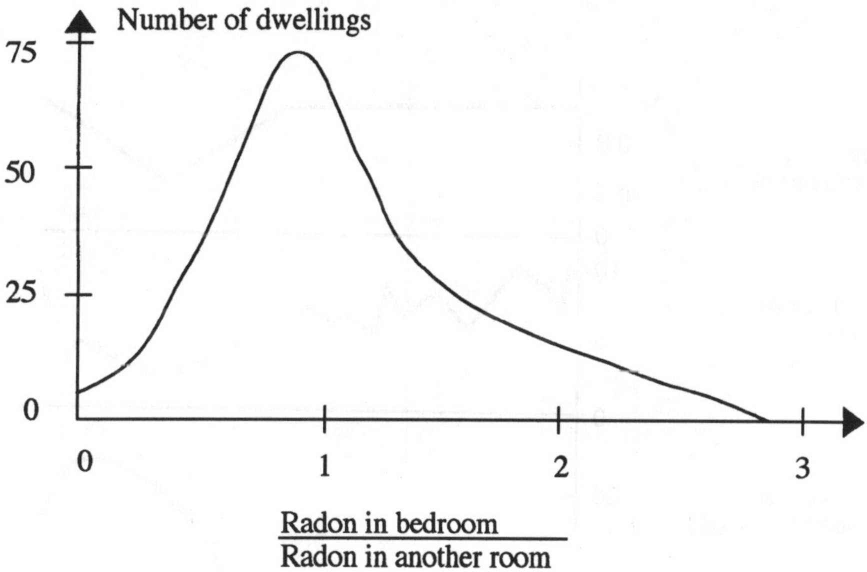

Figure 3.1.11: The distribution of the ratio (radon in bedroom)/(radon in another room) in Swedish houses where the building material is the main radon source (Jönsson, 1988).

3.1.8 Short-Time Variations of the Indoor Radon Levels

As radon is a gas, the ventilation in a dwelling is an important factor affecting the indoor radon level. In houses with self-ventilation, weather conditions outside mean that the ventilation rate indoors is changeable. A negative pressure or a pressure gradient may exist in one part of the hous connection between wind velocity outside, the air exchange rate, and radon level is illustrated in Fig. 3.1.12. All three parameters are studied for over three days in elf-ventilated house. When day no. 3 starts, the wind velocity is very low, which m ns a reduced air exchange and an increase of the radon level.

High indoor radon levels caused by soil radon depend on how people behave in the dwelling: whether the doors are open or kept closed; whether the kitchen fan is running; whether the windows are open; and so on. Open fire in oil-burning system, or petroleum gas burning, need fresh air: which normally means an under-pressure in the house. If soil air is coming into the indoor air as a consequence of this under-pressure, the radon level is probably raised indoors. The short-time variation during a day may be extreme. Fig. 3.1.13 shows how the indoor radon concentration changes rapidly in a self-ventilated house, where the ground under the house is the main radon source. The convective air movements are the main reason for this variation.

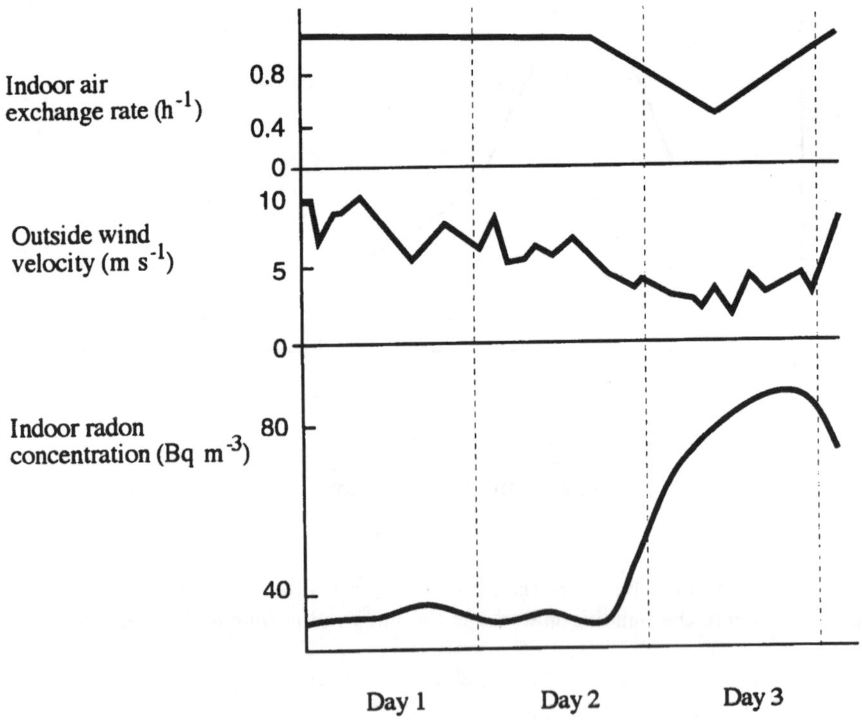

Figure 3.1.12: The indoor radon levels are affected by the outside wind velocity in a house with self-ventilation (Jönsson, 1992).

One lesson from Fig. 3.1.13 is that it is no use letting a window remain open to the fresh air for only half an hour, because the radon level in the room rapidly rises again after closing the window.

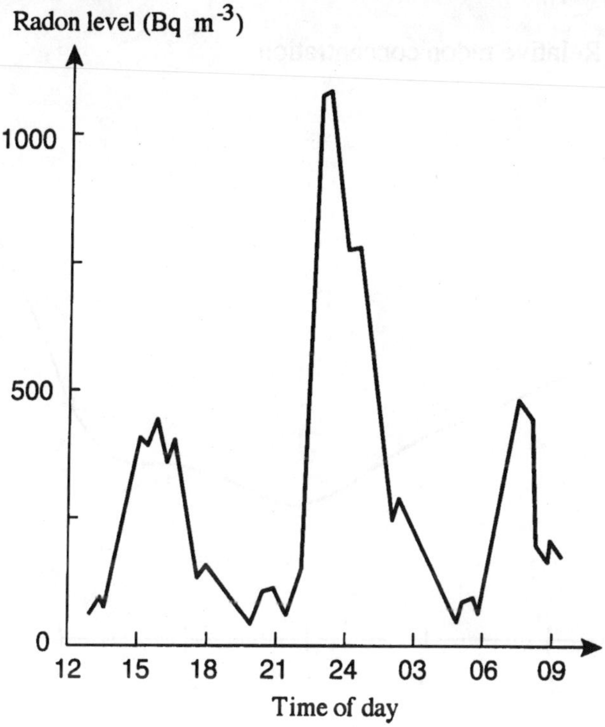

Figure 3.1.13: The indoor radon concentration changes rapidly in a self-ventilated house (Jönsson, 1992). The measurement is made in the winter season in Sweden.

3.1.9 Seasonal Variations of the Indoor Radon Levels

There is a seasonal variation of the indoor radon levels especially in self-ventilated houses. In summertime, when the outdoor temperature is as high as or higher than the indoor temperature, the convective movements of the indoor air are negligible. If the temperature is, say 10°C higher or more indoors, the convective forces are working and there is a small under-pressure, maybe 10 Pa, indoors. If the ground is the chief radon source, it may be a drawback to have an under-pressure indoors at least if this under-pressure exceeds 2-5 Pa. If the building materials or the house-hold water are the radon sources the under-pressure is an advantage. A functioning self-ventilation gives a faint under-pressure.

Concerning the radon daughters, the relative humidity indoors contributes to the aerosol density and keeps the radon daughters in the indoor air. The humidity changes from one season to another, at least in the Scandinavian countries.

The radon exhalation from indoor building materials increases in the summertime, when the mean temperature of the building materials is higher than in winter times (Stranden *et al.*, 1984).

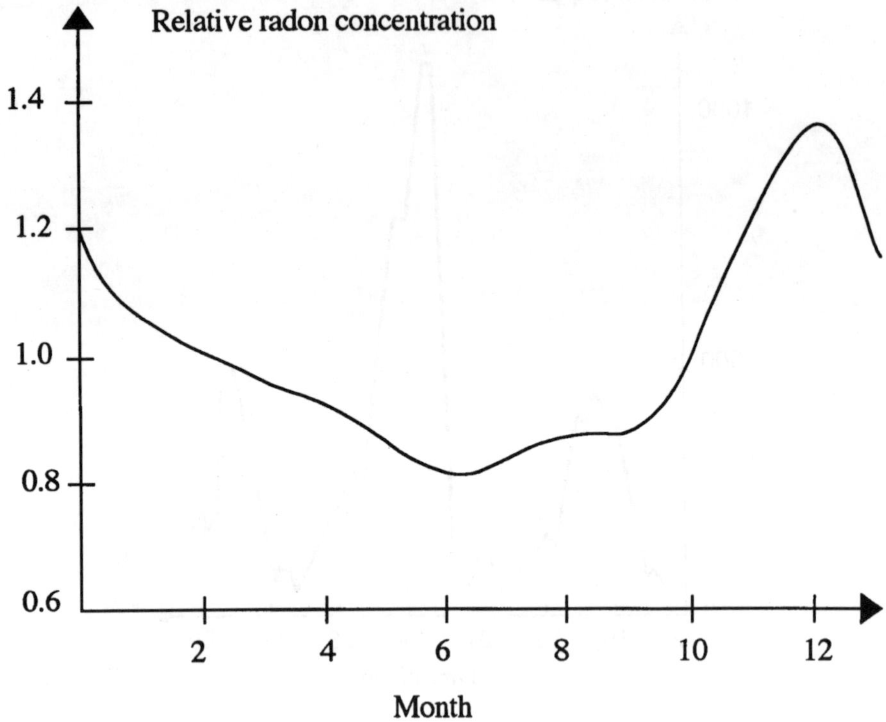

Figure 3.1.14: Yearly variation of the indoor radon concentration in Germany with a mixed radon source (Heller and Poulheim, 1992).

Table 3.1.12: Radon concentration in single-unit houses (A) and in apartment houses (B) at different seasons in Sweden (Jönsson, 1988).

Indoor radon concentration (Bq m^{-3})		< 140		140-400		> 400		Season
Number of dwellings		Percentage						
A	B	A	B	A	B	A	B	
317	156	33	70	53	29	14	1	Summer
612	261	55	83	42	17	3		Winter
59	34	56	82	37	18	7		Spring
43	9	28	67	63	33	9		Autumn

Table 3.1.12 shows the results of a study of the seasonal variations of the indoor radon levels in Swedish houses where the building material is the main radon source. The outside absolute humidity in this study is about 3 g m^{-3} in winter months and 9 g m^{-3} in summer months. The study shows that the radon levels are lower in the winter time, which is due to the higher ventilation rate during winter, caused by the chimney effect.

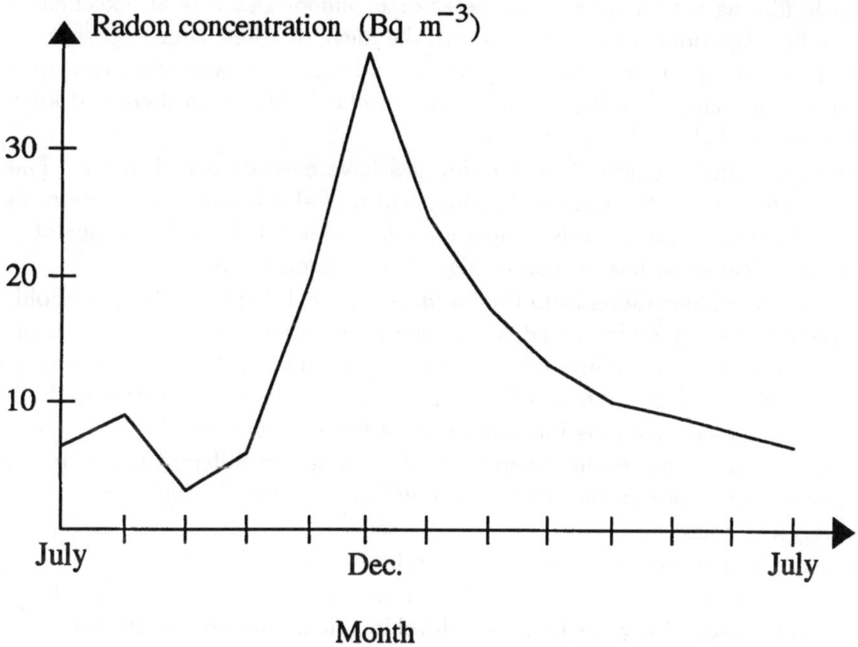

Figure 3.1.15: Yearly variation of indoor radon levels in India (Ramachandran *et al.*, 1990)

Figs. 3.1.14 and 3.1.15 show the variation of the radon concentration during one year from one German and one Indian investigation. There is a rise of the radon level in the autumn and a decrease in spring down to the minimum level in summer. No preferential radon source is given. Referring to the Figs. 3.1.14 and 3.1.15, it is normal to have more open windows in summertime and consequently a lower indoor radon level. In countries with frozen soil in the winter the chimney-effect in houses with self-ventilation may increase the amount of soil radon streaming into the indoor air.

The conclusion is that the seasonal variation of the indoor radon levels depends on several parameters:
- type of house,
- radon source,
- living habits of the inhabitants,

- ventilation system of the house,
- heating of the house,
- outside climate.

3.1.10 Concluding Remarks

The plastic film as a time-integrating detector of indoor radon is an excellent tool in large-scale investigations. The detectors give the mean of a long-time exposure of radon which is important in order to estimate the risk of lung cancer and other cancers for the inhabitants of a house. The benefit of the use of plastic film is an increased knowledge of the behaviour of the radon gas indoors.

Some of the characteristics of the radon gas have been discussed above. The basis of the characteristics is the source of radon, which is also important in the mitigation process of the indoor radon levels. The knowledge achieved of the behaviour of the gas can give an indication of how to reduce high indoor radon levels.

High levels of indoor radon from the building material and from the household water may be reduced simply by increased ventilation rate. In reducing the high indoor levels of radon from the soil by the use of increased ventilation rate, the underpressure indoors must be very small. The methods of radon mitigation are much like the methods used to reduce other gases streaming into the building from the ground.

Another characteristic is the different levels of radon in different rooms in a house; which means that it is not enough to use only one detector for the estimation of the radon levels in a house.

The seasonal variation of the indoor radon levels means that the radon level may be a factor of two higher during one season in comparison with another season. This implies that the mean value of the measured radon level in a house has an uncertainty not caused by the technique of measurement but by the nature of radon. This uncertainty may be around 50% of the mean value.

It is also evident that each house has its own specific "life" of radon. Therefore it is not possible to generalize and say that it is enough to measure the radon level in one house and believe that the neighbouring houses have the same radon level.

References

Åkerblom G. (1994) Radon in water. *Environment and Health* **3**, 30-35 (In Swedish).

Asikainen M. and Kahlos H. (1980) Natural radioactivity of drinking water in Finland. *Health Phys.* **39**, 77-83.

Heller H. and Poulheim K. F., eds. (1992) Die Exposition durch Radon und seine Zerfallsprodukte in Wohnungen in der Bundesrepublik Deutschland und deren Bewertung. *Veröffentlichungen der Strahlenschutzkommission*, Band 19, Gustav Fischer Verlag, Stuttgart (In German).

Humar H., Šutej T., Skvarč J., Mljač L., Radež M. and Ilić R. (1992) Indoor and outdoor radon survey in Slovenia by etched track detectors. *Radiat. Prot. Dosim.* **45**, 549-552.

ICRP Publication 50 (1987) Lung cancer risk from indoor exposures to radon daughters. *Annals of the ICRP* **17**, 1-60.

Jönsson G. (1987) Indoor radon gas and its detection with Kodak plastic film. *Nucl. Tracks Radiat. Meas.* **13**, 85-91.

Jönsson G. (1988) Indoor ^{222}Rn measurements in Sweden with the solid-state nuclear track detector technique. *Health Phys.* **54**, 271-281.

Jönsson G. (1991) Solid state nuclear track detectors in radon measurements indoors and in the soil. *Nucl. Tracks Radiat. Meas.* **19**, 335-338.

Jönsson G.(1992) *About Radon - where, when and how?* Studentlitteratur Publ. Comp., Lund, 1-112 (In Swedish).

Kulich J., Möre H. and Swedjemark G. A. (1988) *Radon and Radium in Household Water.* Swedish Radiation Protection Institute, SSI-Report 88-11 (In Swedish).

Lide D. R., ed. (1993) *Handbook of Chemistry and Physics.* 74th edition, 6.3-6.4, CRC Press, Florida.

Möre H. (1985) *Radioactive Substances in Building Material.* Swedish Radiation Protection Institute, SSI-report 85-08 (In Swedish).

Ramachandran T. V. and Subba Ramu M. C. (1994) Variation of equilibrium factor F between radon and its short-lived decay products in an indoor atmosphere. *Nucl. Geophys.* **8**, 499-503.

Reiners Chr., Steffer C. and Messersmidth O. (1992) *Strahlenrisiko durch Radon.* Gustav Fischer Verlag, Stuttgart (In German).

SOU 1990:94 The environment in western Scania. 109-116 (In Swedish).

Stranden E., Kolstad A. K. and Lind B. (1984) Radon exhalation: Moisture and temperature dependence. *Health Phys.* **47**, 480-484.

Swedish radon commission: *Radon in Dwellings.* SOU 1983:6 (In Swedish).

Swedjemark G.A. (1983) The equilibrium factor F. *Health Phys.* **45**, 453-462.

Tanner A.B. (1964) Radon mitigation in the ground, a review. In *Natural Radiation Environment* (eds. J. A. S. Adams and W. M. Lowder), University of Chicago Press, pp. 161-190.

WHO (1988) Man-made mineral fibres and radon. *IARC Monographs on the Evaluation of Carcinogenic Risks to Humans* **43**, 173-259.

Winter M. and Wicke A., eds. (1993) Environmental radioactivity, radiation ecology, radiation effects. *Progress in Radiation Protection*, Band I., Verlag TÜV, Rheinland, (In German).

G. Jönsson[1]

3.2 Soil Radon Surveys

3.2.1 Radon Levels in Outdoor Air

Radon gas exhaling from the upper soil is normally so diluted by air that its concentration is negligible at knee-height. Levels of 8-14 Bq m^{-3} are usually quoted. There are exceptions confirmed by measurements. Special meteorological conditions may favour the stay of the radon gas *in situ* in the free air. The radon gas is then spread only by the diffusion of its atoms. Inversion conditions mean that the radon gas stays in the air above the point of exhalation, giving radon concentrations of the order of 100 Bq m^{-3}. The dilution of the radon gas means that the presence of the radon daughters in the air is very capricious.

3.2.2 Depth Dependence of Radon Concentrations in the Soil Air

Radon concentration has a depth dependence. The concentration increases with depth, and may reach a maximum at a depth of about 2 m in the soil. Normally there is a break in the increase of the radon concentration at a depth of about 2 m. The transport upwards of the radon gas in soil is not only a process of diffusion but also of forced transport, which explains the long-distance transport of radon (Kristiansson and Malmqvist, 1982; Mogro-Campero *et al.*, 1980; Monnin and Seidel, 1991). One possibility is a carrier gas, like bubbles moving upwards through water-filled cracks. Another is a pumping effect by compression and decompression in the ground, possibly in connection with earthquakes. The radon concentration $C(z)$ at depth z in the soil can be described theoretically by the equation

$$C(z) = C_0(1 - e^{kz}) \tag{3.2.1}$$

where C_0 is a constant and

$$k = \frac{v}{2D\varepsilon} - \sqrt{\frac{v^2}{4D^2\varepsilon^2} + \frac{\lambda}{D}} \tag{3.2.2}$$

with ε as the porosity of the ground medium, v as the flow velocity, D as the diffusion coefficient, and λ as the radioactive decay constant.

The complexity of the radon transport is illustrated by Fig. 3.2.1, which shows the radon concentration versus depth in sand (Kristiansson and Malmqvist, 1982). The

[1]University of Lund, Lund, Sweden

measured radon may be produced both by a local source and by a deep-lying source. It appears that the increase in radon concentration with depth takes off.

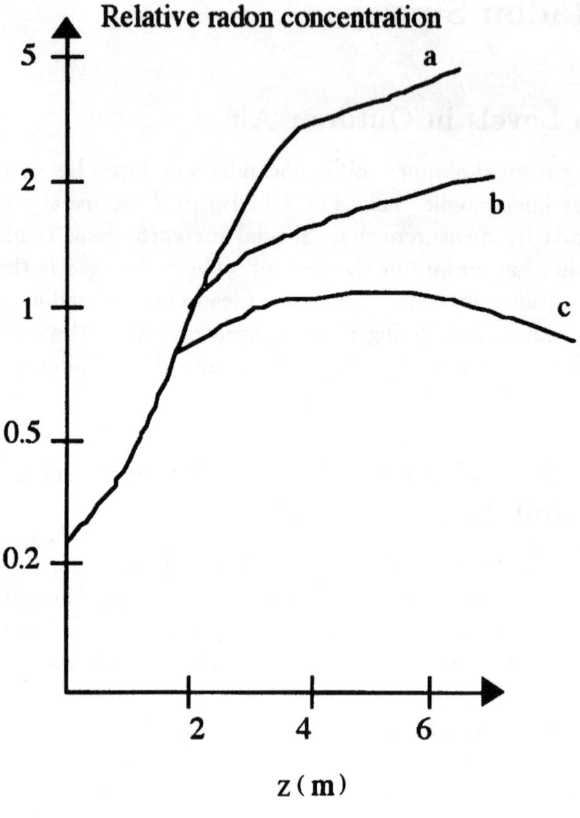

Figure 3.2.1: Depth dependence of soil radon in sand (Kristiansson and Malmqvist, 1982), where z is the depth. a: Theory. Radon diffusion from a deep source. b: curve fitted to measurements. c: Theory. Radon diffusion from a local source.

Table 3.2.1: Approximate values of porosity of some types of soil (Åkerblom, 1993). The porosity is the proportion of a volume filled with air.

Soil type	Sand	Gravel	Morain	Packed Morain	Blasted rock
Porosity	0.40	0.40	0.30	0.25	0.40

Generally, the depth dependence of the radon concentration is different in various types of soil. The porosity is an important parameter because it differs form one type of soil to another. Table 3.2.1 gives approximate values of the porosity of some types of soil. The porosity of clay is missing because of its dependence of the water content of the clay. A second important parameter is the diffusion coefficient. It is given in Table 3.2.2 for some media. A third important parameter affecting the soil radon levels is the moisture content of the soil. This is illustrated by Fig. 3.2.2.

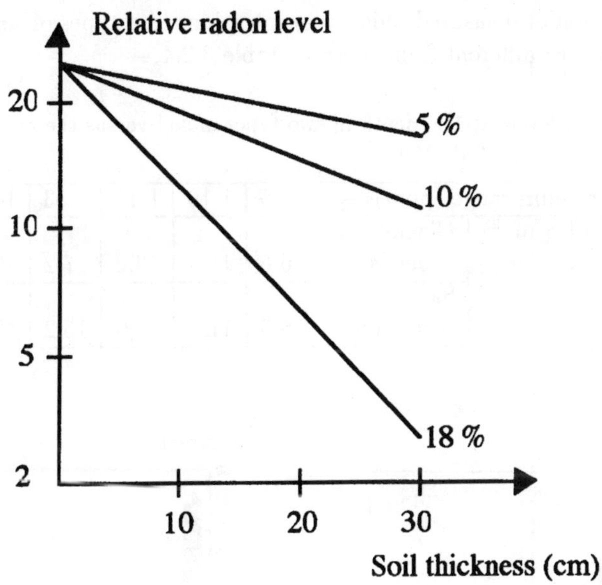

Figure 3.2.2: Variation of the relative radon concentration with depth for different moisture contents, shown as % values (Shweikani *et al.*, 1995).

Table 3.2.2: Reported values of radon diffusion coefficient D (Tanner, 1964).

D(10^{-2} cm^2 s^{-1})	Medium
4.5	Deposits of granodiorite
1.8	Metamorphic rocks
1.5	Deposits of granite
8.0	Loams
7.0	Varved clay
2.0-5.7	Mud

The uranium content of the soil is important for the locally produced radon gas. The U content is another important parameter alongside the porosity and diffusion coefficient. Table 3.2.3 shows the soil radon levels as a function of the uranium content of the soil in a south Swedish community. An increasing range of the uranium content means a higher soil radon level. The uranium content of the soil given in Table 3.2.3 is based on a biogeochemical mapping method (Tell *et al.*, 1994). In another Swedish community with relatively low and homogeneious uranium content of the soil (1-25 ppm), the radon levels for different types of soil are given in Table 3.2.4. Typical for this type of measurement is the wide spread of the radon levels. The measured radon levels given in Table 3.2.4 are only representative of the area of measurement, which is confirmed by soil radon measurements in another community in south Sweden according to Fig. 3.2.3, which gives the distribution of measured radon concentrations in 4 types of soil. The means of the concentrations are different from those in Table 3.2.4.

Table 3.2.3: Radon levels at 1 m depth in two types of soil versus the range of the uranium content of the soil (Jönsson *et al.*, 1991).

Uranium range (ppm) →		1-5	1-10	1-17	1-53	1-69
Rn (kBq m^{-3})	Glaciofluvial deposits	6.6	11.9	13.2	17.2	19.4
	Sandy morain	6.5	11.0	11.9	15.2	43.3

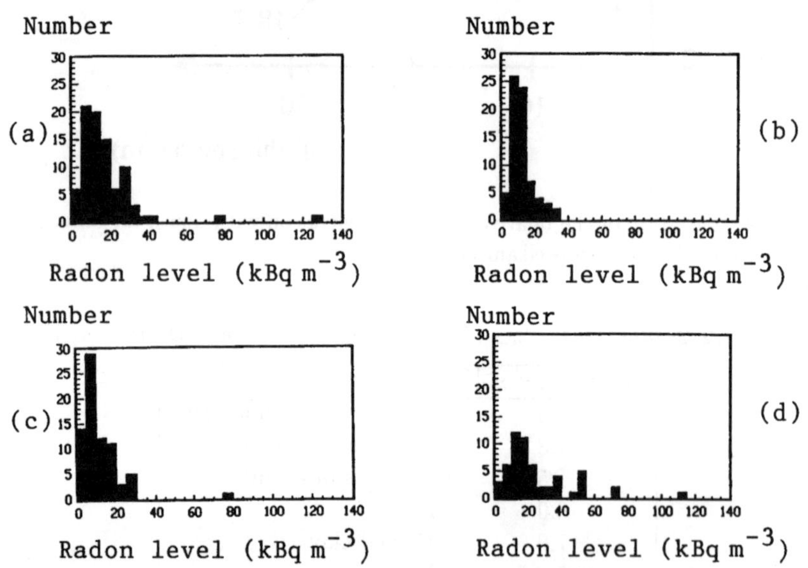

Figure 3.2.3: Distribution of measured soil radon levels in four types of soil in South Sweden. The depth of measurement is 1 m (Jönsson *et al.*, 1991). (a) Morain, 85 measurements, Mean: 17.3±16.3 kBq m^{-3}; (b) Glaciofluvial deposits, 71 measurements, Mean: 13.1±13.1 kBq m^{-3}; (c) Clay, 51 measurements, Mean: 22.7±19.7 kBq m^{-3}; (d) Sand, 75 measurements, Mean: 11.4±10.2 kBq m^{-3}.

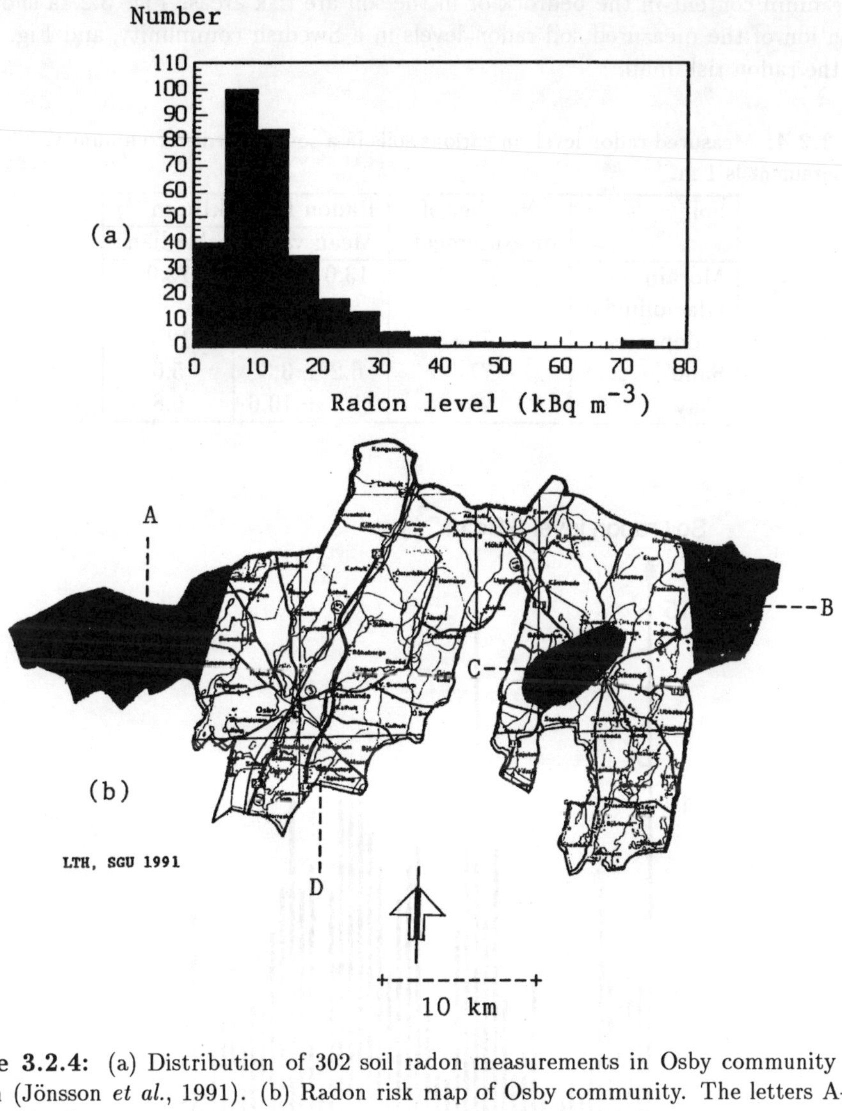

Figure 3.2.4: (a) Distribution of 302 soil radon measurements in Osby community South Sweden (Jönsson *et al.*, 1991). (b) Radon risk map of Osby community. The letters A-D are explained in the text.

3.2.3 Large-Scale Surveys for Risk Assessment

Maps of soil risk areas are required mainly for two reasons. One reason is to identify houses with enhanced indoor radon levels. The other reason is special building regulations. The radon gas must be prevented from entering the building from the ground, and the constructor of a new house must think of that. In large-scale surveys of soil radon, many geological and other data are needed; but there is no doubt that areas with high uranium content in the bedrock or in the soil are risk areas. Fig. 3.2.4a shows the distribution of the measured soil radon levels in a Swedish community, and Fig. 3.2.4b shows the radon risk map.

Table 3.2.4: Measured radon levels in various soils in a south Swedish community. The depth of measurement is 1 m.

Soil	Number of measurements	Radon level (kBq m^{-3}) Mean value	Median
Morain	136	13.0 ± 7.9	12.0
Glaciofluvial deposits	75	13.5 ± 6.1	12.7
Sand	77	6.2 ± 3.3	5.6
Clay	28	11.3 ± 10.6	6.8

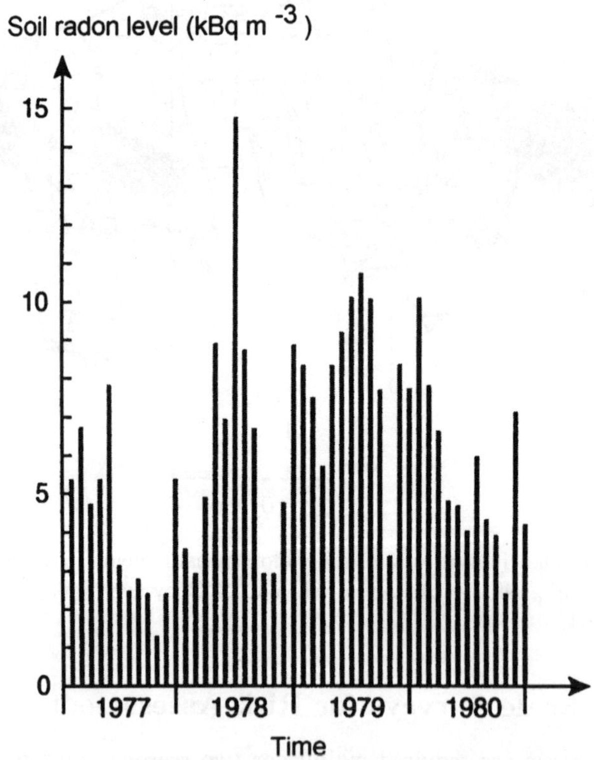

Figure 3.2.5: Soil radon levels at the same point of measurement during 4 years. Monthly means.

The distribution of Fig. 3.2.4a has a tail of radon levels exceeding 25 kBq m^{-3}. These are found in areas where the soil or the bedrock has enhanced uranium content ($>$ 50 ppm). Fig. 3.2.4b shows three black areas. Area A has soil radon levels less than 10 kBq m^{-3}, because neither the bedrock nor the soil has uranium content exceeding 10 ppm. Area B has soil radon levels exceeding 40 kBq m^{-3}, because the bedrock is uranium-rich granite ($>$ 50 ppm). Area C has soil radon levels of 25-40 kBq m^{-3}, because the soil is uranium-rich ($\geq$ 10 ppm). The soil components are collected and brought by the ice at the last glaciation from the uranium-rich granite. Area D has soil radon levels of 10-25 kBq m^{-3}.

3.2.4 Time Variations of Soil Radon Levels

The time-dependent variations of the soil radon levels are mainly of two types: Long-time variations and short-time variations. The long-time variations are conditioned by parameters of seasonal character. Such parameters are ground-water levels and temperature. Fig. 3.2.5 illustrates the soil radon levels at the same measuring point during 4 years. It seems that the radon level is higher during one part of the year and lower during the other part of the year in this study. One possible parameter affecting the measured radon level is the ground water level, which may be higher in the autumn or in the winter. Another possibility is the effect of lower temperature.

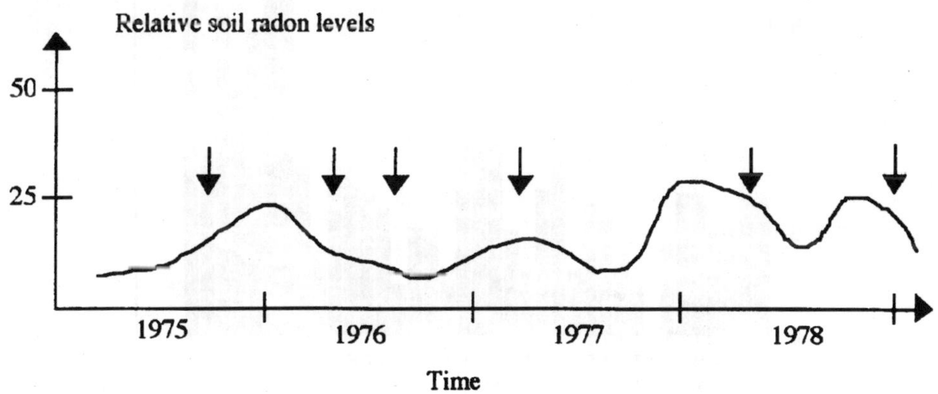

Figure 3.2.6: Soil radon levels during 4 years in seismic active areas. The arrows mark the time of an earthquake of about magnitude 4 (Birchard and Libby, 1980).

For permeable soils the groundwater level has less influence; but for clay this influence may be of importance. Table 3.2.5 shows a comparison of radon concentrations in clay at two different times (during a dry summer and during a wet winter).

Table 3.2.5: Radon levels in clay at 1 m depth. In October-November 1988 the clay was wet and saturated with water. In August-September 1989 the clay was dry and full of cracks (Jönsson, 1995). The levels are mean values over 2-3 weeks.

Point no.		1	2	3	4	5	6	7	8	9	10	11
Radon level (kBq m^{-3})	Oct-Nov 1988	7	32	16	16	7	18	5	15	12	12	10
	Aug-Sept 1989	14	111	31	25	6	71	20	13	3	17	6

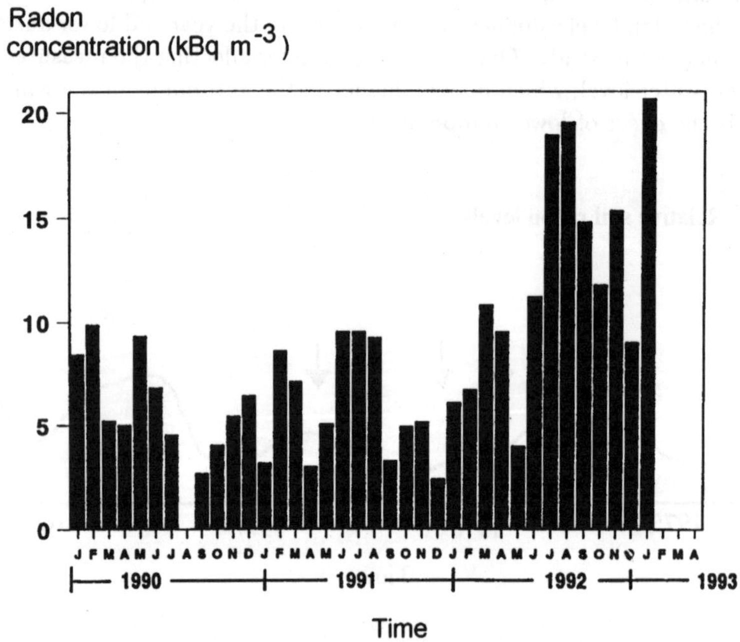

Figure 3.2.7: Soil radon concentration versus time in a seismically active area in Mexico (Segovia *et al.*, 1995).

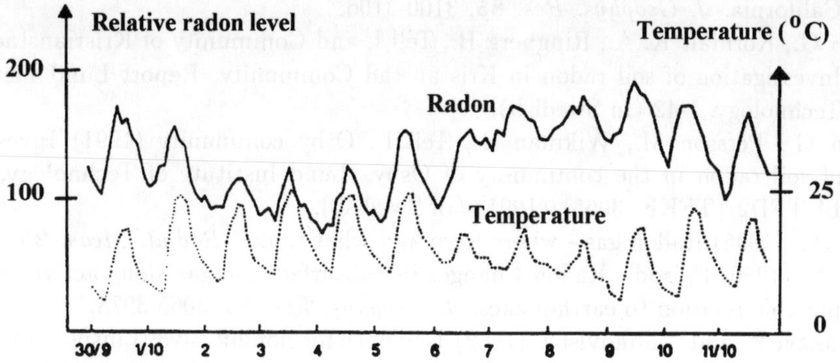

Figure 3.2.8: Temperature-dependent radon variations (Monnin and Seidel, 1991). Upper curve: relative radon level; Lower curve: temperature. Time units are days.

The seasonal pattern is disturbed if the radon gas is associated with seismic activity as shown schematically in Fig. 3.2.6. Fig. 3.2.7 shows the radon concentration in the soil versus time in a seismically active area in Mexico. There are big variations from one month to another also in this case.

Long-term variations are modulated with daily fluctuations. These daily varations are the effect of the atmospheric temperature. Consequently, soil radon measurements must be made at a depth where the radon levels is less influenced by this temperature effect. Fig. 3.2.8 shows laboratory experiment in a tower in order to simulate soil radon levels at shallow depths.

3.2.5 Concluding Remarks

The soil radon level depends on, or is affected by, several parameters. Some of these parameters and their effects have been discussed above. Many of the parameters have a geological or a meteorological origin. Sometimes one parameter influences the radon level in soil very much at one point of measurement, but only a few metres away this influence may be negligible (Kristiansson and Malmqvist, 1982). This fact makes it difficult to specify or predict the soil radon concentration, especially if one or two important parameters are not known. It is then easy in a given measurement to wrongly indict the soil radon detector for giving the wrong response.

References

Åkerblom G. (1993) Ground radon-monitoring procedures in Sweden. *Geoscientist* **4**, 21-27.

Birchard G. F. and Libby W. F. (1980) Soil radon concentration changes preceeding and following four magnitude 4.2-4.7 earthquakes on the San Jacinto fault in Southern California. *J. Geophys. Res.* **85**, 3100-3106.

Jönsson G., Kornfält K. A., Ringberg B., Tell I. and Community of Kristianstad (1990) Investigation of soil radon in Kristanstad Community. Report Lund Institute of Technology, 1-42 (In Swedish).

Jönsson G., Persson M., Wikman H., Tell I., Osby community (1991) Investigation of soil radon in the community of Osby. Lund Institute of Technology, Report LUTFD2/(TFKF- 3065)/(1991) (In Swedish).

Jönsson G. (1995) Radon gas - where from and what to do? *Radiat. Meas.* **25**, 537-546.

King C-Y. (1980) Episodic Radon Changes in Subsurface soil gas along active faults and possible relation to earthquakes. *J. Geophys. Res.* **85**, 3065-3078.

Kristiansson K. and Malmqvist L.(1982) Evidence for nondiffusive transport of ^{222}Rn in the ground and a new physical model for the transport. *Geophysics* **47**, 1444-1452.

Mogro-Campero A., Fleischer R-L. and Likes R. S. (1980) Changes in subsurface radon concentration associated with earthquakes. *J. Geophysical Res.* **85**, 3053-3057.

Monnin M. and Seidel J. L. (1991) Radon and geophysics- recent advances. *Nucl. Tracks Radiat. Meas.* **19**, 375-382.

Shweikani R., Giaddui T. G. and Durrani S. A. (1995) The effect of soil parameters on the radon concentration values in the environment. *Radiat. Meas.* **25**, 581-584.

Segovia N., Mena M., Tamez E., Pena P., Salazar S., Monnin M. and Seidel J. L. (1995) Radon mapping in hazardous zones in Mexico. In *Gas Geochemistry* (ed. C. Dubois), Science Reviews, Northwood.

Tanner A. (1964) Radon in the ground. A review. In *Natural Radiation Environment* **1** (eds. J. A. S. Adams and W. M. Lowder), Univ. of Chicago Press, Chicago, pp. 161-190.

Tell I., Bensryd I., Rylander L., Jönsson G. and Daniel E. (1994) Geochemistry and ground permeability as determinants of indoor radon concentrations in southernmost Sweden. *Appl. Geochem.* **9**, 647-655.

J. F. Pineau[1]

3.3 Radon Dosimetry and Monitoring in Mines

3.3.1 Introduction

In most mines, the presence of radioactive materials such as uranium or thorium in the surrounding rocks, as well as the confinement of the atmosphere underground, may lead to an increased concentration of radon isotopes (which are all uranium and thorium decay products, escaping from the rocks mainly by convective transport) and of their decay products created in the mine atmosphere. The discovery at the end of the last century of the diseases (mainly lung cancers) appearing in the presence of radon led to the development of dosimetry and prevention techniques, allowing, just after the second world war, the implementation of radiation protection structures in uranium mines all over the world. Some non-uranium mines, showing high radon concentrations, have also been involved.

The use of Solid State Nuclear Track Detectors (SSNTDs) to detect the alpha-particle emissions of the radon itself or of its decay products in mining environments has led to two types of rugged heavy-duty measuring instruments:

- passive integrating detectors, which are mainly used to perform radon gas monitoring,
- active integrating dosimeters, allowing the measurement of the indicator of risk, which is, as proposed by the International Commission on Radiation Protection (ICRP), the Potential Alpha Energy Concentration (PAEC) of the Radon Decay Products (RnDPs) existing in the measured mine atmosphere.

This section of the book will describe the two types of measuring devices, and will give information allowing a choice to be made, depending upon the objectives of the measurement. It will also give information about the data which can be extracted from the raw measurements, allowing, for example, the assessment to be made of the quality of the ventilation or the optimisation of the radiation protection in a mine. The special case of the non-uranium mines is discussed, too.

The hazard due to radon and to its decay products was identified as early as the 17th century, and has been recognised since the last century as being responsible for increased lung cancer incidence among exposed miners. With the increased demand for uranium after the second world war, the need for a survey of the atmosphere of the underground uranium mines, and of dosimetric monitoring of the miners involved in this particular

[1]ALGADE, Bessines, France

activity, appeared at the end of the nineteen-forties. As an example, in French uranium mining operations, the first systematic radon samplings appeared as early as 1948, and the first dosimetric data record sheets were made available to mining operators in 1955. Even at that early time, the dosimetric records already took into account the combination of the three sources of exposure arising in a mine, that is to say:

- external exposure,
- radon decay products internal exposure,
- internal exposure by long-lived aerosols.

The knowledge gained from the increased number of measurements (up to 40,000 radon measurements a year for a typical mining site) led to the development between 1972 and 1982 of an active individual dosimeter, using a SSNTD, for the assessment of the exposure to radon decay products, and to its use on a routine basis (Bernhard, 1983a; Chapuis, 1981; Duport 1979b), since 1983 in the French as well as in Canadian and African uranium mines. A derivative of this individual dosimeter has also been built, using the same measuring assembly, but with a pump ensuring a greater air flow, for the monitoring of the mining environment. As well as this implementation, some studies have been conducted with passive dosimeters to estimate their behaviour in the rough mining conditions, and also in the mining environment. The whole set of data gained since 1983 (≈200,000 man-months) in many different mines has led to the development of specific dosimetry concepts (Pineau, 1982a). A brief overview of these concepts is given in this section.

3.3.2 The Atmosphere in Underground Mines

The atmosphere of underground mines contains, in a gaseous form or in the form of aerosols, all the natural radionuclides of the decay chains of uranium and thorium. As a result of their confinement, the concentration of some of them, in particular radon and its decay products, may reach high levels which are potentially dangerous to the health of workers, and which therefore must be systematically evaluated by the mine operators (ICRP, 1981, 1986). We deal mainly, in this section, with radon and its decay products, substances which are respectively gaseous and in the form of solid aerosols, which may, by being inhaled by workers, cause internal "irradiation" of the organism.

3.3.2.1 Origin of the radioactivity of the atmosphere in underground mines

The source of the radionuclides present in the atmosphere of a mine is made up of:

1. for radon
   - the mineralised working faces,
   - all the walls of the tunnels and the workings;

The quantity of mobilisable radon depends mainly on the: (i) pressure gradient through the source rock, (ii) radium content of that rock, and (iii) degree of any water saturation of the rock.

2. for decay products of radon
   - the radon having diffused and aged in the atmosphere of the mine,
   - originating in the above-mentioned sources.

3.3.2.2 Main characteristics of the atmosphere of mines

The atmosphere of a mine may be characterised mainly by the following physical parameters: temperature, relative humidity, particle size distribution of aerosols, volume concentration of radon, age of the ventilation air, and volume concentration of radon decay products.

3.3.2.2.1 Temperature. The temperature of the atmosphere of a mine may vary between -5°C and +50°C; it has generally no great influence on the characteristics of the atmosphere itself, but may have a strong influence on the functioning of the devices used to quantify it, either by the phenomenon of frost formation for the lowest temperatures, or, for the highest temperatures, by the phenomenon of erasure of latent tracks on the detectors used.

3.3.2.2.2 Relative humidity. Relative humidity is a characteristic of great importance; in fact it modifies the conditions of attachment of the free fraction of the aerosols by changing the total number of condensation nuclei; a high level of humidity (often close to 100% in underground mines) thus reduces the proportion of the free fraction. It may also modify the behaviour of the measuring devices by reducing the path of the alpha particles in the air and by causing the uncontrolled clogging of the sampling or input filters, either by the deposition of droplets of water, or by the phenomenon of condensation.

3.3.2.2.3 Particle size distribution of the aerosols. A knowledge of the particle size distribution of the aerosols is vital for the assessment of the exposure to potential alpha energy. It has, in fact, an effect on the conversion factor which allows us to go from exposure to the dose; This knowledge is necessary to ensure that a relevant sampling device is functioning correctly.

3.3.2.2.4 Volume concentration of radon. This parameter depends principally on two variables:

- the magnitude of the source of radon (surface of walls, and of cracks and fissures in contact with the air; pressure gradient; water saturation),

- the rate of ventilation, which also defines the dwell time in the atmosphere in contact with the source.

It may be considered: (i) either as a simplified (and conservative) indicator of the radioactive risk (e.g. in the case of the global evaluation of the risk of exposure for non-uranium mines, where a measurement of this parameter allows the mine to be classified as "to be monitored" or as "not requiring monitoring") in relation to the constraints of the regulations in force. The precautions for the use of this parameter (the least favourable situation must systematically be assumed to be the case) do not allow the slightest optimisation procedure to be envisaged using this indicator; or (ii) as an indicator of the quality of the ventilation.

The measurement of radon alone does not allow reliable information to be obtained concerning the indicator of the risk, which is exposure to Potential Alpha Energy (PAE). In fact, the eminently variable aspect (Tsivoglou *et al.*, 1953) of the parameters which define the atmosphere of a mine (in particular the multiplicity of sources and the variability of the dwell time of the air in contact with them, as well as the variation in the number and in the particle size distribution of the aerosol) makes it impossible to determine the 'Equilibrium Factor' from the measurement of the radon alone. It is therefore necessary, in order to respect the constraints of the regulations in force, to measure directly the exposure to PAE. The radon measurements taken with this in mind must be carried out over periods of the order of half a day for the daily monitoring of the ventilation quality, and of the order of a week for the definition of criteria for the monitoring of a mine.

3.3.2.2.5 Age of the ventilation air. This parameter also allows the quality of the ventilation of a mine to be qualified; it is directly linked to operating costs (size of the ventilation system) and to the age of the mine (size of old workings which increase the total volume to be scanned).

3.3.2.2.6 Volume concentration of radon decay products. This depends on only two main parameters:

- the volume concentration of radon,
- the age of the air concerned.

Generally speaking, this parameter is replaced by the parameter "volume concentration of potential alpha energy" (PAEC); it corresponds then to the indicator of the real radioactive risk, as defined by the ICRP (ICRP, 1981). This parameter then must be measured. The parameter PAE is a concept of health physics, which designates the risk indicator used in the radiation protection of persons exposed to radon and its decay products. Because of its gaseous form, radon - which does not stay in the lungs (the main target organ) - is not normally concerned in this concept. It is its short-lived solid decay products which represent a potential health risk because, present in the atmosphere in the form of aerosols, they may be deposited in the respiratory tract and leave their energy there, thus causing cellular lesions.

The PAE of an atom corresponds to the total alpha energy emitted by that atom during its decay down to ^{210}Pb and ^{208}Pb respectively for the chains of ^{238}U and ^{232}Th.

The volume concentration of potential alpha energy of ^{222}Rn ($PAEC_{222}$) and ^{220}Rn ($PAEC_{220}$) should always be calculated or measured separately, for the PAE as a health risk indicator is directly linked to the Annual Limit of Intake (ALI) whose values are different for ^{222}Rn decay products ($RnDP_{222}$) and ^{220}Rn decay products ($RnDP_{220}$).

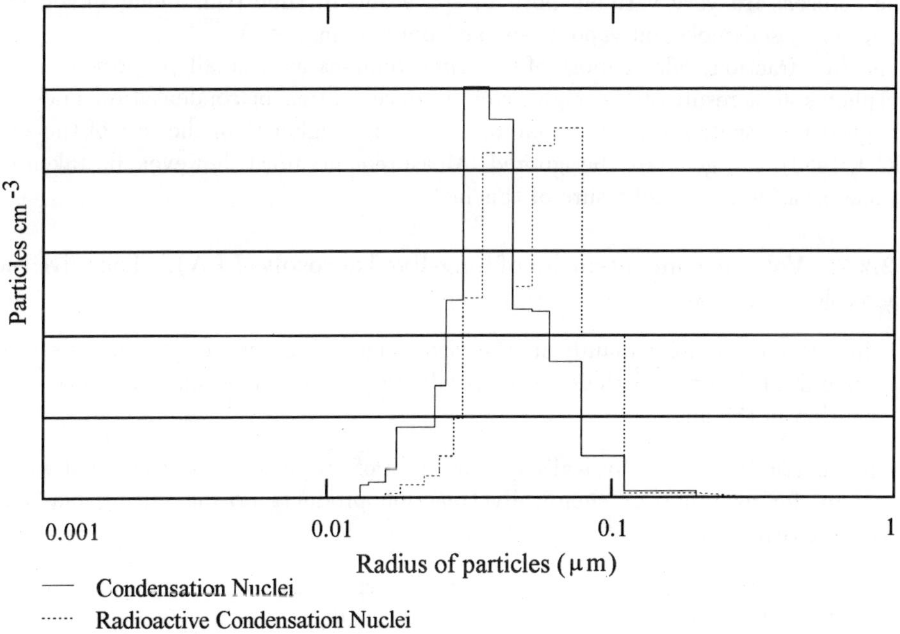

Figure 3.3.1: Typical particle size distribution in the atmosphere of an underground mine.

For radioactive risk prevention operations, PAEC measurements must be carried out over periods of the order of half a day; whereas in order to carry out staff dosimetry, a period of measurement of one month is generally required by regulations. The decay products of radon are constituted for the most part of isotopes of lead, bismuth, thallium and polonium, all of which are solid substances at ambient temperatures. These elements are therefore found in the atmosphere studied in the form of aerosols (Strong, 1887), and may be classified in two categories:

1. Radioactive condensation nuclei, made up of "condensation nuclei": inert particles which serve as seeds, onto which are fixed the radioactive elements which are decay products of radon. The size of this aerosol usually falls within the range of dimensions between 0.05 and 0.3 μm. As an example, Fig. 3.3.1 (Duport, 1976) gives a typical particle size distribution for the atmosphere of an underground mine.

2. The radioactive particles (short lived) formed by the decay of the radon or of one of its short-lived decay products. These particles, which are not fixed on condensation nuclei, remain: either (i) free: this part of the spectrum of the aerosol is then called the "free fraction", or (ii) gather together as clusters of a few atoms, and then form the category of ultrafine particles which, when they are ionised, may take the name of "large ions".

In the particular case of mines, the particle size distribution and the electrical charge of the aerosols are very variable both in space and in time (this being linked to the presence of diesel smoke, oil vapour, smoke from blasting, etc.).

The free fraction, which most of the time remains as a small proportion in mine atmospheres as a result of the high level of concentration of condensation nuclei; and although it represents a conversion factor Sv/J much higher than the rest of the aerosol (OECD, 1983), can generally be ignored. Measurements must, however, be taken on at least one occasion so as to be sure of this fact.

3.3.2.2.7 Volume concentration of long-lived aerosols (LLA). Long-lived aerosols are aerosols made up of:

- dust (large particles mainly in the size range 1-100 μm) from minerals bearing uranium, thorium and their respective decay products, mechanically placed in suspension in the air,
- fine particles formed physically (evaporation of droplets of water charged with Radium, for example) or chemically (reaction products arising during blasting by explosives),
- condensation nuclei bearing long-lived isotopes which are decay products of radon (^{210}Pb, ^{210}Po, etc).

3.3.2.3 Order of magnitude of the volume concentrations to be measured

The volume concentrations of radon that may be encountered in underground mines range from 100 to 10,000 Bq m^{-3} in non-uranium mines, and from 100 to 500,000 Bq m^{-3} in uranium mines. The order of magnitude of the PAEC that may be encountered ranges from 5 to 50,000 nJ m^{-3} in non-uranium mines and from 0.1 to 1,000 μJ m^{-3} in uranium mines.

3.3.3 Dosimetry: Application to Miners

The dosimetric monitoring of miners must be carried out according to the regulations in force in the country concerned. These national regulations are generally based on the recommendations of the ICRP. Such a dosimetry must be carried out by adding up the three radioactive risks to which miners are exposed:

- external exposure due to ambient gamma radiation,

- internal exposure due to long-lived aerosols,
- internal exposure to potential alpha energy due to short-lived aerosols which are decay products of radon.

The measurement of all these risks must be done (Bernhard, 1992) for each exposed worker if the total exposure is liable to be over 0.3 of the annual limit; it may be carried out for groups of workers if total exposure is between 0.1 and 0.3 of the annual limit. If it can be shown that exposure cannot in any circumstances be more than 0.1 of the annual limit, no personal dosimetry is required; but a dosimetry based on the monitoring of the physical parameters of the atmosphere is recommended. Generally speaking the constraints limiting quarterly exposure are added to the constraints of the annual limits.

3.3.3.1 Dosimetry of miners in France

In France, an Integrated Dosimetry System designed around a dosimeter providing for the measurement of these three risks was developed between 1972 and 1983, at which date (i.e. 1983) its use was on routine basis: and after a short delay time, its use was made obligatory by the French regulations concerning the monitoring of miners exposed to a radioactive risk, in uranium mines or in any other mineral mine. As a result of its being designed specifically for use in mines, the dosimeter is also used by numerous mining companies throughout the world (Bernhard, 1995a).

3.3.3.1.1 Integrated dosimetry system. This system (Bernhard, 1986, 1995c; Pineau, 1980b, 1982a) comprises the following elements:

- a contactless system for recharging the accumulator of the dosimeters,
- a special flow meter,
- a housing containing a sampling pump and its electrical and electronic supply for charging and switching,
- a measuring head, optimised for the measurement of the three above-mentioned risks.

The dosimeter itself is made up of the two latter items.

3.3.3.1.2 Measuring head. The measuring head (Chapuis, 1972; Duport, 1978; Pineau, 1980a, 1986b) comprises a body made of an aluminium alloy containing a filter holder, the aerosol sampling filter, a part with air inlets and collimators, a set of absorbers for the discrimination of the energy of the alpha particles emitted by the RnDPs collected on the filter, and a cellulose nitrate detector film (LR 115 type II from Dosirad, France). The sampling filter is a membrane filter of 1.2 μm pore size. The pump ensures a nominal sampling rate of 5 l h^{-1}. The air inlets are made up of three holes, 1.5 mm in diameter placed at 120° around the upper part of the filter holder. They are situated in the centre of wide grooves machined in the body of the head. The body of the dosimeter also has a wide groove at the level of the air inlets in order to limit to a minimum the loss of aerosols (Pineau, 1982b, 1993), without compromising the solidity of the unit which is necessary for resisting the extreme conditions of use met in mines. Figure 3.3.2 illustrates the composition of the measuring head.

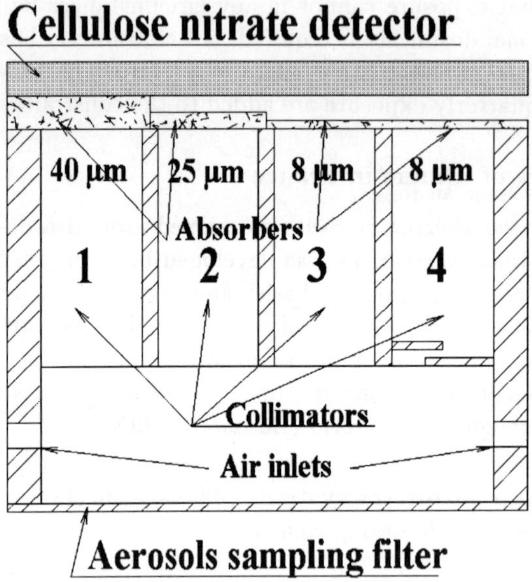

Figure 3.3.2: Composition of the measuring head.

In such a measuring head, the measurement of exposure to the potential alpha energy of the decay products of radon is carried out with the aid of a solid state nuclear track detector, as shown above. The measurement of exposure to long-lived aerosols is done by analysing in the laboratory the sampling filter after a cooling time needed to allow for the complete decay of the RnDPs. The measurement of external exposure is carried out with the aid of two thermoluminescent pellets made of lithium fluoride placed in the lid of the head.

3.3.3.1.3 Unit for the detection and measurement of exposure to potential alpha energy. Samples of the aerosols bearing the radon decay products are taken, at the nominal flow rate of 5 l h^{-1}, through the three air inlets, on the sampling filter where they are trapped. The RnDPs, short lived alpha emitters ($\tau_{1/2} < 10.64$ h), decay on the filter and isotropically emit their alpha radiation. The set of collimators selects the alpha particles emitted by these radionuclides, as they emerge near-normally from the filter and thus have an energy very close to, or equal to, their maximum energy. The absorbers carry out a selection of energy (mechanical spectrometry), thus allowing the discrimination and the identification of the radioactive elements. As the LR 115 type II cellulose nitrate detector film responds to an energy range limited to 2 to 4 MeV for the detection of alpha particles, it is therefore possible to selectively record the different radionuclides, in four detection zones defined by the position of the four collimators (Pineau, 1986c). Fig. 3.3.3 (Duport, 1979a) specifies the conditions in which each of the radionuclides is recorded by the solid state track detector.

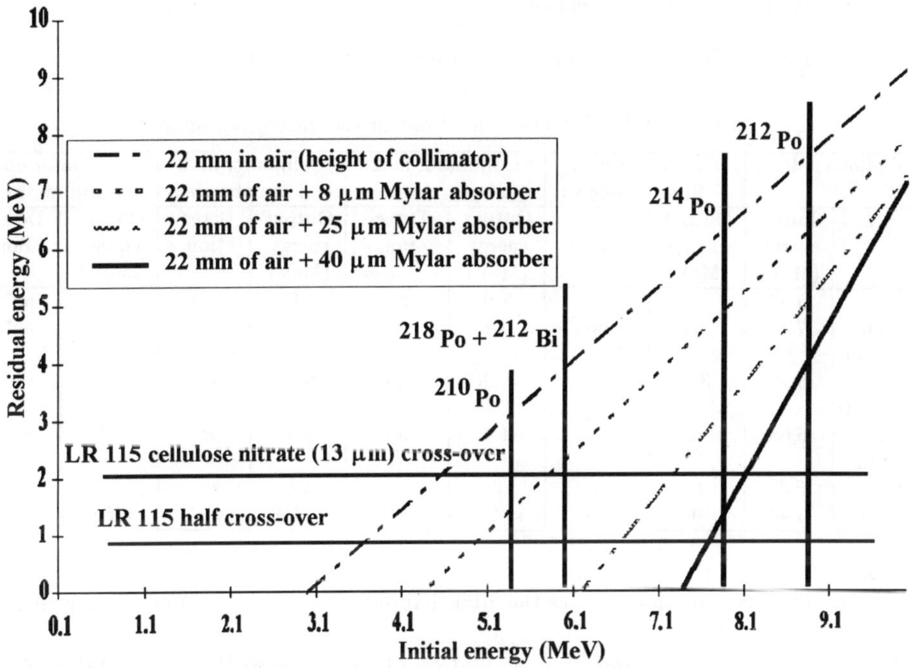

Figure 3.3.3: Residual energy of alpha particles after passing through collimators and absorbers as a function of initial energy (the vertical lines indicates the alpha energy of the radionuclides concerned).

Table 3.3.1 represents synthetically the way the detection functions in the measuring head. It may be seen from this Table and from the curves in Fig. 3.3.3 that in the measuring head a separate detection is obtained of the different radionuclides present and contributing notably to the total exposure due to RnDPs:

- the potential ^{218}Po and ^{214}Po (we will include in the term potential ^{214}Po: all the ^{214}Pb, ^{214}Bi and ^{214}Po detected, after total decay, by the alpha emission of the ^{214}Po) are recorded respectively in channels 3 and 2,

- the potential ^{212}Po is recorded in channel 1,

- the ^{212}Bi is recorded in channel 3 and is subtracted from the total number of tracks in this zone by applying the equation: $^{212}\text{Bi} = 0.56 \times {}^{212}\text{Po}$ (branching ratio),
- the ^{222}Rn is recorded in channel 4, which is closed by a baffle at the bottom; this arrangement ensures that no alpha emitted from the filter may reach the detector in zone 4, only the atoms of ^{222}Rn decaying inside the collimator being detected. Since the alpha particles emitted by the ^{222}Rn atoms present in collimators 2 and 3 have an influence on the measurement of ^{218}Po, ^{212}Bi and ^{214}Po, the values given by the channel 2 and 3 detection zones have to be corrected using the information given by the channel 4 detection zone.
- the ^{210}Po only leaves non-etched out tracks on the detector.

Table 3.3.1: Detection functions in the measuring head.

Radionuclide		Collimator no.1 + 40 μm absorber		Collimator no.2 + 25 μm absorber		Collimator no.3 + 8 μm absorber		Collimator no.4 + 8 μm absorber	
	Initial energy (MeV)	Output energy (MeV)	Detection	Output energy (MeV)	Detection	Output energy (MeV)	Detection	Output energy (MeV)	Detection
^{218}Po	5.99	0	no	1.0	no	2.6	yes		no**
^{214}Po	7.7	1.0	no	3.1	yes	5.05	no		no**
potential ^{212}Po	8.78	3.4	yes	5.3	no	6.4	no		no**
potential ^{212}Bi	6.08	0	no	1.1	no	2.6	yes		no**
potential ^{210}Po	5.3	0	no	0	no	1.5	no		no**
^{222}Rn	5.49	0	no	≈ 2.0	yes*	1.7-4	yes*	≈2-4	yes*

* in part only (tail of the spectrum)
** this zone of the detector does not see the filter, but only the internal volume of the collimator.

Although it appears simple, the calculation of exposure to PAE must take into account, in particular for the calculation of error, the different interactions existing between the zones of detection, the efficiency of the sampling unit as far as the correct collection of aerosols is concerned, and the efficiency of the automatic optical system used for the counting of tracks left in the detector by the passage of the alpha particles. The simplified block diagram in Fig. 3.3.4 presents an overview of all the parameters connected with the measurement of exposure to PAE. In this figure the diamonds represent the coefficients which are subject to statistical uncertainty and may introduce uncertainties in the final result. It may be seen that the causes of uncertainty are many, and that it is therefore necessary to limit their individual values to the strict minimum if we wish the system to be of the highest quality in relation to what is required of it for the purposes of meeting regulations. In order to be as aware as possible of the different causes of uncertainty, the dosimeter equipped with its measuring head has undergone tests, in real mining conditions and in the laboratory, in order to find out the response overall to different:

- sizes of aerosols (Pineau, 1993; Rannou, 1985; Tymen, 1983),
- temperature and humidity conditions,
- ventilation conditions,
- concentrations of radon and RnDPs.

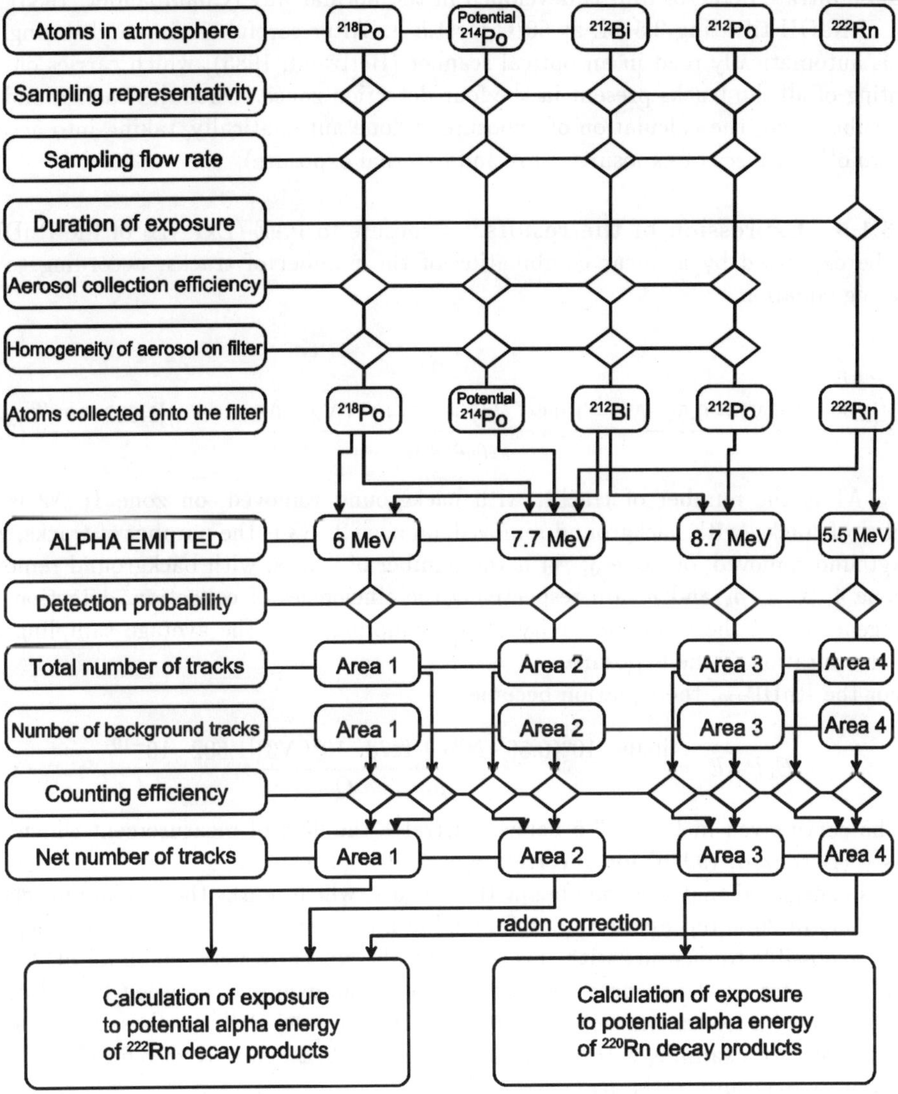

Figure 3.3.4: Operating diagram of the measuring head for potential alpha energy exposure assessment.

These tests have allowed the dosimeter to be qualified for use in the different types of atmosphere which may be encountered in a mine.

3.3.3.1.4 Treatment and reading of the detector films. At the end of the month's use imposed by the regulations, the complete measuring head, or just the detector film, is sent to the dosimetry processing laboratory. After being removed, the cellulose nitrate detector film is developed in the normal way (Chambaudet, 1985) in a bath of NaOH titrating 2.5 M, at 60°C for 1.5 h. After careful rinsing and drying, the film is automatically read in an optical scanner (Bertrand, 1985), which carries out the counting of all the tracks present in the four detection zones. Once the background has been subtracted, the calculation of exposure is done automatically, taking into account the two other causes of exposure (LLA and external exposure).

3.3.3.1.5 Expression of the results. Exposure to PAE (ExPAE) of the $RnDP_{222}$ may be expressed by a linear combination of the number of tracks, according to the following equation

$$E_xPAE = \\ = \frac{[5.99 \cdot 10^6(N1 - 0.56 \cdot N3 - 1.099 \cdot N4) + 7.7 \cdot 10^6\ (N2 - 0.045 \cdot N4)]1.602 \cdot 10^{-19}}{\rho_c \rho_g \rho_h \rho_r Q} \quad (3.3.1)$$

where $N1$ is the number of tracks, with background removed, on zone 1; $N2$ is the number of tracks, with background removed, on zone 2; $N3$ is the number of tracks, with background removed, on zone 3; $N4$ is the number of tracks, with background removed, on zone 4; ρ_c, ρ_g, ρ_h and ρ_r are respectively the efficiencies of collection, detection, the homogeneity and the representativity of the sampling; Q is the average sampling rate over the duration of the exposure.

For the $RnDP_{220}$, the equation becomes

$$E_xPAE = \frac{[6.08 \cdot 10^6(0.56 \cdot N3) + 8.78 \cdot 10^6(N3)]1.602 \cdot 10^{-19}}{\rho_c \rho_g \rho_h \rho_r Q} \quad (3.3.2)$$

The curve given in Fig. 3.3.5 below illustrates the field of measurement which may be used (Bernhard, 1995b) for uncertainty of $\pm$ 20% (1 σ).

It is considered that the zone below the line 0.2, which marks the maximum relative uncertainty of 20%, represents the useful field of measurement. It may be seen therefore that it is possible to measure with this set, monthly exposures between 50 μJ m^{-3} h and 60 mJ m^{-3} h for an atmosphere free of ^{220}Rn; or, as an example, between 140 μJ m^{-3} h and 60 mJ m^{-3} h for an atmosphere containing 700 nJ m^{-3} of ^{212}Po; these values are to be compared, for example, with the derived limit for PAE exposure of 3.6 mJ m^{-3} h for the monthly exposure to the $RnDP_{222}$.

The number of tracks obtained in the different zones of detection, moreover, allow the following information to be obtained:

- the equilibrium factor may be estimated following modelling based on the ratio

$$\frac{N2}{N1 - 0.56 \cdot N3 - 1.099 \cdot N4} \tag{3.3.3}$$

- the ageing of the air may be estimated from the ratio

$$\frac{N3}{N1 + N2} \tag{3.3.4}$$

Furthermore, the simultaneous measurement of potential ^{218}Po and ^{214}Po ensures a high level of redundancy of the dosimetric information.

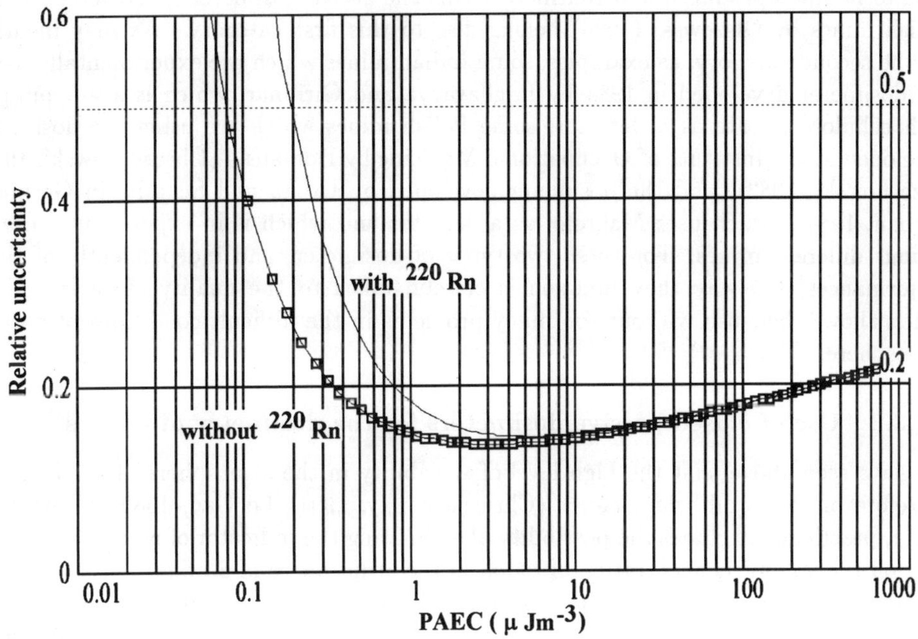

Figure 3.3.5: Relative uncertainty vs Potential Alpha Energy Concentration, PAEC.

3.3.3.2 Other examples of operational dosimetry

Different mining operators have chosen to use other types of dosimeters to carry out the dosimetry of their staff. These other types may be divided into two categories:

(i) active dosimeters including a device for pumping and sample taking on filters, functioning according to a principle similar to that of the French dosimeter; and (ii) passive dosimeters similar to those used for houses or workplaces; here dosimeters are almost always of the "closed" type, and include a diffusion membrane permeable only by radon gas. The principles according to which they function are amply described in other parts of the book.

In the first category, we may place the dosimetric unit built in South Africa for the Chamber of Mines, by Strydom (Leuschner, 1992; Strydom, 1992), which includes a detection device using the polycarbonate Makrofol DE from Bayer, Germany, as an SSNTD, in association with a gravimetric dust sampling device. This active dosimeter however measures only potential ^{214}Po, since the material chosen as a detector does not allow the carrying out of spectrometry. The active dosimeter built in Czechoslovakia in the 1970s, used cellulose nitrate as a detector, with spectroscopic separation of potential ^{218}Po and ^{214}Po. This device, used in Czech mines, was one of the few devices of this type to be mass produced and routinely used. The active dosimeter constructed for the Polish mines by Olszewski (1996) belong also to this first category. We may mention for the second category, as examples, some Indian mines which are experimentally using the dosimeter developed in 1984 by Raghavayya and Giridhar, which is also equipped with cellulose nitrate as a detector; some Polish mines which are using the dosimeter developed at the Institute of Occupational Medicine by Domanski (Chruscielewski, 1982; Orzechowski, 1982); and the dosimeter developed by Urban and Schmitz in Germany (Urban, 1984), which uses Makrofol as a detector and which is in experimental use in various different mines. For these two types of dosimeter, and independently of their performances, the way they function is closely linked to the quality of their design, which allows their use without too many problems in the difficult conditions of a mine atmosphere.

3.3.3.3 Use of closed passive dosimeters for the dosimetry of miners

We have seen above that the high level of variability in the atmosphere of a mine, and therefore of the equilibrium factor (Chruscielewsky, 1991; Le Gac, 1980; Skowronek, 1991), practically excludes the possibility of estimating the indicator of risk (PAE) based only on the volume concentration of radon in the atmosphere of a mine.

Different intercomparisons between passive dosimeters and the French active dosimeters have been carried out over the last few years. The documents cited as references (Bertrand, 1990; Pineau, 1985, 1986a) show all the factors of no, or very low, correlation between the exposures estimated using passive radon dosimeters and the real exposures measured with the Integrated Dosimetry System. It is therefore not very realistic to expect to be able to carry out reliable dosimetry in terms of the qualification of individual exposures using passive dosimeters, and this is even more so because the latter do not permit the measurement of LLAs, which play an important part in the risk. These dosimeters may, however, if they are used for the purpose for which they were designed, be extremely useful for the monitoring of physical parameters of the atmospheres.

3.3.4 Monitoring of Physical Parame[...]ospheres

The monitoring of physical parameters of the atmos[...] field of application of dosimeters, and in particular closed passive dosim[...]g SSNTDs.

3.3.4.1 Qualification of non-uranium mines

The measurement of radon gas may give a valuable indication of the overall quality of the air in a mine (Bernhard, 1981, 1983b). This is the reason why in France, for example, the classification of non-uranium mines and underground quarries as "to be monitored" or as "not requiring any particular monitoring" is done by annually measuring the volume concentration of radon, averaged out over at least one week, of the air returns. This measurement can be efficiently carried out, using, amongst other measuring devices, passive dosimeters.

The volume concentrations, when they are significant, do in fact allow an acceptable degree of precision to be obtained using this type of device. Careful attention must be paid however to the limiting conditions for the use of these dosimeters, such as the relative humidity of the air, the conditions of the forced circulation of ventilated air around the device, or the presence of ^{220}Rn in non-negligible quantities compared with the quantity of ^{222}Rn. The dosimeters will have to have been calibrated taking into account these particular conditions of use. However, if the measurements taken according to these principles show that the mine must "be monitored", then the case is the same as that of a uranium mine, and the regulations concerning functional or individual dosimetry must be applied, depending on the levels of exposure measured.

3.3.4.2 Monitoring of the environment of mining sites

The monitoring of the environment of mining sites (Fourcade, 1987) is necessary to ensure that the mining operations do not generate radon in such quantities that the exposure added (anthropogenic) to the natural exposure (geogenic) of the site, for the public in the neighbourhood of the site, exceeds the limits imposed by the regulations. The latter can be checked in a useful way only by using dosimeters which allow the simultaneous measurement of the $RnDP_{220}$ and the $RnDP_{222}$, organised in a monitoring network.

The ^{212}Pb in the ^{220}Rn decay chain may in fact be considered as a tracer (Zettwoog *et al.*, 1996) of the local vertical stability of the atmosphere, being directly linked with its ^{220}Rn parent (which, due to its very short half life, is not able to come from far away). The knowledge of the equilibrium factor, as well as that of the concentration of ^{222}Rn (obtained with the aid of the information on potential ^{218}Po and ^{214}Po), allows the "regional background" measured by some of the dosimeters in the network to be defined relatively easily. By simple subtraction, the value of the concentration of anthropogenic radon measured by the other stations of the network can be obtained (Bernhard, 1994). To our knowledge, only the site dosimeter developed in France, which uses the same measuring head as the individual dosimeter, but with a pump with a higher sampling rate, allows this type of measurement to be made. It, in fact, measures ^{212}Po (and allows the calculation of ^{212}Bi and ^{212}Pb), simultaneously with the decay products of ^{222}Rn.

The setting up of a netw[illegible] such site dosimeters around a mining site may then allow the operator to know the [illegible]ortion of local radon which is exhaling from his mine, and thus to respond to the criti[illegible]n or attacks coming, for example, from opponents of his mining operation.

3.3.4.3 Optimisation of radiation protection using the dosimetric data

The high level of accuracy of the measurements carried out by individual dosimeters allows the distribution of miners according to their exposure to be studied for a given mine, and the different mines to be classified according to the statistical distribution of the exposures they present (Pradel, 1982). These are in fact log-normal distributions, characterizable by their average and their standard deviation. The operator may thus follow, with a view to optimisation, the evolution of collective doses and the quality of his operations from a "radiation protection" point of view, with the aid of just these two parameters, traced for example in the form of Henry lines, in a log-Gossip reference mark. It is easy to observe, for example, that the closing of old workings always leads to a reduction in the standard deviation (reduction of the slope of the Henry line) of the distribution of the exposures concerned.

The knowledge of the ratio $^{212}Po/^{218}Po$, using a model giving access to the ratio $^{220}Rn/^{222}Rn$, in turn allows the relative contributions of the mineralised workings and the sterile tunnels to the total exposure received by the miner to be assessed (Zettwoog, 1980). This information may allow the cost of ventilation to be reduced, by allowing the engineer in charge of ventilation to better adjust the distribution of the ventilation power between primary and secondary ventilation.

The above few examples show the wealth of information that may be obtained by a separate individual dosimetry of the different RnDPs for the health officer responsible for radiation protection, if the operator wishes to put into place an effective and optimised policy of radiation protection in his mines.

3.3.5 Concluding Remarks

The choice of dosimetry for the monitoring of radiation in a mine depends on: (i) the constraints imposed by the regulations applying to the operator, (ii) his desire to improve the protection of his workers, and (iii) the amount of money he can reasonably give over to such actions (the cost of radiation protection is a problem which cannot be evaded).

The choice between the different technical solutions which exist will depend on these different parameters, and on the nature of the risk that must systematically be assessed by a preliminary study, including in non-uranium mines or underground quarries. The technical criteria of choice have been described in this section; the bibliography will help the reader to usefully complete this overview of 40 years of studies on radiation protection in mines carried out by the staff of the French Atomic Energy Commission (CEA), the Center for Radiation Protection in Mines (CRPM) and now ALGADE, in constant collaboration with experts all over the world.

References

Bernhard S., Le Gac J., Seguin S. and Zettwoog P. (1981) Radon levels and radon daughter exposures of workers in non uranium mines of the E.C. In *Proc. Int. Conf. on Radiation Hazards in Mining,* Golden, Colorado, pp. 625-628.

Bernhard S., Pineau J.F., Rannou A. and Zettwoog P. (1983a) One year of individual dosimetry in French mines. In *Proc. Int. Conf. Occupational Radiation Safety in Mining,* Toronto, pp. 526-540.

Bernhard S., Le Gac J., Seguin S. and Zettwoog P. (1983b) Identification of nonuranium mines, which present a radon risk, through measurements of exhaust air. In *Proc. Int. Conf. Occupational Radiation Safety in Mining,* Toronto, pp. 307-311.

Bernhard S. and Pineau J.F. (1986) Techniques de dosimétrie. In *Séminaire sur la radioprotection des travailleurs dans les mines et usines de traitement des minerais d'uranium,* Niamey, Niger, édition CRPM.

Bernhard S. and Zettwoog P. (1992) Optimizing dosimetry strategies in an uranium mine. In *Proc. Int. Conf. Radiation Safety in Uranium Mining,* Sakatoon, Saskatchewan.

Bernhard S., Miller W. and Zettwoog P. (1994) Recent developpment of instrumentation and methods for the assessement of the radon 222 pathway in the vicinity of uranium ore mining and milling wastes. In *HLW, LLW, Mixed Wastes and Environmental Restoration - Working Toward a Cleaner Environment,* WM'94, Tucson, Arizona, **3**, pp. 2261-2269.

Bernhard S., Bertrand C., Pineau J.F. and Sarradin F. (1995a) Personal dosimetry in uranium mines and processing plants of COGEMA and its African subsidiaries. In *IRPA Int. Conf. Radioactive Protection and Radioactive Waste Management in the Mining and Minerals Processing Industries*, Johannesburg.

Bernhard S., Bertrand C., Pineau J.F. and Sarradin F. (1995b) Integrated measurement of potential alpha energy (PAE) due to short lived decay products of radon 220 and 222. In *Int. Symp. on the Natural Radiation Environment*, Montreal, p. 217.

Bernhard S., Bertrand C., Pineau J.F. and Sarradin F. (1995c) Le systéme intégré de dosimétrie individuelle dans les mines d'uranium. In *L'Instrumentation en radioprotection des travailleurs, Journées SFRP*, Saclay.

Bertrand C. and Pineau J.F. (1985) Readout and data processing of the film used in the CEA integrated dosimeter. In *13th International Conference on SSNTD*, Roma.

Bertrand C., Gibaud C., Janot M. and Pineau J.F. (1990) *Results of the 2nd European Intercomparison of Radon and Radon Daughters Measurement Methods and Devices in an Underground Mine Atmosphere.* CCE B16/0340/F.

Chambaudet A., Felleman D., From M., Gibaud C. and Pineau J.F. (1985) The processing, on routine basis, of the film used in the CEA integrated dosimeter. In *13th Int. Conf. on SSNDT*, Roma.

Chapuis A.M. and Dajlevic D. (1972) Détecteur de particules alpha. *French Patent 72,25,550.*

Chapuis A.M., Fourcade N., Le Gac J., Pineau J.F. and Zettwoog P. (1981) State of the art of the alpha individual dosimetry in France. In *Proc. Int. Conf. Radiation*

Hazards in Mining, Golden, Colorado, pp. 321-331.

Chruscielewsky W., Orzechowski W., Domanski T. and Swiatnicki G. (1982) Measurement of exposure to radon and its progeny using Kodak LR 115 Type II foil: II Calibration of the detector. In *Proc. of the Specialist Meeting on the Assesment of Radon and Daughter Exposure and Related Biological Effects*, Roma, 1980, RD press, Radiobiology Division, Univ. of Utah, Salt Lake City, pp. 30-38.

Chruscielewsky W., Domanski T., Klusczynski D. and Olszenski J. (1991) Measurement systems and their impact on the estimation of individual miners exposures. In *5th Int. Symp. on the Natural Radiation Environment*, Salzburg.

Duport Ph. (1976) L'aérage et les caractéristiques de l'atmosphère d'une mine d'uranium laboratoire: Application à la répartition des particules radioactives dans les voies respiratoires. *Thèse de Doctorat* Faculté des Sciences de Brest.

Duport Ph., Chapuis A.M. and Pradel J. (1978) Appareil individuel pour la dosimétrie des descendants du radon. *Coll. Int. sur les progrès du contrôle radiologique aux fin de radioprotection*, Stockholm, In *Advances in Radiation Protection Monitoring*, IAEA, Vienna, SM-229/90, 1979, pp. 435-443.

Duport Ph., Madelaine G., Zettwoog P. and Pineau J.F. (1979a) Enregistrement des rayonnements alpha dans les dosimètres individuel et le dosimètre de site du CEA. In *Proc. 10th Conf. on Solid State Nuclear Track Detectors* (eds. H. Francois *et al.*), Lyon, Pergamon Press, pp. 609-615.

Duport Ph., Chapuis A.M. and Pineau J.F. (1979b) Six mois d'expérience en dosimétrie individuelle des descendants du radon dans une mine d'uranium. In *Proc. 10th Conf. on Solid State Nuclear Track Detectors* (eds. H. Francois *et al.*), Lyon, Pergamon Press, pp. 617-622.

Fourcade N., Pineau J.F., Robe M.C., Schmitz J., Urban M. and Zettwoog P. (1987) Impact of radon 222 on the environment near the uranium or extraction and processing site in the Lodève Department (Hérault). *IAEA Technical Committie for the Technological Aspects of Uranium Extraction*, Montpellier.

Le Gac J., Pineau J.F. and Zettwoog P. (1980) Evaluation of the radon risk to uranium miners, an experimental comparison of an ambiant dosimetry method and an individual dosimetry method. *CEA/SPT Internal Report n° 221.*

Leuschner A.H., Strydom R., Steyn A., van Staden J.A. and Rolle R. (1992) Performance evaluation and calibration of an integrating personal radon daughter monitor. In *Proc. Int. Conf. on Radiation Safety in Uranium Mining*, Sakatoon, Saskatchewan.

OECD/NEA (1983) Dosimetry aspects of exposure to radon and thoron daughter products. *Report by a Group of Experts of the OECD Nuclear Energy Agency*, OECD, Paris.

ICRP (1981) Limits for inhalation of radon daughters by workers. *Annals of ICRP, ICRP publication 32, Vol. 6, n° 1*, Pergamon Press, Oxford.

ICRP (1986) Radiation Protection of Workers in Mines. *Annals of ICRP, ICRP Publication 47*, Pergamon Press, Oxford, Vol. 16, n°1.

Orzechowski W., Chruscielewski W. and Domanski T. (1982) Measurement of exposure to radon and its progeny using Kodak LR 115 Type II foil: I Laboratory investigations of the detection response. In *Proc. Spec. Meeting on the Assesment of Radon and Daughter Exposure and Related Biological Effects*, Roma, 1980, RD Press, Radiobiology Division, Univ. of Utah, Salt Lake City, pp. 20-29.

Olszewski J., Chruscielewski W. and Skalski H. (1996) Polish active dosimeter for measurement of individual exposure of miners to radioactive radon daughter products. In *Proc. of the Ninth International Congress of the International Radiation Protection Association*, Vienna, 1996, vol. 4, pp. 445-447.

Pineau J.F. (1980a) Système intégré de dosimétrie individuelle pour les mineurs d'uranium. *Session d'étude sur la protection contre les rayonnements lors de l'exploitation et du traitement des minerais d'uranium.* Exposé n° 10, tome II, Vassivière Edition CRPM.

Pineau J.F. (1980b) Instrumentation en dosimétrie et en radioprotection dans les mines d'uranium. *Session d'étude sur la protection contre les rayonnements lors de l'exploitation et du traitement des minerais d'uranium*, Exposé n° 17, tome II, Vassivière, Edition CRPM.

Pineau J.F. (1982a) Individual integrated dosimetry in uranium mines: consequences on preventing methods. *Radioprotection* **17**, 2, 133-136.

Pineau J.F. (1982b) Efficacité de collection des têtes des appareils de prélèvement d'aérosols radioactifs en fonction de leur granulométrie. *2ème séminaire sur la radioprotection dans les mines d'uranium*, Vassivière.

Pineau J.F. (1985) Results of the 1st European intercomparison of radon and radon daughters measurement methods and devices in an underground mine atmosphere. CCE 1424/1/84.

Pineau J.F., Schmitz J., Urban M. and Zettwoog P. (1986a) Experiences with differents techniques for routine dose assessment of uranium miners. In *Proc. 4th European Congress and 13th Regional Congress of IRPA*, Salzburg, pp. 493-500.

Pineau J.F. (1986b) Routine use of cellulose nitrate films for the dosimetry of underground miners in France. *Nucl. Tracks* **12**, 1-6, 725-729.

Pineau J.F. (1986c) Détection et mesure des rayonnements. *Séminaire sur la radioprotection des travailleurs dans les mines et usines de traitement des minerais d'uranum*, Niamey, Niger, édition CRPM.

Pineau J.F. (1993) The behaviour of the individual integrated dosimeter in relation to the aerosol size: application to long lived aerosols dose asessments. In *First Int. Symp. on Rad. Prot. in the Mining, Milling, and Downstream Processing of Mineral Sands*, Bunbury, Western Australia.

Pradel J. and Zettwoog P. (1982) Application of the methodology on optimisation and dose limitation to the cas of underground uranium mines. In *Conf. Int. Sur l'expérience acquise dans le domaine nucléoénergetique*, IAEA, Vienna.

Rannou A. (1985) Contribution à l'étude du risque lié au radon 220 et 222 dans l'atmosphère des habitations, *Thèse de doctorat*, Faculté des sciences de Brest.

Skowronek J. (1991) Les descendants à vie courte du radon dans les mines de charbon en Pologne. *Radioprotection* **26**, 1, 65-74.

Strong J. C. and Tu. K W. (1987) Design and calibradon of a multichannel diftusion battery for the measurement of radon daughter activity-size distributions. In *Proc. of a Conf. Aerosols: Their Generation: Behaviour and Applications*, London Aerosol Society, Loughborough.

Strydom *et al.* (1992) A cost effective technique for integrating personal radiation dose assessement with personal gravimetric sampling. *J. Mine Vent. Soc. of S. Africa* **45**, 1, 7-14.

Tsivoglou E. C., Ayer H. E. and Holaday D. A. (1953) Occurrence of non-equilibrium atmospheric mixtures of radon and its daughters. *Nucleonics* **11**, 9, 40.

Tymen G., Mouden A., Rannou A., Pineau J.F. and Zettwoog P. (1983) Experimental study of the collection efficiency of the CEA individual alpha dosimeter for aerosol particles in the 0.001-0.1 microns size range. *J. Aerosol Sci.* **3**, 280-283.

Urban M. (1984) Dosisbestimmung durch gleichzeitiges Messen der Radon/Thoronkonzentration und der Gleichgewichtfaktoren in Luft mit Hilfe eines passiven Dosimeters. Kernforschungzentrum Karlsruhe, Karlsruhe, *KFK Report n° 3726.*

Zettwoog P. (1980) Modèle théorique des processus de l'irradiation du personnel dans les mines d'uranium. Internal report CEA/SPT n° 187.

Zettwoog P., Lemaitre N., Bernhard S. and Vauzelle Y. (1996) Use of isotopic signature of radionuclides released from uranium mines and mills to discriminate low levels of environmental impact against natural background levels. In *Proc. IAEA/TCM*, Vienna, 1996, in press.

J. C. H. Miles[1] and T. K. Ball[2]

3.4 Mapping the Probability of High Radon Concentrations Occurring in Buildings

3.4.1 Purposes of Mapping

Because radon levels vary widely between apparently identical buildings on the same geological unit, no map can predict the radon level in an individual building. Maps can, however, give the probability that a building in a particular locality is above a threshold of radon concentration such as a reference or action level. The probability may be calculated for a particular building type or for a mixture of building types. In the latter case the probability is in effect an estimate of the proportion of buildings above the threshold level. Alternatively, maps can provide estimates of the mean radon levels in buildings by area.

Maps showing the geographical variation in the probability that new or existing buildings will exceed a radon reference level are particularly useful for directing action to prevent excessive human exposures to radon. The information provided by the maps may be used in various ways, such as to target information campaigns encouraging measurement of radon levels in homes or to allow building regulations to be altered in particular areas to prevent new buildings from having radon problems.

The data which are used to provide the estimates of the proportion of buildings above a threshold may be the results of radon measurement in a sample of buildings, or may be indirect indicators such as soil type, radium concentrations and permeability measurements. The uses and relative advantages of different types of data are discussed below.

Various different types of geographical boundary are used in the analysis of data and presentation of maps, such as administrative boundaries, geological boundaries, or arbitrary divisions such as 10 km squares. Use of administrative areas has the advantage of simplifying any subsequent administrative action, and it is usually easier to determine the administrative area in which a building falls rather than its map coordinates. However, it has the disadvantages that the areas have shapes and sizes which may not be suitable for the amount of data available, and the variation in shape and size obscures the underlying pattern of variation in radon levels. Use of grid squares allows an appropriate size of area to be chosen and simplifies the analysis, but can make it difficult to apply administrative action. Geological boundaries delineate differences in radon potential much more closely than other types of boundary, but are often complex and difficult to determine.

[1]National Radiological Board, Chilton, UK
[2]British Geological Survey, Nottingham, UK

3.4.2 Mapping Based on Radon Levels in Buildings

Since the purpose of maps of radon-prone areas is to indicate radon levels in buildings, maps based on actual measurements of radon in buildings are preferable to those based on other data. It is also relatively cheap to map by this method since passive radon detectors can be distributed by post.

3.4.2.1 Lognormal distribution of radon levels in buildings

There is a long chain of factors that influence the radon level found in a building, such as radium content and permeability of the ground below it, and construction details of the building. Variations in these factors between buildings produce the wide range of radon levels measured. The distribution of radon levels in buildings is usually found to be lognormal, whether the whole of a country or a small area is considered. The reasons for radon concentrations following this distribution can be understood from statistical considerations.

When a parameter depends on the addition of a large number of independent factors, it can be shown that the measured values of that parameter will follow the 'normal' (bell-shaped) distribution. In the case of radon in buildings, there are many independent factors, but they are multiplicative rather than additive. The radon concentration indoors can be represented by the equation

$$Rn_{in} = Rn_{out} + A \cdot B \cdot C...... \qquad (3.4.1)$$

where Rn_{in} is the radon concentration indoors, Rn_{out} is the radon concentration out of doors, and *A, B, C,...* are factors such as radium concentration in the ground, permeability of the ground, number and size of entry routes, underpressure in the building and ventilation of the building. These factors determine how much of the radon from the source enters the building and how long it remains. The model assumes that these factors are largely independent and are multiplicative. The equation can be rewritten as

$$ln(Rn_{in} - Rn_{out}) = ln(A) + ln(B) + ln(C) + ... \qquad (3.4.2)$$

This equation conforms to the requirements for normal distributions of being a sum of independent terms: hence if there are sufficiently many independent terms, each randomly distributed, then $ln(Rn_{in} - Rn_{out})$ should be normally distributed. Equivalently, $(Rn_{in} - Rn_{out})$ can be said to be lognormally distributed.

This distribution has been widely observed in practice in surveys of radon levels in buildings when whole countries are considered (for instance, Gunby *et al.*, 1993). Nero *et al.* (1994) showed that results grouped by US county also conformed to the lognormal distribution. Miles (1994) showed that the same applied to UK radon results grouped by 5 km grid square. Given data which conform in this way, it is simple to estimate the proportion of the distribution above a threshold from the geometric mean (GM) and geometric standard deviation (GSD) for each area by subtracting average outdoor radon concentration, taking natural logarithms and calculating mean and standard deviation.

Recent computer spreadsheets have a function which will calculate the proportion of a normal distribution above a threshold; an example is NORMDIST in Microsoft Excel.

The requirements for mapping radon-prone areas using the results of radon measurements in buildings are:

1. Accurate and representative radon measurements made using a consistent protocol,
2. Centralised data holding,
3. Sufficient data evenly spread,
4. Location of addresses of measurements on the map.

The first three requirements depend on financing and organising a survey of radon levels. Making a survey properly representative of the building stock is a point that is often neglected. One approach to the problem is demonstrated by Wrixon *et al.* (1988). Determining the exact location of addresses can be problematic in some countries, while others have central sources of data that allow the conversion to be carried out automatically from postal codes to map coordinates.

3.4.2.2 Standardisation of results of radon measurements

Radon measurements in buildings are usually made for purposes other than mapping radon potential, and are often carried out in ways which make them unsuitable for this use without correcting the results first. For instance, measurements are sometimes carried out over a few days in a basement for screening purposes. The result is difficult to relate to the average level over a year in the occupied parts of the building. However, where data are sparse it is important to make the best use of whatever information is available, so techniques have been developed to incorporate such results into databases for mapping.

Before standardising results, it is necessary to decide the parameter which is to be mapped. Usually this will be the annual average radon concentration in a standard type of building or mixture of building types, or a related parameter such as the proportion of buildings above a threshold. Variations in radon levels caused by building parameters such as house type, floor of measurement and presence of double glazing were considered by Gunby *et al.* (1993), using the results of the UK national survey. They showed that these parameters affected radon levels as independent multiplying factors. Verger *et al.* (1994) carried out a similar analysis using French data, and used the results of the analysis to estimate the radon results that would have been obtained if all the houses had been of a standard design.

If measurements have been carried out over a period of months instead of a full year then it is necessary to correct for seasonal variation in radon levels. Measurements of average seasonal variations in large numbers of houses have been made by Majborn (1992) and Miles *et al.* (1992) among others.

Where it is necessary to combine short-term measurement results made under untypical conditions with more representative data, the statistical problems are more complex.

Price (1995) and Price *et al.* (1996) showed that a Bayesian statistical approach can be used to improve the estimates of long-term average radon levels under these circumstances.

3.4.2.3 Mapping on the basis of sparse data

If the lognormal model is to be used for radon mapping, it is necessary to estimate the GM and GSD for areas in which few results are available. Miles (1994) examined the variation of GSD as a function of GM in 5 km grid squares with 100 or more results each. The results showed that although the GMs of the distributions varied from 18 Bq m^{-3} to 250 Bq m^{-3}, the spread on GSD values was small, with a mean of 2.5 and 90% of values between 2 and 3. Similarly, Nero *et al.* (1994) showed that radon results in a set of US counties were consistent with the hypothesis that all counties had the same GSD of 2.1, while GM varied widely. These results imply that the same GSD may be used for all similar sized areas within a region. Surveys tend to produce large numbers of results in some areas even if there are few in other areas, so the mean GSD determined from those areas with many results may be used for all.

In order to estimate the GM for an area where few results are available a different strategy must be applied. The GM varies greatly between areas (this is the reason for mapping) so it is not possible to use the mean of the GMs in different areas. However, there is considerable uncertainty in the estimate of GM if only a few results are available, and it is desirable to reduce this uncertainty. One approach to this problem is to use information on the variation, between areas, of other parameters which may be related to radon levels, such as soil type, permeability and gamma ray dose rate. This approach has been taken by Nero (1993), Kunz *et al.* (1994), and Price *et al.* (1996). In areas where sufficient results of indoor radon measurements are available, these results are used to estimate the variation in radon with the other parameters. In areas where few or no radon results are available, the other parameters are used to estimate what radon levels would have been measured.

An alternative approach to estimating GMs in areas where few measurements have been made is to smooth values between adjacent areas. This technique is applicable only to grid square mapping. If the GMs in adjacent squares are correlated, it may be possible to improve the estimate of GM by smoothing between squares. This was investigated by Miles (1994). To derive the smoothed estimate for a grid square a weighted mean was taken of the values for the square and the eight surrounding squares. Using a standard smoothing algorithm, a weight of 7 was applied to the central square, weights of 3 to vertically or horizontally adjacent squares, and weights of 2 to diagonally adjacent squares.

To discover whether this procedure improved the estimate of the GM in cases where there were few results, further tests were carried out using a subset of the data. These showed that smoothing GMs where results are sparse does indeed improve the estimated GM. For this data set it was shown that the optimum weight on the central square

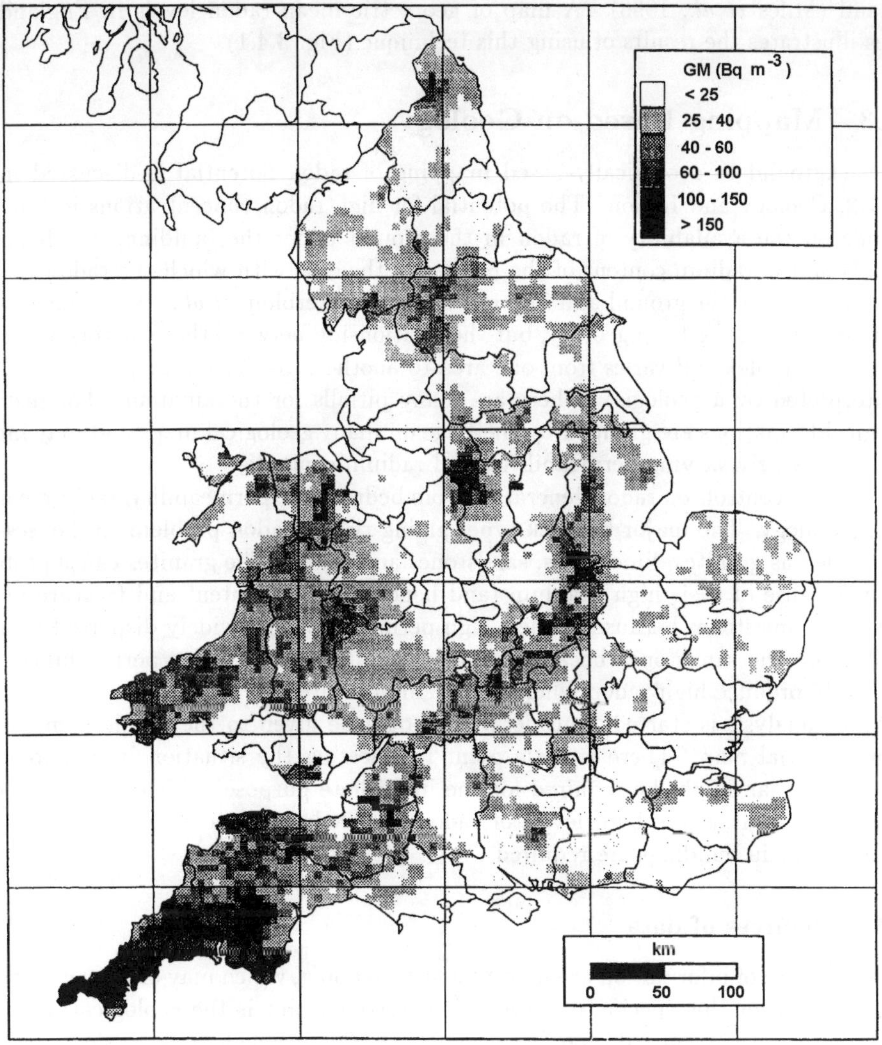

Figure 3.4.1: Estimated geometric mean radon levels in each 5 km grid square of England and Wales, based on lognormal modelling of radon measurements in houses.

depended on the number of measurements, and was given by the function

$$Weight = 3.5 + 1.7n \tag{3.4.3}$$

where n is the number of measurements in the central square. Where $n > 20$, smoothing does not significantly increase the accuracy of the estimated GM. This refinement of the smoothing technique has been used to map radon-prone areas across the whole of England (Miles *et al.*, 1996). A map of geometric mean radon levels in England and Wales illustrates the results of using this technique (Fig. 3.4.1).

3.4.3 Mapping Based on Geology

The background to geologically-based mapping of radon potential is discussed in section 1.2, Geology and Radon. The potential for high radon concentrations in buildings depends on the availability of radon in the ground under the building, which in turn depends on the radium content of the rocks and the ease with which the radon released can move through the ground and into the building (Åkerblom *et al.*, 1984). Such factors clearly depend on the local geology, but the relationship between these factors and radon indoors is complex and varies from one area to another. Geological maps should always be interpreted by a geologist - there are many pitfalls for the amateur. For instance, geological formations are grouped by their age on many geological maps and may include rocks with widely varying permeabilities and radium contents.

The main controls on radon generation from bedrock are permeability, radium content and mineralogy. The major bedrock types giving rise to radon problems in houses may be classified as granites, limestones, sandstones and shales. The granites cause problems mostly because of their high uranium (and hence radium) content and fracture permeability; the limestones because of their high permeability and widely dispersed uranium distribution; the sandstones because of their fracture and granular permeability. The shales only produce high radon where the uranium content is high.

Before analysis is started, consideration should be given to the requirements of the radon potential map. There is little point in assessing the situation in a more rigorous manner than might be required for the immediate purpose, which is often to draw attention to areas where high levels of radon are likely to issue from the ground and measurements in buildings are required.

3.4.3.1 Sources of data

Various sources of information are used to assist mapping, which may or may not include data collected for this specific purpose. The starting point is the geological map. For many countries the standard mapping scale is 1:50,000, although larger scale maps (such as 1:10,000) may be available for certain areas: usually highly urbanised zones or districts of economic importance. The largest scale maps available should be used. Many recent maps are digitised and may be used with a Geographical Information System (GIS) to allocate data such as gamma ray dose rates or house radon results to the geology automatically. Older maps, however, require manual processing. Problems arise with early maps, which although spatially accurate may have units of the geology assigned using different criteria to the modern map, so that often a high degree of interpretation is required.

The prerequisite for using a geological map for estimating radon potential is that the map should be reasonably accurate. Clearly the accuracy of a geological map can vary as a result of the nature of any soil or overburden cover, the scale of the mapping and the complexity of the geology. Geological maps are based generally upon secure knowledge obtained from natural exposures or artificial ones such as quarries and boreholes. Between such sites the information is interpolated, often supported by feature mapping, geometrical considerations or shallow excavating and augering, but the uncertainty increases with distance from the well-defined location. Where there is thick overburden cover and low relief, the bedrock map is more uncertain than in relatively well exposed (usually hilly) areas.

Geological maps summarise the surface existence of rock types which are normally easily recognisable in the field. They represent the bedrock geology at the surface or sub-surface and usually some idea of the three-dimensional configuration is given by incorporating structural information. Information on unconsolidated overburden such as glacial drift or alluvium is either included with the solid geology or, if the overburden is thick or complex, may be presented on a separate map. Mappable units are called "formations" if they are of sedimentary origin. Each formation may be distinguished by a number of related or unrelated combinations of properties. Lithology, colour index, chemistry, fossil content and other factors all play a part in their identification.

Additional information may be derived from, for example, systematic drainage geochemical surveys or airborne radiation surveys. Information which has been collected specifically for radon mapping includes measurements of uranium or radium contents of rocks, gamma-ray dose rates and radon concentrations in soil gas, among other parameters. In the UK the radon concentrations in soil gas have been found to have the closest correlation with radon in buildings (Ball *et al.*, 1992); but great care has to be exercised in developing a prior understanding of the behaviour of the soil gas before this easily measured parameter is used. Ball *et al.*; found that radon in soil gas measurements were reasonably reproducible in summer conditions when transpiration and evaporation normally exceed precipitation, and during periods of settled weather. Variability increased under winter conditions. Local climatic conditions will determine the most suitable time for soil gas surveys.

Superficial deposits must be taken into account as well as bedrock. The presence of unconsolidated overburden can have a substantial effect upon the radon emanation from bedrock. Some overburdens derive from the glaciations that affected much of the northern hemispheric land masses during the Pleistocene. Others result from periglacial conditions, for example soliflucted head. In many areas there are marine or estuarine alluvial deposits, blown sand or other light materials. Residual deep weathering profiles sometimes remain, and these can also change the nature of emanation from the bedrock. Gundersen *et al.* (1992), for example, note that residual soils over limestone are normally enriched in uranium and radium, and conclude that this factor can cause the large amounts of radon measured in such areas. Ball and Miles (1993) concluded that deep weathering profiles over brittle slates and sandstones could result in up to 10% of houses having high radon levels even when the uranium content was equal to the crustal average in the UK (2.7 ppm uranium).

The effects of overburden composition and thickness have been dealt with extensively in the USA and Swedish literature and there are several examples of the relationship between thick glacial deposits and high house radon, the controlling factors being primarily the composition, emanation coefficients and permeability of the drift. Eskers with their granitic provenance, high permeability and great thickness have been highlighted especially in Sweden (see section 1.2 of this book).

In upland areas, or close to glacial limits, drift deposits are seldom so thick that they can be regarded as the sole providers of radon to the exclusion of the bedrock. In these cases the nature of the bedrock has to be considered, especially owing to the complex interplay of permeability and geochemistry.

The nature of this relationship indicates that the overburden and bedrock combination is important and should always be assessed in the production of radon potential maps. Boulder Clay generally and consistently reduces the indoor radon signature from the bedrock. It should be noted that despite the sometimes marked reduction caused by such deposits, over very permeable rocks the radon risk class could still be high. Glacial Sands and Gravels have a variable effect, generally enhancing the signal where the bedrock substrata are permeable and with a reduction in affected houses over less permeable strata. Experience has shown that there are many and variable superficial deposits, but they must be considered in relation to the bedrock where possible.

3.4.3.2 Classification procedures

The manner in which radon potential is classified varies from country to country. In Sweden the radon potential is estimated and mapped using a three-fold radon risk classification based upon geology, permeability and soil gas radon (Åkerblom, 1987). These criteria are used at a mapping scale of 1:50,000, in conjunction with airborne gamma spectrometry surveys to produce provisional radon risk maps. Where the overburden cover is less than 2 metres thick over bed-rock, soil gas radon measurements are routinely recommended even if other data indicate low radioactivity. Detailed radon surveys are usually made before erecting buildings (Åkerblom, 1994). Maps are produced by the Swedish Geological Company for local communities.

Table 3.4.1: Radon risk classification used in the Czech Republic.

Class	Radium (Bq kg^{-1})	Soil gas radon (kBq m^{-3})	Permeability
Very high	>500	>250	high
High	100-500	50-250	high
Normal	30-100	10-50	medium
Negligible	<35	<30	low

The broad relationship between radon emission potential, radon concentration in soil gas and rock/soil permeability in the Czech Republic has been demonstrated by Barnett (1991). Information on rock and soil permeability was obtained from hydrogeological and pedological maps. A radon risk classification is based upon the classes given in Table 3.4.1.

The US Geological Survey classify radon potential on the basis of a scoring system which considers five factors in a Radon Index Matrix: indoor radon, radioactivity of the soil or overburden, geology, soil permeability and house architecture (Gundersen *et al.*, 1992). These are used to produce regional summaries for the Environmental Protection Agency. The summaries give a detailed evaluation of geological radon potential for each state.

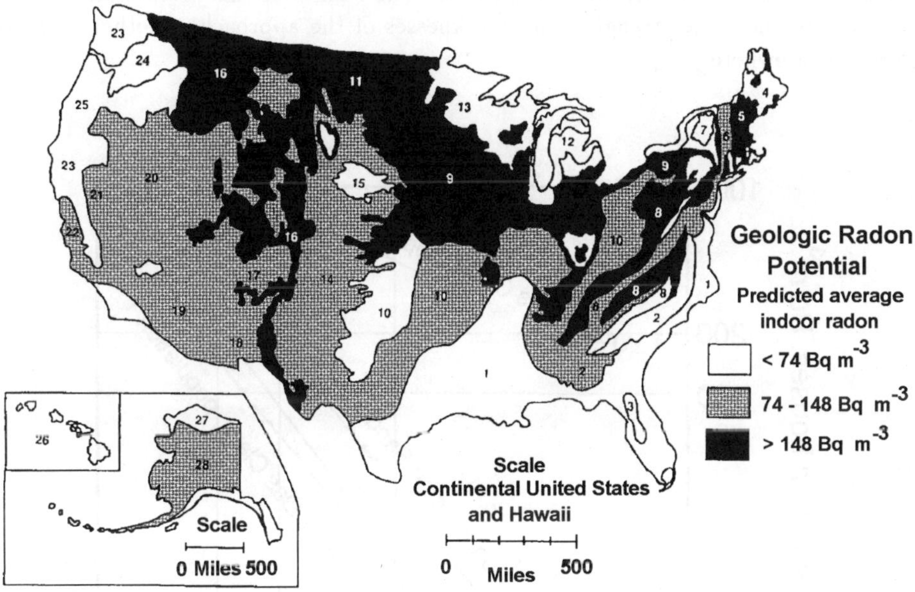

Figure 3.4.2: Predicted indoor average radon level in the United States of America, based on geology (from Gundersen *et al.*, 1992).

There is a comprehensive aeroradiometric survey of the USA resulting from the National Uranium Evaluation Program (NURE). Catafe (1989), Cui (1990), and Duval and Otton (1990) have produced assessments of radon potential using a combination of radiometric results and soil permeability. The general procedure is to use the NURE data to produce a radium concentration map, assign soil permeability parameters to soils, and analyse the correlation between indoor radon levels and other parameters. Duval *et al.* (1989) and Duval and Otton (1990) concluded that there was a linear relationship between average indoor radon levels and surface radium levels for soils with low to moderate permeability. Similar procedures were used to assess the potential risks from radon in houses across Canada (Cocksedge *et al.*, 1993). Gundersen *et al.* (1992) used class intervals of <74 Bq m^{-3}, 74-148 Bq m^{-3} and >148 Bq m^{-3} to provide a map of the USA with classes based upon predicted average indoor radon (Fig. 3.4.2).

3.4.4 A Combined Approach

An alternative mapping technique is to group indoor radon measurements by geological unit. This approach has the potential to combine the advantages of grid square mapping of indoor radon results (accurate estimates of the numbers of buildings affected) with those of geologically based mapping (fine detail). Miles and Ball (1996) have tested this approach in an area of England in which the outcropping of rock types at the surface makes a complex and rapidly varying pattern. The results of this investigation are discussed here to show the strengths and weaknesses of the approach, whether applied in this area or elsewhere.

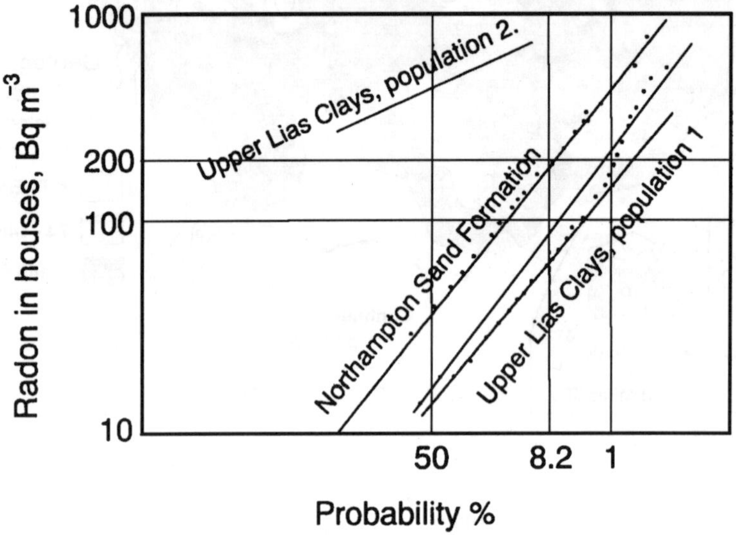

Figure 3.4.3: Cumulative probability distribution of indoor radon levels over two geological formations, showing a single lognormal distribution (Northampton Sand Formation) and a distribution which can be separated into two lognormals (Upper Lias Clays).

A geological map may be regarded as a series of irregular polygons. House radon results can be related to these polygons in a similar manner to grid squares and the statistical distributions tested in the same way. The radon measurements for this study provided by surveys carried out by NRPB on behalf of the UK Department of the Environment. Each house was measured using two detectors exposed over a three month period, and the results corrected for seasonal variations. For this exercise, results from measurements made in the areas of two British Geological Survey (BGS) sheets were used. The survey sheets were the adjacent Northampton and Wellingborough survey sheets, each covering an area 29 km x 19 km. Intensive surveys of radon in homes have been carried out in this area, resulting in 35,000 results being available. Such large numbers of measurements are not required for the production of radon potential maps, but allow the advantages and drawbacks of the technique to be explored in detail.

Once the results have been assigned to a geological feature, statistical treatment of the results is then undertaken by correcting for outdoor atmospheric background and testing the results for lognormality. This may be carried out entirely by machine, but we have found it useful to examine the data visually using log-probability plots, as this helps to identify the presence of sub-populations, perhaps resulting from contamination from other factors, and distributions not following the lognormal rule may be easily recognised. The advantage of this approach is that it enables an accurate description of the data to be made, it means that small data sets may be used and it reduces the effect of outliers. It also enables the recognition of more than one population in the data set, thereby testing for possible contamination, variability due to geological changes or building styles. Log-probability plots for two formations are given in Fig. 3.4.3, one distribution conforming well to the lognormal and one which can be separated into two lognormal distributions.

Table 3.4.2: Comparison of actual percentages of houses in England above the Action Level (200 Bq m^{-3}) with percentages estimated using the lognormal distribution. Two adjacent geological survey sheets; houses on bedrock with no superficial deposits. Geometric mean (GM) radon levels (expressed in Bq m^{-3}) and geometric standard deviation (GSD) are also shown.

	Northampton sheet				Wellingborough sheet			
Bedrock	GM	GSD	Actual % above 200 Bq m^{-3}	Estimated % above 200 Bq m^{-3}	GM	GSD	Actual % above 200 Bq m^{-3}	Estimated % above 200 Bq m^{-3}
Blisworth Limestone	23	2.4	0.8	1.2	35	2.6	3.6	4.5
Upper Estuarine Series	18	2.3	0.4	0.5	28	2.7	3.1	3.0
Lower Estuarine Series	22	2.7	2.3	2.3	48	3.0	11.3	12.0
Northampton Sand Formation	41	3.0	8.4	8.2	82	3.0	21.5	24.0
Upper Lias Clays	20	2.5	1.9	1.8	22	2.4	1.0	0.9

The thickness or composition of rock types in geological formations may vary laterally. This is particularly common in the Northampton area and the effect of the subtle changes in geology on indoor radon values is one of the factors investigated. The locations of the buildings were identified from the postcode, which can generally give the grid reference of a building with an uncertainty of a few hundred metres or less in towns, but the uncertainty may be 1 km or more elsewhere. Because the results of measurements are confidential, a system was devised to allow the geology and superficial cover of each measurement point to be identified without loss of confidentiality.

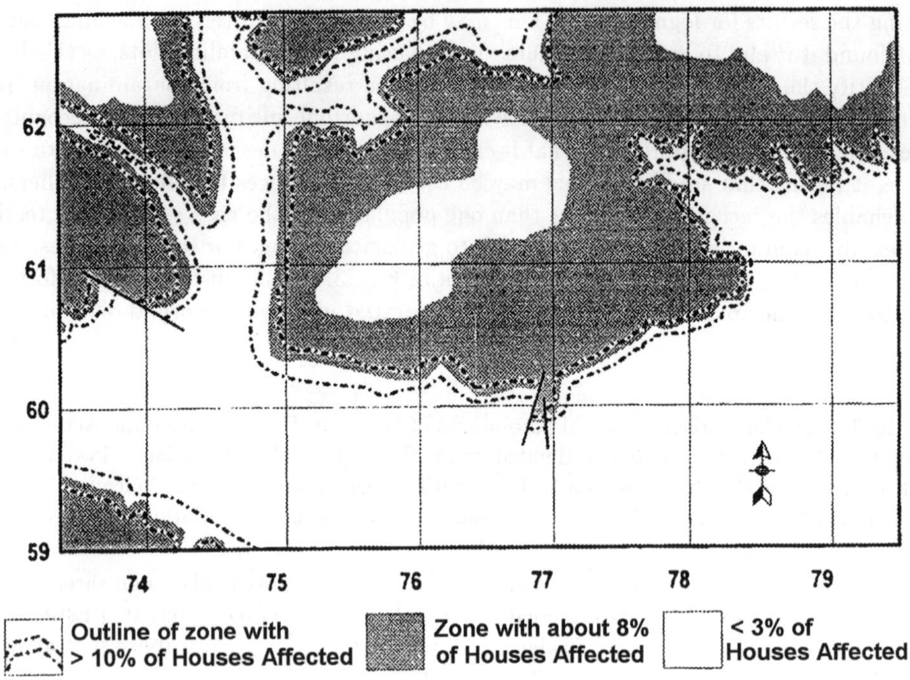

Figure 3.4.4: Estimated percentage of houses above 200 Bq m^{-3} in Northampton, United Kingdom, based on lognormal modelling of radon measurements in homes grouped by geological formation. Each square shown is 1 km^2.

Each set of indoor radon results for a particular combination of geology and superficial cover was analysed separately. The mean UK outdoor radon concentration (4 Bq m^{-3}) was subtracted from each result and the GM and GSD of each set were calculated. These parameters were used to estimate the fraction of homes above 200 Bq m^{-3} (the UK Action Level for radon) on the assumption that the distribution was lognormal. The estimated fraction could then be compared with the actual fraction found above the threshold in each data set. Table 3.4.2 shows the results for rock types which appear on both the Northampton and the Wellingborough sheets, in areas without superficial cover. The measured percentage of homes above the UK Action Level is in all cases close to the percentage estimated using the lognormal model, showing that the results conform well to this model. The results also show the large variation in the percentage of homes above the UK Action Level between different bedrocks, from 0.5% to > 20%. Fig. 3.4.4 shows the results of such mapping. Here features as small as 100 metres in width can be assigned numerical estimates of the proportion of homes above the Action Level.

The GM for a single rock type differs by up to a factor of 2.2 between the two geological survey sheets. There are no differences in housing type or age of houses between the two sheets that could account for this difference, which is therefore attributed to lateral variation in the geological formations. This demonstrates that conclusions about radon potential over individual geological formations cannot necessarily be extrapolated to adjacent areas. This is likely to be particularly true of areas with a high proportion of shelf facies sedimentary rocks.

If the results are grouped by 5 km grid squares, the GSDs found vary from 2.6 to 4.7, significantly higher than the national average of 2.5. This is because each grid square contains a mixture of rock types. When the results are grouped by rock type, the GSDs vary from 2.3 to 3.0. The smaller GSDs found imply that the data sets are more homogeneous when results are grouped by geology. The smallest GSDs were found for rock types where there was little spatial variation in the GM. This suggests that much of the remaining variation between radon levels in homes when the results are grouped by geology is caused by the lateral variation in the geological formations. Analysis of the geographical variation in radon potential between buildings on the same lithology may therefore produce even more detailed and accurate estimates of the fraction of homes exceeding a reference level.

Table 3.4.3: The effects of superficial deposits on radon levels in houses. Northampton geological survey sheet; houses on Northampton Sand Formation with different superficial deposits.

Superficial deposit	GM	GSD	Actual % above 200 Bq m^{-3}	Estimated % above 200 Bq m^{-3}
None	41	3.0	8.4	8.2
Boulder Clay	29	2.9	2.7	6
Glacial Sand and Gravel	39	3.1	9.4	10

Table 3.4.3 shows the effects of superficial deposits on one lithology, the Northampton Sand Formation on the Northampton map sheet. A superficial cover of glacial sand and gravel makes little difference to the GM or the number of homes above 200 Bq m^{-3}. A capping of Boulder Clay, however, leads to a marked reduction in the number of homes above the Action Level. In this case the results do not fit the lognormal model well, but could be a combination of two lognormal distributions. It is possible that this is due to variations in the thickness of the boulder clay. In some cases the foundations of homes will break through the superficial layer, and in others the clay will form an almost impermeable barrier. It is clearly important to take superficial deposits into account in

mapping radon potential.

3.4.5 Concluding Remarks

This exercise has shown that grouping indoor radon results by geological formation and superficial cover can produce radon potential maps which are more detailed than grid square maps and more accurate in estimating numbers of homes affected than mapping based only on geology. However, lateral variations in geological formations mean that conclusions drawn for one area cannot necessarily be extrapolated to the same geological formations in adjoining areas. The results also show that it is important to take superficial deposits into account in mapping radon potential.

Analysis of this type is possible only where there were sufficient results of measurements in homes for each combination of geological formation and superficial cover. In areas where there are insufficient homes to allow this analysis (either because the population density is too low or because the geology varies too rapidly) other information such as radon in soil gas may assist in estimating radon potential. These data will require careful interpretation, particularly where impermeable deposits cover permeable rocks, and may only provide a provisional estimate.

References

Åkerblom G. (1987) Investigations and mapping of radon risk areas. In *Proc. Int. Symp. on Geological Mapping In the Service of Environmental Planning*, Trondheim, 1986, Norges Geologiske UndersAkelse, Oslo, pp. 96-106.

Åkerblom G. (1994) Ground radon-monitoring procedures in Sweden. *Geoscientist* **4**, 4, 21-27.

Åkerblom G., Andersson P. and Clavensjo B. (1984) Soil gas radon - a source for indoor radon daughters. *Radiat. Prot. Dosim.* **7**, 49-54.

Ball T. K., Cameron D. G. and Colman T. B. (1992) Aspects of radon potential mapping in Britain. *Radiat. Prot. Dosim.* **45**, 211-214.

Ball T. K. and Miles J. C. H. (1993) Geological and geochemical factors affecting the radon concentration in homes in Cornwall and Devon, UK. *Environm. Geochem. Health* **15**, 27-36.

Barnett I. (1991) Radon risk mapping in the Czech Republic: Results. In *Radon Investigations in Czechoslovakia* 11 (ed. I. Barnett), *Geological Survey*, Prague, pp. 13-19.

Catafe J.S. (1989) Regional NURE, geology, and soils data as predictors of indoor radon. In *Proc. Symp. Radon and Radon Technology*, Denver, Poster papers, Vol 2, US Environmental Protection Agency Report, EPA 600 9 89 006B.

Cocksedge W., Rankin W., Tostowaryk W., Rahman W., Charbonneau B. W. and Grasty R. L. (1993) National Native Home Radon Survey - maximising resources through radon potential assessment. *26th Health Physics Society Meeting on Environmental Health Physics*, Lake Coeur d'Alene, Idaho.

Cui L. P. (1990) Radiometric methods in regional radon hazard mapping. *Nucl. Geophys.* **4**, 3, 353-364.

Duval J. S. and Otton J. K. (1990) Radium distribution and indoor radon in the Pacific Northwest. *Geophys. Res. Lett.* **17**, **6**, 801-804.

Gunby J. A., Darby S. C., Miles J. C. H., Green B. M. R. and Cox D. R. (1993) Factors affecting indoor radon concentrations in the United Kingdom. *Health Phys.* **64**, 2-12.

Gundersen L. C. S., Schumann E. R., Otton J. K., Dubief R. F., Owen D. E. and Dickenson K. E. (1992) Geology of radon in the United States. In *Geologic Controls on Radon* (eds. A. E. Gates and L. C. S. Gundersen), Boulder, Colorado, Geological Society of America, Special Paper 271.

Kunz C., Schwenker C., Green J. and Kitto M. (1994) Identifying high risk areas in New York State: mapping indoor radon data. *Int. Radon Symp. of the American Association of Radon Scientists and Technologists*, Atlantic City.

Majborn B. (1992) Seasonal variation of radon concentrations in single-family houses with different sub-structures. *Radiat. Prot. Dosim.* **45**, 443-447.

Miles J. C. H. (1994) Mapping the proportion of the housing stock exceeding a radon reference level. *Radiat. Prot. Dosim.* **56**, 207-210.

Miles J. C. H., Green B. M. R. and Lomas P. R. (1996) Radon affected areas: England. Documents of the NRPB, 7, 2, 1-9.

Miles J. C. H., Green B. M. R. and Lomas P. R. (1992) Radon affected areas: Derbyshire, Northamptonshire and Somerset. Documents of the NRPB, 3, 4, 19-28.

Miles J. C. H. and Ball T. K. (1996) Mapping radon-prone areas using house radon data and geological boundaries. To be published in *Environment International.*

Nero A. V. (1993) Methodologies for identifying high-radon areas: A brief review. In *Proc. 6th Int. Conf. on Indoor Air Quality and Climate* (eds. M. Jantunen and O. Seppanen), Publisher Indoor Air '93, Vol. 4, 419-425.

Nero A. V., Leiden S. M., Nolan D. A., Price P. N., Rein K. L., Revzan H. R., Wollenberg H. R. and Gadgil A. J. (1994) Statistically based methodologies for mapping of radon 'actual' concentrations: the case of Minnesota. *Radiat. Prot. Dosim.* **56**, 215-219.

Price P. N. (1995) Joint analysis of long- and short-term radon monitoring data from the Northern USA. *6th Int. Symp. on the Natural Radiation Environment*, Montreal.

Price P. N., Nero A. V. and Gelman A. (1996) Estimation of county mean radon concentrations in Minnesota: a Bayesian analysis. To be published in *Environment International.*

Verger P., Hubert Ph., Cheron S., Bonnefous S., Bottard S. and Brenot J. (1994) Use of field measurements in radon mapping in France. *Radiat. Prot. Dosim.* **56**, 225-229.

Wrixon A. D., Green B. M. R., Lomas P. R., Miles J. C. H., Cliff K. D., Francis E. A., Driscoll C. M. H., James A. C. and O'Riordan M. C. (1988) Natural radiation exposure in UK dwellings. NRPB-R190.

M. Sohrabi[1]

3.5 High Radon Levels in Nature and in Dwellings: Remedial Actions

3.5.1 Introduction

A High Level Natural Radiation Area (HLNRA) may be defined here for working purposes as "an area where at least one of its existing natural radioactivity sites and/or exposures in the environment (soil, water, air, etc.) or in dwellings has potential to lead to internal and/or external exposures of public higher than a predetermined level or a limit with a potentially increased health risk". Many areas in the world meet this definition, such as HLNRAs in Austria (Pohl-Rüling *et al.*, 1982), Brazil (Cullen and Penna-Franca, 1977), China (Wei *et al.*, 1993), India (Mishra, 1993; Sunta, 1993) and Iran (Sohrabi, 1993a,b). There are some other countries which have been subjects of long-term dosimetric and epidemiological studies with a view to estimating the risk factors, which are at present only based on the available data on Hiroshima and Nagasaki (ICRP-60, 1991).

A series of conferences on HLNRAs have been held over the last two decades: the first in Pocos de Caldas, Brazil in 1975 (Cullen and Penna-Franca, 1977); the second in Bombay, India in 1981 (Vohra *et al.*, 1982), the third in Ramsar, Iran in 1990 (Sohrabi *et al.*, 1993a); and the fourth in Beijing, China in 1996. Many high-level radon areas or radon-prone areas and dwellings have also been discovered during national surveys within the last decade. Owing to the importance of radon indoors and its remedial actions, many topical conferences and workshops at national and international levels have been organized; for example the first and second International Workshops on Radon Monitoring in Radioprotection, Environmental and/or Earth Sciences in 1989 and 1991 at the International Centre for Theoretical Physics (Furlan and Tommasino, 1993; Tommasino *et al.*, 1990), the UK National Conference on "Radon 2000" (O'Riordan and Miles, 1992); and the International Conference on "Indoor Radon Remedial Actions" (Campos-Venuti *et al.*, 1994).

According to UNSCEAR (1993), the population-weighted global average radon concentration is 40 Bq m^{-3} indoors and 10 Bq m^{-3} outdoors. By using equilibrium factors of 0.4 for indoors and 0.8 for outdoors, the population-weighted average Equilibrium Equivalent Concentration (EEC) is therefore estimated to be 16 Bq m^{-3} indoors and 8 Bq m^{-3} outdoors. By using the conversion factors recommended by UNSCEAR (1993), i.e. the effective equivalent dose (E) of 9 nSv h^{-1} per 1 Bq m^{-3} for EEC of radon and for both indoors and outdoors, and 1.5 μSv y^{-1} from inhalation of 1 Bq m^{-3} radon dissolved in tissues, the effective dose E from the inhalation of the above-quoted average radon

[1]Atomic Energy Organization of Iran, Tehran, Islamic Republic of Iran

concentrations indoors and outdoors is estimated to be 1.2 mSv y^{-1} by using a simple calculation given in the Appendix.

The problem arises from the fact that many people in the world live in dwellings with radon levels as much as over 100 times the global average, and receive doses higher than the present ICRP-60 dose limit of 20 mSv y^{-1} or than the older limit of 50 mSv y^{-1} for radiation workers (ICRP-60, 1991), as will be discussed later in this section. For example, in one dwelling in Ramsar, Iran, the dose E due to external and internal exposures including radon has been measured to be at least 160 mSv y^{-1}, which is 8 times higher than the present dose limit for radiation workers (Sohrabi, 1993a,b). Therefore, dwellings having such high levels of radiation require implementation of immediate remedial actions. Considering the known health risks from ionizing radiation, the above can easily justify why studies of high level natural radiation and/or radon-prone areas worldwide, as well as mitigation of their radiation levels, are of importance and concern to regulatory authorities and the public. Although many conferences have been held and extensive researches have been carried out during the past two decades, and vast amounts of data have been accumulated on the above subjects, definition of a HLNRA and the criteria for selecting such areas have not yet been well defined and internationally accepted.

In this section, the criteria for HLNRAs and radon-prone areas are defined, followed by a review of some important HLNRAs and radon-prone areas worldwide, action levels, remedial actions and international recommendations with regard to the protection of the public (Sohrabi, 1996).

3.5.2 Definition and/or Criteria of a High Level Natural Radiation or Radon Area

So far, no official definition of a HLNRA, nor criteria for it other than the ICRP system of dose limitation for public due to man-made sources and recommendations of some action levels, have been established (ICRP-60, 1991). A simple working definition of a HLNRA was given by this author in subsection 3.5.1 above. Others have also given other definitions and criteria characterizing a HLNRA or a radon-prone area. Cullen and Penna-Franca (1977) have characterized a HLNRA by one or more of the following conditions:

1. The exposure rate from external terrestrial sources, over extended areas, is greater than 200 mR y^{-1} ($\approx$ 2 mGy y^{-1}).

2. The long-lived alpha activity ingested through the local diet and water is greater than 50 pCi d^{-1} (1.85 Bq d^{-1}).

3. The radon concentration of potable water is greater than 5000 pCi l^{-1} (185 kBq m^{-3}).

4. The ^{222}Rn and ^{220}Rn concentrations in the atmosphere are greater than 1 pCi l^{-1} (37 Bq m^{-3}).

Mishra (1993), based on the above guidelines and the UNSCEAR results on average global external dose from terrestrial sources and by introduction of a population factor, has proposed the following criteria for a HLNRA.

1. The exposure rate from external terrestrial sources, over extended areas, is greater than 4 mSv y^{-1} (10 times the average global value).

2. The long-lived alpha activity ingested with the local diet and water is greater than 2 Bq d^{-1}.

3. The ^{222}Rn concentration of the potable water is greater than 200 kBq m^{-3}.

4. The ^{222}Rn and ^{220}Rn concentrations of the atmosphere is greater than 40 Bq m^{-3}.

5. The population receiving radiation dose from one or more of the above sources is greater than 1,000, so that data obtained has some statistical significance.

Each of the above proposals has some advantages and disadvantages for defining a HLNRA and a radon-prone area. The present author, drawing on the existing experiences on HLNRAs and desiring to establish criteria more dependent on a system of limitation of effective equivalent dose received by public rather than on different sources of potential exposures, and in the light of the definition of subsection 3.5.1 above, proposes the following criteria for natural radiation areas by dividing them into four groups (Sohrabi, 1996):

1. A "Low Level Natural Radiation Area" (LLNRA) or "Normal Level Natural Radiation Area": an area or a complex of dwellings where soil, outdoor air, indoor air, water, food, etc., have exposures and/or radioactivity levels that would lead to an internal and/or external public exposure which falls below or is equal to two times the average global annual effective equivalent dose from natural sources, given for example by UNSCEAR (1993); i.e. 2x2.4 or $\approx$ 5 mSv y^{-1} or a dose level $\leq$ 5 mSv y^{-1}. No remedial action is recommended for such LLNRAs, although some simple measures can always mitigate the national average annual effective equivalent dose.

2. A "Medium Level Natural Radiation Area" (MLNRA): an area or a complex of dwellings where soil, outdoor air, indoor air, water, food, etc., have exposures and/or radioactivity levels that would lead to an internal and/or external public exposure which is higher than that of LLNRA; i.e. 5 mSv y^{-1}, but falls below or is equal to a predetermined level or limit, for example the dose limit of 20 mSv y^{-1} for radiation workers (ICRP-60, 1991). A remedial action is required to be implemented within a time frame to be determined; for example 5 years.

3. A "High Level Natural Radiation Area" (HLNRA): an area or a complex of dwellings where soil, outdoor air, indoor air, water, food, etc., have exposures and/or radioactivity levels which lead to an internal and/or external public exposure which is higher than 20 mSv y^{-1} and falls below or is equal to 50 mSv y^{-1}, the former

ICRP dose limit for radiation workers. A remedial action should be implemented subject to a regulatory control within a time frame to be calculated; for example one year.

4. A "Very High Level Natural Radiation Area" (VHLNRA): an area or a complex of dwellings where soil, outdoor air, indoor air, water, food, etc., have exposures and/or radioactivity levels that would lead to an internal and/or external public exposure higher than the upper limit for a HLNRA; i.e. > 50 mSv y^{-1}. Evacuation is recommended as the first step in the remedial programme to be enforced by a regulatory authority.

The proposed criteria are believed to have some advantages in being applied as a regulatory framework and as a "system of dose limitation" compatible with that of ICRP for natural radiation indoors and outdoors, based on which remedial actions could be accordingly implemented. Yet the above are open to criticism, improvement, and/or approval to be used as international criteria for radiation-prone areas. It should also be mentioned that the term "HLNRA" is used in this section as a general term, not yet applying the above proposed criteria.

So far as radon alone is concerned, the criteria proposed above could also cover that gas. However, some international and national definitions have also been proposed previously. For example, ICRP has defined a "radon-prone area" as one in which more than 1% of dwellings have a radon concentration of more than 10 times the national average value (ICRP-65, 1993). The National Radiological Protection Board (NRPB) in the UK has used the term "radon affected area" rather than "radon-prone area" and has expressed it as an area within which a certain percentage of present and future dwellings might exceed a predetermined action level and from which certain consequences would follow (Webb, 1992). Also Scott (1993), selecting one of the three concepts for radon-prone areas that he has proposed, defines a radon-prone area as an area where the average risk (radon concentration) to the population is high enough to justify an action programme. The derived criterion is arithmetic mean (AM) radon concentration greater than some value; e.g., AM $\geq$ Y Bq m^{-3}.

In view of the above definitions, one can easily conclude that there is no single definition that can be universally used.

3.5.3 High Level Natural Radiation Areas Worldwide

Detailed studies on some HLNRAs in Austria (Pohl-Rüling *et al.*, 1982), Brazil (Cullen and Penna-Franca, 1977), China (Wei *et al.*, 1993), India (Sunta, 1993), Iran (Sohrabi, 1993a,b) and Japan (Iwasaki *et al.*, 1993) as well as in some other countries have led to some interesting results, in particular from epidemiological points of view. Some areas such as Badgestein in Austria (Pohl-Rüling *et al.*, 1982), Ramsar in Iran (Sohrabi *et al.*, 1993a,b), Slovenia (Kobal and Renier, 1987), Tuwa in India (Joshi and Mishra, 1980), and Rudas in Budapest, Hungary (Szerbin, *et al.*, 1994) have thermal waters used as spas. They are usually rich in radium and thus have high radon concentrations indoors, which deliver high doses to staff as well as to the visitors, some of which have been introduced long ago (Eisenbud, 1987).

In Brazil, due to high thorium contents and traces of uranium in minerals, gamma exposure levels of up to 2 mR h^{-1} have been detected, registering a dose from 1 to 32 mGy y^{-1} with an average dose rate to public of 6.4 mGy y^{-1} in Guarapari, built on a Peninsula over monazite sand deposits, and at Araxa, lying over a volcanic intrusive area (Cullen and Penna-Franca, 1977).

On the SW coast of India, in an area with 70,000 inhabitants, a sizable proportion of the population receives exposures exceeding 10 mG y^{-1}, with the highest personal dose rate of 32.6 mGy y^{-1} belonging to a resident of a house that registered 38.4 mGy y^{-1} (Sunta, 1993). Although the dose levels are significant, the results of the demographic survey for dose-genetical effects correlation, and epidemiological studies as well as studies on chromosomal anomalies of human blood cells and plants, have shown no conclusive statistically significant biological effects on the population of HLNRAs in India in comparison to control groups.

In HLNRAs of Ramsar, Talesh Mahalleh has the highest radiation levels, with personal doses registered by personal dosimeters up to 132 mGy y^{-1}, having a mean value of 10.23 mSv y^{-1}, with a maximum personal dose detected for inhabitants of a house in that area (Sohrabi, 1993a,b). The potential gamma exposures indoors and outdoors measured by survey meters range form 0.05 to 9 mR h^{-1}, with annual potential exposures in houses ranging from 0.6 to 360 mGy y^{-1}. One school, in particular, showed potential gamma exposures ranging from 5.6 to 15 mGy y^{-1}. The dose received by the inhabitants in the area from gamma exposures reached up to 132 mSv y^{-1} (Sohrabi, 1993a,b). The gamma exposures in Brazil and India seem to be almost equal, but the level in Ramsar is a few times higher than in these two regions. On the other hand, HLNRAs of India are densely populated, where epidemiological study data can be of interest, unlike in Brazil and in Ramsar with their lower population densities.

In Yangjiang, China, which has been under investigation since 1972, the average effective equivalent dose is 5.4 mSv y^{-1}, which is about three times higher than the mean background of 2 mSv y^{-1}. The results of epidemiological studies over a population of 80,000 have led to a firm conclusion; i.e. no difference was found in mortality from all types of cancer or due to leukaemia in the HLNRAs of Yangjiang in comparison with the control group studied (Wei *et al.*, 1993).

Some areas in Japan, with exposure rates (i) below 7.6 μGy h^{-1}, (ii) between 7.6 and 10.5 μGy h^{-1}, and (iii) above 10.5 μGy h^{-1} respectively, with population sizes of (i) 2,230,300; (ii) 2,885,787; and (iii) 2,790,818, from 39 areas including 28 cities and 11 towns and villages, have been under epidemiological investigations as reported by Iwasaki and co-workers (1993). The results of cancer mortality rate studies of public exposed to natural radiation level in different geographic zones in Japan have shown no detectable increase in cancer mortality.

The HLNRAs discussed above, although they have relatively higher backgrounds, do not necessarily mean that very high radon levels also exist both indoors and outdoors in such areas. There are many areas and/or complexes of dwellings worldwide which have high indoor radon, as recently discovered, and deliver a high dose to public. Some of these are discussed below.

3.5.4 High Level Radon Areas and Dwellings Worldwide

Out of the 2.4 mSv average annual effective equivalent dose from natural sources, about 1.3 mSv is due to radon and its decay products indoors and outdoors, while the rest is due to cosmic rays, cosmogenic radionuclides, terrestrial gamma rays and radionuclides in the body (other than radon). The average dose due to radon, as stated above, corresponds to the average global population-weighted concentration of 40 Bq m^{-3} indoors and 10 Bq m^{-3} outdoors (UNSCEAR, 1993). The dose to public may be even higher in some regions, being elevated to 10 times the average. However, combination of local geological formations, construction materials, and type and ventilation of some houses may give dose rates from radon decay products of up to several hundred times the average.

The outdoor radon comes mainly from the soil, the level of which depends on the type of rock and the radon flux density from the earth, height above the ground and dispersion in the atmosphere: which are all affected by the meteorological conditions. The radon indoors in the basements is largely due to soil gas that enters through any cracks and openings in the structure of the house, and especially its floor. Radon also emanates from construction materials, tap water, etc., where the type of soil or rock plays an important role. For example, rocks pertaining to uranium deposits, phosphate activity and in particular alum shale from Upper Cambrian or Lower Ordovician formations have ^{226}Ra activity concentrations ranging from 600 to 4,500 Bq kg^{-1}; while those from diorite, sandstone and limestone vary from 1 to 60 Bq kg^{-1} (UNSCEAR, 1993). Very high soil gas concentrations of 700 kBq m^{-3} in alum-shale soil in Sweden have led to very high outdoor and indoor radon concentrations in some regions (Åkerblom *et al.*, 1984). Also, high outdoor radon concentrations have been measured in the provinces of Manitoba and Saskatchewan in Canada with respectively 59 and 61 Bq m^{-3}, which are about 6 times higher than the global average (Grasty, 1994). Therefore, depending on the type of rock, the radon level outdoors and indoors varies significantly from place to place, leading to some high levels in some areas and dwellings. Table 3.5.1, reproduced from UNSCEAR (1993), will provide information on the measurements made in a large number of countries, giving year of survey, type of survey, duration of exposure, mean radon concentrations and the maximum values recorded. Some high-level radon dwellings as taken from Table 3.5.1 or from other reports are reviewed below.

One very high-level and important historic area is Saxony and the neighbouring state Thuringen, as reported by Becker (1993), where 40 underground and four open-pit mines have been in operation over the years. More than 200,000 underground miners have been exposed to annual exposures of 30-300 WLM, corresponding to an annual effective equivalent dose of 0.3 to 3 Sv y^{-1}. These levels were reduced later to about 10 to 40 mSv y^{-1} by the implementation of remedial actions. Dumps from the mines cover an area of 10,000 km^2, with radon levels of 1 to 3 kBq m^{-3} in open areas. Houses using the old and new tailings or waste rocks as building materials, and constructions above shallow shafts and adits, have indoor concentrations more than 100 kBq m^{-3} in storage rooms and

Table 3.5.1: Radon concentrations in dwellings in indoor surveys (UNSCEAR, 1993).

Country Province/regions	Year of survey	Type of survey	Duration of exposure	No. of dwell. survey.	Radon concentration[a] ($Bq\ m^{-3}$)			GSD[b]
					AM	GM	MV	
Algeria	1987	Exploratory	60 days	50	32		137	–
Argentina, 3 cities	1990	Preliminary		180	32	31	127	2.0
Australia	1990	National	1 year	3,413	12	8.7	423	2.1
Austria, Salzburg	1980	Local	Grab samp.	729	-	15	190	-
Belgium	1991	National	6 months	450	48	-	4,000	-
Canada	1977-80	National	Grab samp.	13,413	34	14	1,724	3.6
Canada, N. Scotia	1990	Regional	3 months	719	108	-	5,920	3.6
China, 7 provinces	1989	Regional		3,945	24	20	378	2.2
China, Sechuan	1990	Regional		1,967	19	17	170	1.7
China, Shenzhen	1986	Regional		69	16	14	54	2.0
Czechoslovakia	1982	National	Grab samp.	1,200	140[a]	-	20,000	-
Denmark	1985	National	6 months	496	47	29	560	2.2
Egypt	1991	National		329	9.0		24	-
Finland	1982	National	1 month	8,150	90	64	-	3.1
France	1988	National	60 days	3,006	62	41	4,687	2.7
Germany (F.R.)	1984	Regional	3 months	5,970	49	40	-	1.8
Germany, Cottbus	1989	Regional	3 months	67	35	23	153	2.5
Germany, Saxony and Thuringia	1990	Exploratory	3 days	5,000	270	190	115,000	2.4
Germany (F.R.)	1991	Regional	3 days	1,040	57	34	3,100	2,9
Ghana, Legon	1990	Exploratory	9 months	25			340	-
Greece	1988	Exploratory	6 months	73	52		492	-
Hongkong	1991	Regional		140	41		140	-
India	1991	Exploratory	3 months	1,208	57	42	214	2.2
Indonesia	1991	Exploratory		165	12		120	-
Iran, 4 cities	1988	Exploratory	90 days	121	82	-	3,070	-
Ireland	1987	National	6 months	736	-	37	1,700	-
Italy	1991	National	1 year	2,250	80	62		1.9
Japan	1990	National	1 year	6,000	29	23		1.6
Kuwait	1988	Exploratory	1 year	69	41		103	-
Luxembourg	1991	National		2,500		65		-
Netherlands	1982-84	National	1 year	1,000	29	24	118	1.6
New Zealand	1988	National	1 year	717	20	18	94	-
Norway	1991	National	6 months	7,500	60	30		-
Pakistan	1991	Exploratory	2.5 months	50	30		83	-
Poland	1991	Preliminary	1 year	345	38		568	-
Portugal	1991	National	4 months	4,200	81	37	2,795	-
Spain	1991	National		1,700	86	43	15,400	3.7
Sweden	1980-82	National	2 weeks	512	108	62	3,310	-
Sweden	1990-91	National	3 months	1,360	108	56	3,900	-
Switzerland	1991	National	2.5 months	1,600	70	-	3,000	-
Syria, 2 areas	1990	Exploratory	6 months	77	20		72	-
United Kingdom	1991		3 months	96,000	20	-	10,000	-
United States	1991	National	1 year	5,967	46	25		3.1
U. S., New York	1988	Regional	1 year	2,043	42	26	1,420	2.7
Median values					42	30		2.2

a - (AM - Arithmetic Mean; GM - Geometric Mean; MV - Maximum Value), b - (GSD - Geometric Standard Deviation), c - (Derived from radon EEC measurement, using an equilibrium factor of 0.4.)

30 kBq m^{-3} in the living rooms (with a current showing around 150 kBq m^{-3}).

In Schneeberg and Schlema, as reported by Keller (1993) and by Lehman and Czarwinski (1994), houses built on the mining waste materials have median radon concentrations of about 300 Bq m^{-3} with a 7-fold increase compared with the median value in dwellings of the former West Germany and the global average. Further, in 60 houses in Schneeberg, radon concentrations of more than 10 kBq m^{-3}, with maximum values reaching even 80 kBq m^{-3}, have been measured - with maximum values in cellars of about 200 kBq m^{-3} (Keller, 1993). A similar area is Joachimsthal, where houses built of prefabricated blocks of slag concrete with radium concentrations up to 300 kBq kg^{-1} have accumulated unacceptably high radiation and radon levels indoors (Thomas *et al.*, 1993).

Another interesting area is Badgastein in Austria with a population of 6,500, where 20 thermal springs originate in the centre of town. The major part of 5 million liters of water from the springs, with a mean concentration of about 1.5 kBq m^{-3} of ^{222}Rn, flows to 120 hotels and treatment centers distributed over the town (Pohl-Rüling *et al.*, 1982). The air activities on the periphery, which range from 37 to 185 Bq m^{-3} indoors and 3.7 to 55 Bq m^{-3} outdoors, are lower than in the center where the springs originate and where concentrations range from 74 to 555 Bq m^{-3} indoors and 18 to 130 Bq m^{-3} outdoors. The mean level is even higher in the bathrooms, with a concentration of about 3.3 kBq m^{-3}; and it is the highest in the "Thermal Gallery". This is a former gold mine near Badgastein, used as a natural inhalation facility for therapeutic purposes with a mean ^{222}Rn concentration of over 100 kBq m^{-3}, where the short-lived products are 70 to 80% in equilibrium. The mine is a treatment house with a medical examination room in the entrance, followed by therapy and recreation areas. The employees work in the house in an environment with a mean radon concentration of about 0.3 to 5.5 kBq m^{-3}. The scientific development of the former gold mine to the thermal facility ("Thermal Gallery") has been recently reported by Pohl-Rüling (1993).

In HLNRAs of Ramsar (Sohrabi, 1993a,b), ^{222}Rn concentrations measured in some 473 rooms of about 350 houses in different villages as well as in the town of Ramsar showed arithmetic mean values of 615 Bq m^{-3} in Talesh Mahalleh; 326 Bq m^{-3} in Chaparsar; 258 Bq m^{-3} in Schools of Ramsar; 246 Bq m^{-3} in Ramak; 111 Bq m^{-3} in Ramsar City; 50 Bq m^{-3} in Sadat Mahalleh and Katalom; 49 Bq m^{-3} in Tonekabon; 27 Bq m^{-3} in Talesh Mahalleh of Katalom; and 90 Bq m^{-3} in old, and 50 Bq m^{-3} in new, Ramsar Hotels. The maximum value of 3,070 Bq m^{-3} was observed in Talesh Mahalleh. In one house, for example, the effective equivalent dose due to external and internal exposures including radon has been measured to be 128; 161; 167; and 11.2 mSv y^{-1} respectively for the father (53 y); housewife (47 y); son (17 y); and daughter (22 y) (Sohrabi, 1993a,b; Sohrabi *et al.*, 1993b). The results have been easily justified by their residence times indoors and outdoors. In Ramsar areas, 9 hot springs used as spas have shown ^{226}Ra contents ranging from 1 to 146 kBq m^{-3} (Sohrabi, 1993a,b; Sohrabi *et al.*, 1993a).

There are also mineral and thermal springs used as spas for therapeutic purposes in Hungary, some of which are located at the foot of Mount Gellért in Budapest and have been the subject of some studies. High radon concentrations of up to 7.15 kBq m^{-3} have been detected (Szerbin *et al.*, 1994). The effective equivalent doses were determined to be from 14 to 41 mSv y^{-1} for staff, and 4 mSv y^{-1} for visitors with 2-3 visits per week, (104 hours per year). Also there are thermal spas in Tuwa, India, with ^{226}Ra concentrations ranging from 400 to 900 Bq m^{-3} leading to radon concentrations indoors in thermal water up to 40 kBq m^{-3} and in dwellings up to 420 kBq m^{-3} (Joshi and Mishra, 1980). In Slovenia, ^{226}Ra concentrations in thermal drinking water of about 140 Bq m^{-3} while having a ^{222}Rn concentration of about 16.7 kBq m^{-3}, and relatively low indoor radon concentrations of up to 190 Bq m^{-3} in the air have been reported (Kobal and Renier, 1987).

In a village of 2,600 inhabitants in an Alpine region of Western Tyrol, Austria, very high levels of radon have been reported indoors (Ennemoser *et al.*, 1994). The measurements made indoors, for example during the period January-April 1992, showed radon concentrations of 20 to 88,000 Bq m^{-3} with a median value of 1,180 Bq m^{-3} on the ground floor of 346 houses, and 21 to 210,000 Bq m^{-3} with a median value of 3,750 Bq m^{-3} in the basement. In 208 houses (63.5%), the mean annual radon concentration exceeded the Austrian action level of 400 Bq m^{-3} (Ennemoser *et al.*, 1995).

There are many other areas around the world where a certain percentage of houses screened also shows high or very high level radon concentrations, some examples of which will be reviewed below. Measurements were made in 55,000 randomly selected houses in 38 states divided into 225 regions in the Unites States, to identify houses with screening level of radon (Alexander *et al.*, 1994). Twenty four regions were identified as having the highest concentrations: with 78.4% above 74 Bq m^{-3}; 57.3% above 148 Bq m^{-3}; 31.7% above 296 Bq m^{-3}; and 8.6% above 740 Bq m^{-3}. An extremely high radon level, exceeding 410 kBq m^{-3}, has been measured in the basement of a house in Prescott in the state of Arizona, which had a well opening in the basement with the extremely high radon concentration of 3.5 MBq m^{-3} in the well water (Kearfott, 1989).

In a survey made in Belgium, it has been estimated that about 100,000 houses have radon concentrations above 150 Bq m^{-3}; 10,000 above 400 Bq m^{-3}; and 1,000 above 4,000 Bq m^{-3} (Eggermont, 1990). Also, in a survey made in the massif of Visé in Belgium, known to geologists for its special geological features, of several hundreds of houses surveyed, 160 houses (about 2%) showed radon concentrations above 400 Bq m^{-3}, and three houses showed concentrations above 3,000 Bq m^{-3} (Poffijn *et al.*, 1994).

In a survey made in Sweden, about 1% of houses have shown concentrations above 800 Bq m^{-3}, with the highest being 40,000 Bq m^{-3}. The studies made by Åkerblom *et al.* (1984) on the effects of soil gas radon concentrations in 105 Swedish houses showed indoor radon variations from 20 to 20,000 Bq m^{-3}, and soil gas radon levels from 5,000 Bq m^{-3} in sandy soil to 700,000 Bq m^{-3} in alum-shale rich soils. Also, radon surveys in other countries show high radon levels in certain dwellings; for example the highest level of 5,920 Bq m^{-3} in Nova Scotia in Canada; 20,000 Bq m^{-3} in former Czechoslovakia; 4,687 Bq m^{-3} in France; 3,070 Bq m^{-3} in Iran; 15,400 Bq m^{-3} in Spain; 10,000 Bq m^{-3} in UK, etc. as taken from Table 3.5.1 (quoted from UNSCEAR, 1993).

3.5.5 Indoor Radon Action Levels

3.5.5.1 Action levels

Action level is defined by the IAEA Basic Safety Standards (1994) as "the level of dose rate or activity concentration above which remedial actions or protective actions should be carried out in chronic exposure or emergency exposure situation". The choice of an action level is complex, depending not only on the level of exposure, but also on the likely scale of action, which has economic implications for the community and for the individuals (ICRP-65, 1993). Since radon occurs in both workplaces and residences, regulatory limits are established for radiation workers, while action levels are set for dwellings. Such levels are established by radiation protection regulations implemented by a national regulatory authority. However, to set such a level, reasonable amounts of data on radon measurements indoors and outdoors as well as in building materials are required to enable the policy makers to formulate regulatory measures to control radon. Nevertheless, establishing a health and safety culture in a country has an important role.

Radon concentration usually differs from house to house, region to region, and country to country. Therefore, a level should be established in the context of cost-effectiveness for any regulatory action based on the data obtained in a given country. Although radon measurements have been well advanced worldwide, the action level is not yet well established in many countries. Even in some advanced countries with well established regulations, government strategies and high levels of social welfare, the action levels have had to be changed with time owing to the onset of new social, economic, and health-related problems.

Table 3.5.2 shows the action levels established in Australia, Austria, Canada, Denmark, Germany, Ireland Sweden, United Kingdom and United States, as well as the levels recommended by ICRP, IAEA, WHO and CEC. The values are heterogeneous; and even the levels in the EC countries are different from each other and from those recommended by the CEC (Olast and Sinnaeve, 1992). The action level in Germany is 250 Bq m^{-3}, about five times the nationwide mean concentration both for existing and new houses, with a time-frame for remedial action defined on the basis of lifetime (60 years) cumulative exposure of 15,000 Bq m^{-3} y. In Australia and the United Kingdom, the action level is 200 Bq m^{-3} with a time-frame for remedial action defined on the basis of lifetime cumulative exposure of 1,500 Bq m^{-3} y, which is 10 times lower than that in Germany (Olast and Sinnaeve, 1992). The ICRP-60 (1991) has given a very thoughtful advice or recommendation: "The best choice of an action level may well be that level which defines a significant, but not unmanageable, number of houses in need of remedial work; it is then not to be expected that the same action level will be appropriate in all countries". Moreover, it states that for owner-occupied dwellings, general guidance may be adequate, leaving the final decision to be made by the occupier, on behalf of the occupants; but in countries with substantial numbers of rented dwellings, it may be desirable to establish firm national action levels (ICRP-65, 1993). Also WHO have given recommendations on risks to health and on risk to management, as reviewed by Steinhäusler (1994).

Table 3.5.2: Some national and international recommended radon action levels (Ahmed, 1993; CEC,1990; IAEA-BSS, 1994; ICRP, 1991; Olast and Sinnaeve, 1992; Snihs, 1992; WHO, 1993; Steinhäusler, 1994).

Country	Action Level: Old Buildings ($Bq\ m^{-3}$)	Action Level: New Buildings ($Bq\ m^{-3}$)	Remarks and/or Recommended Time for Remedial Action
Australia	200	200	
Austria	400	400	
Canada	800	-	
Denmark	200	200	
Germany	250	250	- A time-frame band on the basis of a lifetime (60 y) cumulative exposure of 15,000 $Bq\ m^{-3}$ y; 10 times higher than the UK (NRPB) level.
Ireland	200	200	
Sweden	200	70	- Between 70 to 200 should be reduced by simple measures if possible.
United Kingdom	200	200	- A time-frame band on the basis of a lifetime (60 y) cumulative exposure of 1,500 $Bq\ m^{-3}$ y; 10 times lower than the level in Germany.
United States	150	150	- For 150-750 $Bq\ m^{-3}$, a few years; for 750-7,500 $Bq\ m^{-3}$, within a few months; above 7,500 $Bq\ m^{-3}$, immediate action or evacuation.
ICRP-65	200-600	200-600	
IAEA-BSS	200-600	200-600	
CEC	400	200	
WHO	200-300	200-300	

One parameter of importance is also the time allowed to mitigate the level which strongly depends on the radon concentration indoors. Therefore, it seems necessary to establish different action levels based on which a remediation time can be determined. The criteria developed by the present author in subsection 3.5.2 above are believed to be justified criteria, and to represent a good approach not only for the definition of different natural radiation areas, but also for implementation of remedial actions during a time-frame for its execution. For example, by defining a time-frame concept based on cumulative exposure X $Bq\ m^{-3}$ y, the time for each category of action level can be determined.

3.5.5.2 Indoor radon remedial actions

Remedial action as defined by IAEA Basic Safety Standards (1994) is an action taken when a specified action level is exceeded, to reduce radiation doses that might otherwise be received, in an intervention situation involving chronic exposure. To achieve this, the radon level indoors should be mitigated to below a well-established action level, and as low as reasonably achievable, by applying a proper remediation technique. The techniques applied should be economically efficient and reasonable. The goal to implement such a programme can be divided into two different strategies: (1) to mitigate the mean radon concentration at a national level, and (2) to mitigate radon levels only in those dwellings that have radon concentrations above the action level. By comparing the two strategies, it can be justified that mitigating an extremely high domestic level of radon is more reasonable than meeting a goal to reduce average radon level in the country (Clarke, 1995; Mossman and Sollitto, 1991). This is the same philosophy as recommended by ICRP Publication 65 (ICRP, 1993), in which the intervention is applied to reduce the risks to those most likely to be exposed rather than trying to reduce the collective dose in a whole country. However, the first strategy can also be pursued by the personal and voluntary efforts of the home owners using some simple mitigation techniques, by which the national mean concentration of radon will also be automatically reduced.

The mitigation of radon level indoors in dwellings is also a regulatory problem. An interesting article on the regulatory control of indoor radon has been put forward by Mossman and Sollitto (1991). So far only some EC countries and the United States of America have been more serious in radon control and remedial actions, and they have a number of well-studied radon mitigation techniques.

One general protocol which can be applied in all cases consists of five phases, as concluded from Wilson *et al.* (1994):

Part 1. Pre-diagnosis by performing short radon tests of all major locations in a dwelling.
Part 2. Active diagnosis by taking the information from part 1 and performing active radon mitigation diagnosis, and then designing a mitigation system for the dwelling.
Part 3. Mitigation by installing the remediation system.
Part 4. Operation of the mitigation system and performing post-mitigation radon testing.
Part 5. Improvements, if necessary, based on information received from part 4.

The mitigation techniques can generally be divided into two categories: (1) techniques for existing houses, and (2) techniques for new houses. The Proceedings of the First International Workshop on Indoor Remedial Actions (Campos-Venuti *et al.*, 1994) can be a good reference to explore many techniques and their effectiveness in mitigation of radon levels indoors. Excellent reviews dealing with different techniques have been presented by Henschel (1994) for existing houses, and by Najafi *et al.* (1995) for new houses, as well as techniques in specific buildings: such as in large buildings (Craig, 1994); in non-residential buildings (Wilson *et al.*, 1994) including schools (Fisher *et al.*, 1994); in dwellings with timber floors (Welsh and Stephen, 1994), etc. Considering the techniques for existing houses, Henschel (1994) for example has given an excellent review of the full range of techniques (along with their advantages and disadvantage and their effectiveness) that have been installed in existing US houses. These include:

A. Techniques intended to prevent soil gas entry into living areas:

- Active soil depressurization techniques (ASD); active sub-slab and drain-tile depressurization (SSD and DTD); active block-wall depressurization (BWD); and active sub-membrane depressurization (SMD).
- Active soil pressurization techniques; active sub-slab and drain-tile pressurization; active block-wall pressurization.
- Passive soil depressurization techniques: passive SSD, DTD, and SMD.
- Pressure adjustments in house or crawl space: basement pressurization, and crawl-space depressurization; crawl space pressurization, and natural crawl-space ventilation.
- Sealing radon entry routes; seal slab and wall openings in houses on slabs; seal floor between crawl space and living area; seal barrier membrane over exposed soil in crawl space.

B. Techniques intended to remove radon after entering into a living area:

- Ventilation of the living area; forced ventilation with heat recovery.
- Air cleaners: particle (progeny) removal devices, radon gas sorption devices.

Table 3.5.3: Guide to the cost and effectiveness of various measures for buildings (ICRP-65, 1993).

Method	Cost	Effectiveness*
Soil depressurization	Moderate	High
Floor sealing	Moderate	Moderate
Water treatment	Moderate	High
Subsoil removal	High	High
Increased ventilation	Moderate	Low
Increased air movement	Low	Low

* The effectiveness is judged in terms of the effect on that part of the concentration of radon progeny to which the remedial measure applies.

Najafi *et al.* (1995) have also given an excellent review of the new constructions in the United States based on the work carried out by three research programmes applying active and passive approaches. Passive approaches include placement of a vapor barrier, sealing of plumbing penetration, mixing of floor slab concrete with superplasticizers, reinforcing of slab at re-entrant corners, and proper slab curing and loading. Active approaches include a fan used to depressurize sub-slab, thus making the sub-slab pressure less than the indoor pressure. The effectiveness of the techniques applied in 47 houses has also been discussed. Similar techniques have also been applied in special buildings. However, the efficiency of the techniques for both existing and new houses might not be the same when applied to buildings in other countries, having different construction principles.

ICRP Publication 65 (1993) has also given the remedial and protective measures and the principal methods for reducing high radon concentrations indoors. In particular, it has tabulated a qualitative summary of the costs and effectiveness of the various options for radon remedial work, as given in Table 3.5.3.

3.5.5.3 International recommendations

The problem of health risks from radon in dwellings has been of serious concern to international organizations and commissions such as ICRP (1993), IAEA (1994), WHO (1993), UNSCEAR (1993) and CEC (1990). Although the concept of radon-prone areas has been given some thought and definitions, the subject of determination of criteria for HLNRAs has not been considered for detailed investigation. Even the action levels recommended by different organizations are different from each other, as given in Table 3.5.2. The ICRP, by using the dose convention and various conversion coefficients, obtains a radon gas concentration of 600 Bq m^{-3} as the figure where action is almost certain to be justified, and it expects optimization to suggest a figure no lower than 200 Bq m^{-3}; and it gives no lower target level for new constructions (Clarke, 1995). The same principles have also been followed in the IAEA Basic Safety Standards (IAEA, 1944). However, the CEC recommends an average radon gas concentration of 400 Bq m^{-3} for the existing houses and 200 Bq m^{-3} for new houses, while that by WHO is 200-300 Bq m^{-3} (Steinhäusler, 1994).

The ICRP (1993) recommends that all countries should carry out radon surveys to find radon-prone areas and to find houses with the most risk for the occupants; the national authorities should establish action levels; and guidance should be given as to the values using a philosophy similar to giving advice on intervention after a radiological accident; and they should advise residents whose homes have higher radon concentrations to initiate remedial measures.

3.5.6 Concluding Remarks

Defining a radiation-prone area and devising criteria that would apply to all types of exposures and areas, and to search for other areas not yet explored, need to be prioritized. However, radon-prone areas are a matter of immediate intervention in all countries for applying a reasonable and effective approach. Although studies have advanced on different aspects of measuring radon in dwellings, the importance of the problems invites more extensive research.

Appendix

Based on the conversion factors given by the UNSCEAR (1993) and stated in section 3.5.1, it is simple to calculate the effective dose E from radon and its decay products. For example, E due to the population-weighted average radon concentration of 40 Bq m^{-3} indoors and 10 Bq m^{-3} outdoors using equilibrium factors of 0.4 for indoors and 0.8 for outdoors, is estimated to be 1.2 mSv y^{-1} as calculated below:

a. For radon decay products outdoors:
E_{out} = 8 Bq m^{-3} (EEC) x 9 nSv h^{-1} per Bq m^{-3} (EEC)
x 0.2 (occupancy) x 8760 h y^{-1} = 0.13 mSv y^{-1}

b. For radon decay products indoors:
E_{in} = 16 Bq m^{-3} (EEC) x 9 nSv h^{-1} per Bq m^{-3} (EEC)
x 0.8 (occupancy) x 8760 h y^{-1} = 1.0 mSv y^{-1}

c. For radon inhaled indoors and outdoors:
$E_{in\&out}$ = [(10 Bq m^{-3} x 0.2) + (40 Bq m^{-3} x 0.8)]
x 1.5 μSv y^{-1} per Bq m^{-3} = 0.051 mSv y^{-1}

By adding the above doses, the total average annual effective equivalent dose E is estimated to be about 1.2 mSv:
$E = E_{out} + E_{in} + E_{in\&out} = 0.13 + 1.0 + 0.05 = 1.18$ mSv y^{-1}
Based on this procedure, the value E can be simply calculated in any dwelling with known radon concentration, ocupancy factors, etc.

References

Ahmed J. U. (1993) Radon overview in natural environment and dwellings. In *Proc. Int. Conf. on High Levels of Natural Radiation* (eds. M. Sohrabi *et al.*), Ramsar, 1990, IAEA, Vienna, pp. 267-279.

Åkerblom G., Anderson P. and Clavens B. (1984) Soil gas radon - a source for indoor radon daughters. *Radiat. Prot. Dosim.* **7**, 49-54.

Alexander B., Rodman N. White, S. B., and Phillips J. (1994) Areas in the United States with elevated screening levels of ^{222}Rn. *Health Phys.* **66**, 50-54

Becker K. (1993) Environmental and radon dosimetry: national and international standards. In *Proc. Int. Conf. on High Levels of Natural Radiation* (eds. M. Sohrabi *et al.*), Ramsar, 1990, IAEA, Vienna, pp. 281-291.

Bennet B. G. (1993) Natural background radiation exposure world-wide. In *Proc. Int. Conf. on High Levels of Natural Radiation*, (eds. M. Sohrabi *et al.*), Ramsar, 1990, IAEA, Vienna, pp. 17-27.

Campos-Venuti G. (1993) Present and future radioprotection recommendations within the European Community with special regard to radon. In *Proc. 2nd Workshop on Radon Monitoring in Radioprotection, Environmental and/or Earth Sciences* (eds. G. Furlan and L. Tommasino), ICTP, Trieste, 1991, World Scientific, Singapore.

Campos-Venuti G., Jänssens A., Olast M., Piermattei S., Sinnaeve and Tommasino L., eds. (1994) *Proc. 1st Int. Workshop on Indoor Radon Remedial Action.* Rimini, 1993, *Radiat. Prot. Dosim.* **56**, 1-4.

Clarke R. H. (1995) ICRP recommendations applicable to the mining and minerals processing industries and to natural sources. *Health Phys.* **69**, 454-460.

Commission of European Communities (1990) Recommendation of 21 February 1990 on the protection of the public against indoor exposure to radon. *Official Journal of the European Communities* **L80**, 26-28.

Craig A. B. (1994) A review of radon mitigation in large buildings in the US. *Radiat. Prot. Dosim.* **56**, 29-32.

Cullen T. L. and Penna-Franca E., eds. (1977) *Proc. Int. Symp. High Natural Radioactivity, Pocos de Caldas*, Academia Brasileria de Ciencias, Rj.

Eggermont, G. (1990) Recommendation of the Belgian Council on Science Policy. Paper presented at the *1st Int. Seminar on Managing the Indoor Radon Problem*, Mol.

Eisenbud M. (1987) *Environmental Radioactivity.* 3rd. Edition, Academic Press, San Diego.

Ennemoser O., Ambach W., Auer T., Brunner P., Schneider P., Oberaigner W., Purtscheller F. and Stingl V. (1994) High indoor radon concentrations in an Alpine region of Western Tyrol. *Health Phys.* **67**, 151-154.

Ennemoser O., Oberdorfer E., Brunner P., Schneider P., Purtscheller F., Stingl V. and Ambach W. (1995) Mitigation of indoor radon in an area with unusually high radon concentrations. *Health Phys.* **69**, 277-232.

Fisher G., Ligman B., Brennan T., Shaughnessy R., Turk B. H., and Snead B. (1994) Radon mitigation in schools utilizing heating, ventilating and air conditioning. *Radiat. Prot. Dosim.* **56**, 51-54.

Furlan G. and Tommasino L., eds. (1993) *Proc. 2nd Workshop on Radon Monitoring in Radioprotection, Environmental and/or Earth Sciences.* ICTP, Trieste, 1991, World Scientific, Singapore.

Grasty R. L. (1994) Summer outdoor radon variations in Canada and their relation to soil moisture. *Health Phys.* **66**, 185-193.

Henschel D. B. (1994) Analysis of radon mitigation techniques used in existing US houses. *Radiat. Prot. Dosim.* **56**, 21-27.

International Commission on Radiological Protection (1991) *1990 Recommendations of the International Commission on Radiological Protection.* ICRP Publication 60, Pergamon Press, Oxford.

International Commission on Radiological Protection (1993) *Protection Against ^{222}Rn at Home and at Work.* ICRP Publication 65; Ann. ICRP **23**, 2.

International Atomic Energy Agency (1994) *International Basic Safety Standards for Protection Against Ionizing Radiation and for the Safety of Radiation Sources.* Safety Series No. 115-1, IAEA, Vienna.

Iwasaki T., Minova M., Hashimoto S., Hayashi M. and Murata M. (1993) Cancer mortality rates by geographic distribution and levels of natural radiation dose in Japan. In *Proc. Int. Conf. on High Levels of Natural Radiation* (eds. M. Sohrabi *et al.*), Ramsar, 1990, IAEA, Vienna, pp. 503-514.

Joshi L. U. and Mishra U. C. (1980) Studies on geochemical behavior of uranium isotopes in Tuwa springs and Khari River sediments. *J. Radioanalytical Chem.* **59**, 367-372.

Kearfott K. J. (1989) Preliminary experiences with ^{222}Rn gas in Arizona homes. *Health Phys.* **56**, 169-179.

Keller G. (1993) Radiological aspects of former mining activities in the Saxon Erzgebirge, Germany. *Environment International* **19**, 449-454.

Kobal I., and Renier A. (1987) Radioactivity of the atomic spas at Podcetrtek, Slovenia, Yugoslavia. *Health Phys.* **53**, 308-310.

Lehman R. and Czarwinski R. (1994) Influence of shallow mine-working on the radon concentrations in houses: a problem of old mining regions. *Radiat. Prot. Dosim.* **56**, 41-44.

Mishra U. C. (1993) Exposure due to high natural radiation background and radioactive springs around the world. In *Proc. Int. Conf. on High Levels of Natural Radiation* (eds. M. Sohrabi *et al.*), Ramsar, 1990, IAEA, Vienna, pp. 29-47.

Mossman K. L. and Sollitto M. A. (1991) Regulatory control of indoor Rn. *Health Phys.* **60**, 169-176.

Najafi F. T., Lawani L. and Li W. G. (1995) Radon entry control in new house construction. *Health Phys.* **69**, 67-74.

National Radiological Protection Board (1990) Documents of the NRPB. Vol. 1, No. 1, NRPB, Chilton.

Olast M. and Sinnaeve J. (1992) The radon exposure programme of the Commission of the European Communities. *Radiat. Prot. Dosim.* **42**, 185-189.

O'Riordan M. C. and Miles J. C. H., eds. (1992) Radon 2000. *Proc. of a Conf.*, London, 1992, *Radiat. Prot. Dosim.* **42**.

Poffijn A., Eggermont G., Hallez S., and Cohilis P. (1994) Radon in Belgium: mapping and mitigation in the affected area of Visé. *Radiat. Prot. Dosim.* **56**, 77-80.

Pohl-Rüling J. (1993) The scientific development of a former gold mine near Badgastein, Austria, to the therapeutic facility "Thermal Gallery". *Environment International* **19**, 455-465.

Pohl-Rüling J., Steinhäusler F. and Pohl E. (1982) Radiation exposure and resulting risk due to residence and employment in a radon spa. In *Proc. 2nd Special Symp. on Natural Radiation Environment*, 1981 (eds. K. G. Vohra *et al.*), Wiley Eastern Limited, pp. 107-113.

Scott A. G. (1993) Comparison of criteria to define radon-prone areas. *Health Phys.* **64**, 435-436.

Snihs J. O. (1992) Swedish radon programme. *Radiat. Prot. Dosim.* **42**, 177-184.

Sohrabi M. (1993a) Recent radiological studies in high level natural radiation areas of Ramsar. In *Proc. Int. Conf. on High Levels of Natural Radiation* (eds. M. Sohrabi *et al.*), Ramsar, 1990, IAEA, Vienna, pp. 39-47.

Sohrabi M. (1993b) High level natural radiation areas with special regard to Ramsar. In *Proc. 2nd Workshop on Radon Monitoring in Radioprotection, Environmental, and/or Earth Sciences* (eds. G. Furlan and L. Tommasino), ICTP, Trieste, 1991, World Scientific, Singapore, pp. 98-108.

Sohrabi M. (1996) World High Level Natural and/or Radon-Prone Areas with Special Regard to their Classification, Invited paper, *4th Int. Conf. High Levels of Natural*

Radiation, Beijing, in press.

Sohrabi M., Ahmed J. U. and Durrani S.A., eds. (1993a) *Proc. of the Int. Conf. on High Levels of Natural Radiation.* Ramsar, 1990, IAEA, Vienna.

Sohrabi M., Zainali H., Mahdi Sh., Solaymanian A.R., Salehi, M. (1993b) Determination of ^{222}Rn levels in houses, schools and hotels of Ramsar by AEOI passive radon diffusion dosimeters. In *Proc. Int. Conf. on High Levels of Natural Radiation* (eds. M. Sohrabi *et al.*), Ramsar, 1990, IAEA, Vienna, pp. 365-375.

Sohrabi M. and Solaymanian (1988) Indoor radon level measurements in some regions in Iran. *Nucl. Tracks Radiat. Meas.* **15**, 613-616.

Steinhäusler F. (1994) A risk-based approach to health criteria for radon indoors- Report on a WHO initiative. *Radiat. Prot. Dosim.* **56**, 355-356.

Sunta C. M. (1993) A review of the studies of the high background areas of the S-W coast of India (1993). In *Proc. Int. Conf. on High Levels of Natural Radiation* (eds. M. Sohrabi *et al.*), Ramsar, 1990, IAEA, Vienna, pp. 71-86.

Szerbin P., Köteles Gy., and Str, D. (1994) Radon concentration in Rudas thermal Bath, Budapest. *Radiat. Prot. Dosim.* **56**, 319-321.

Thomas J., Mouka L. and Salava J. (1993) New houses with high radiation exposures. In *Proc. Int. Conf. on High Levels of Natural Radiation* (eds. M. Sohrabi *et al.*), Ramsar, 1990, IAEA, Vienna, pp. 177-182.

Tommasino L., Furlan G., Khan H. A. and Monnin M., eds. (1990) *Proc. 1st Int. Workshop on Radon Monitoring in Radioprotection, Environmental and/or Earth Sciences.* ICTP, Trieste, 1989, World Scientific, Singapore.

United Nations Scientific Committee on the Effects of Atomic Radiation (1993) Sources and effects of ionizing radiation. United Nations, New York.

Vohra K. G., Mishra U. C., Pillai K. C. and Sadasivan S., eds. (1982) *Proc. 2nd Special Symp. on Natural Radiation Evironment*, Bombay, 1981, Wiley Eastern Limited.

Webb G. A. M. (1992) Exposure to radon. *Radiat. Prot. Dosim.* **42**, 191-195.

Wei L., Zha Y., Tao Z., He W., Chen D. and Yuan Y. (1993) Epidemiological investigation in high background radiation areas in Yangjiang, China. In *Proc. Int. Conf. on High Levels of Natural Radiation* (eds. M. Sohrabi *et al.*), Ramsar, 1990, IAEA, Vienna, pp. 523-547.

Welsh P. and Stephen R. (1994) Radon remediation of dwellings with suspended timber floors-case studies from the building research establishment (UK). *Radiat. Prot. Dosim.* **56**, 45-50.

Wilson D. L., Dudney C. S., and Gammage R. B. (1994) An interim protocol for the mitigation of radon in non-residential buildings. *Radiat. Prot. Dosim.* **56**, 9-11.

World Health Organization (WHO) (1993) Working group summary report. WHO Doc. ICP/CEH 108 (S).

C. R. Muirhead[1]

3.6 Radon-Induced Health Effects

3.6.1 Lung Cancer

3.6.1.1 Studies of miners

The main epidemiological evidence for the role of radon exposure in the induction of lung cancer comes from a multitude of studies of miners of uranium, tin and iron ores. Details of the main studies are summarised in Table 3.6.1. The total number of men included in these studies is over 60,000, with an aggregated follow-up of approximately one million person-years. The mean cumulative exposure varies widely between studies: from a few Working Level Months (WLM) at the Radium Hill mine in Australia to just over 800 WLM on the Colorado Plateau in the USA (although, for the latter study, it is possible that some of the higher exposures have been over-estimated). The number of lung cancers observed in each study is substantial - generally at least 50 - and is raised to a statistically significant extent above expectation in each instance; the ratio of observed to expected values varies from just under 1.5 to about 5. Furthermore, these studies show that the lung cancer risk increases with increasing cumulative exposure and, in general, they appear to be consistent with a linear relationship between exposure and risk. The exposures at which some of the studies yield a significantly raised risk, i.e. between 20 and 50 WLM, are only about a factor 2-5 above the mean lifetime indoor exposure received in countries such as the US (BEIR IV, 1988) and the UK, and are less than the lifetime exposure in dwellings at the UK action level (i.e. 75 WLM) (NRPB, 1990). A parallel analysis of 11 miner studies has recently been performed by Lubin *et al.* (1994, 1995a, b).

In regard of whether the results have been influenced by the presence of any other carcinogens in the mines, Radford and St Clair Renard (1984) showed that variation in dust levels over time has not been reflected in the miners' lung cancer rates. Also, associations between radon and lung cancer have been seen at mines (e.g. Malmberget, Sweden) where respiratory carcinogens such as arsenic were absent. While findings from the Chinese (Xuan *et al.*, 1993) and West Bohemia (Tomasek *et al.*, 1994) studies suggest that adjusting for arsenic exposure may decrease the radon-related risks, results from the Ontario study (Kusiak *et al.*, 1993) show much larger risks for a given level of arsenic exposure among miners with relatively high radon exposures than among those with low radon exposures. In addition, it is difficult to conceive of a confounding factor that is so well correlated with radon exposure, such that it would be able to explain the exposure-response relationships seen in the different miner studies as artifacts. Further supporting

[1]National Radiological Protection Board, Chilton, UK

Table 3.6.1: Mortality from lung cancer among miners exposed to radon.

Mine(s) (Follow-up period)	Number of men	Mean WLM	Person - years	No of lung cancer deaths Observed	Expected
Colorado Plateau, US (1951-82)[a]	3,346	821	73,642	256	59.1
Ontario, Canada (1955-81)[b]	13,469	30		152	67.6
Beaverlodge, Canada (1950-80)[c]	8,487	13	114,170	65	34.2
Port Radium, Canada (1950-80)[d]	2,103	144	52,930	57	24.7
West Bohemia, Czech Republic (1953-90)[e]	4,320	219		702	138
Malmberget, Sweden (1951-76)[f]	1,294	94	26,567	51	13.1
New Mexico, US (1977-85)[g]	3,469	111	59,000	68	17.0
Newfoundland, Canada (1950-90)[h]	1,744	383	42,417	139	31.1
Yunnan Province, China (1976-87)[i]	17,143	217	175,143	981	267
Cornwall, UK (1941-86)[j]	3,010	~100		105	66.6
Radium Hill, Australia (1952-87)[k]	1,429	7		32	23.1
France (1946-85)[l]	1,785	70	44,995	45	21.1

Notes

(a) Hornung and Meinhardt (1987); BEIR IV (1988)
(b) Kusiak *et al.* (1993)
(c) Howe *et al.* (1986)
(d) Howe *et al.* (1987)
(e) Tomaášek *et al.* (1994)
(f) Radford and St. Clair Renard (1984)
(g) Samet *et al.* (1991)
(h) Morrison and Villeneuve (1995)
(i) Xuan *et al.* (1993)
(j) Hodgson and Jones (1990)
(k) Woodward *et al.* (1991)
(l) Tirmarche *et al.* (1993)

evidence for a causative effect comes from animal studies in the US and France, which show similar patterns of increased risks of lung cancer in rats and dogs exposed chronically to radon (BEIR IV, 1988). The International Agency for Research on Cancer (IARC) has concluded that there is sufficient evidence to classify radon as a carcinogen in humans (IARC, 1988).

3.6.1.2 Estimates of lifetime risk associated with indoor radon exposure

Several committees have attempted to estimate the risk over a lifetime for a general population exposed in an indoor environment; in particular, the International Commission on Radiological Protection (ICRP, 1993) and the US Committee on the Biological Effects of Ionising Radiation (BEIR IV, 1988). One possible approach would involve applying a quality factor to the risks derived from low-LET studies. However, there are difficulties owing to uncertainties concerning both the conversion of exposures to lung doses and the value of the radiation weighting factor w_R. In particular, the practical applications of the improved dosimetric model for the respiratory tract adopted by ICRP (1994) are still being developed. Consequently risk estimates such as those of ICRP (1993) have been based generally on studies of radon-exposed miners, possibly with supplementary information from the low-LET studies. Among the points that are important in producing such an estimate are the projection of doses from a mining to an indoor environment and the projection of risks from the period of follow-up in the miner studies to an entire lifetime. On the former point, the US National Research Council (NRC) concluded that the dose per unit exposure in homes is likely to be 70-80% of that in mines (NRC, 1991), based on a review of factors such as aerosol characteristics and differences in breathing patterns. However, ICRP concluded that there is insufficient justification for adopting a risk coefficient for the public different from that for workers (ICRP, 1993).

In projecting risks over a lifetime, both ICRP and BEIR IV utilised models under which the relative risk decreases with time since exposure. The BEIR IV model, which was derived by a joint analysis of data from the Colorado Plateau, Ontario, Beaverlodge and Malmberget cohorts, incorporates a lower relative risk beyond 15 years following exposure than that at 5-14 years post exposure, together with a lower relative risk at higher attained ages. ICRP considered a model based on analysis of Czech data for which the relative risk decreases exponentially with time since exposure (beyond 12 years) and also decreases with age at exposure.

Using idealised baseline rates that correspond to a world population, ICRP (1993) calculated the probability of lung cancer death due to indoor radon exposure as $3 \cdot 10^{-4}$ WLM^{-1}. The corresponding value in the BEIR IV report based on a similar risk model and applied to US baseline lung cancer rates (which are higher than the world average) was $3.5 \cdot 10^{-4}$ WLM^{-1}. The parallel analysis of Lubin *et al.* (1994) arrived at a similar value for indoor exposure. While the results of these analyses are in close agreement, there are a number of factors affecting the calculation of these estimates for which some uncertainty exists.

3.6.1.3 Factors that may affect risk estimates

3.6.1.3.1 Sex and age at exposure. The miner studies are, by their nature, restricted to providing information on the exposure of adult males. There is therefore the question of how the risks obtained from these studies should be used to predict risks for a general population containing women and children. In assessing the risks for females, the BEIR IV committee (1988) made the assumption that the relative risk is the same for both sexes. The basis for this approach was the idea that differences in the baseline lung cancer rates between sexes could be explained by differences in smoking habits and that a multiplicative model might be used to assess the joint effect of radon and smoking (see paragraph 3.6.1.3.2). Under this approach, the probability of radon-induced lung cancer is higher among males than among females, reflecting the difference in baseline rates. ICRP (1993), on the other hand, decided to use the same absolute risk coefficient for males and females, based largely on results for smoking and low-LET radiation exposure in the Japanese atomic bomb survivors (Kopecky *et al.*, 1986). However, ICRP recognised that the risk factor for females may be an over-estimate.

Regarding the effect of exposure to radon early in life, the only results available are from a study of Chinese tin miners, some of whom were aged less than 13 years when first exposed (Xuan *et al.*, 1993). The relative risks for this group were slightly higher than those for first exposure at ages greater than 20 years, although the study investigators judged that the variation in risk with age was not consistent. It may well be the case that information from epidemiological studies of indoor radon exposures (see subsubsection 3.6.1.4) will be required in order to obtain a better understanding of the effects of both sex and age at exposure.

3.6.1.3.2 Joint effect of radon and smoking. The examination of this topic by the BEIR IV Committee (1988) was based largely on data collected for the Colorado Plateau uranium miners. Their analysis showed that a model under which the joint effect of radon exposure and smoking on the lung cancer risk is additive was inconsistent with the data, whereas a multiplicative model for the joint effect of radon and smoking was consistent with the data. The best fit to the data was an intermediate model, i.e. predicting a supra-additive but sub-multiplicative effect. However, because of its consistency with the Colorado Plateau data, the BEIR IV Committee adopted the multiplicative model. The consequence of this model is to predict radon-induced lung cancer risks that are about an order of magnitude greater for smokers than for non-smokers.

Since the BEIR IV report was published, several other publications on this topic have appeared. Roscoe *et al.* (1989) looked at only non-smokers among the Colorado Plateau uranium miners - whose mean exposure was 720 WLM - and noted an approximately 10-fold increase in the lung cancer risk; this value appears to be higher than the relative risk for the overall cohort of smokers and non-smokers. The study by Ševc *et al.* (1989) of a subset of Czech uranium miners indicated a relationship that was intermediate between additive and multiplicative, although based on only three lung cancer deaths among non-smokers. Data on Chinese tin miners (Xuan *et al.*, 1993) also suggested a sub-multiplicative but supra-additive model for the effect of radon and smoking. However, the study by Samet *et al.* (1991) on New Mexico uranium miners was consistent with a multiplicative model, as was a case-control study of Beaverlodge miners in Canada (L'Abbé *et al.*, 1991), although the latter authors advised caution in interpretation in view of the small numbers (46 cases, 95 controls). Among Newfoundland fluorspar miners (Morrison *et al.*, 1995) the data were again too few to permit a definitive conclusion on this issue.

Taken overall, the evidence from the miner studies indicates that the joint effect of radon exposure and smoking on the risk of lung cancer is greater than additive. However, while some studies are consistent with a multiplicative effect, other studies suggest that a sub-multiplicative model may be more appropriate. A combined analysis by Lubin *et al.* (1994, 1995a) of six miner cohorts with data on both radon and smoking suggested that these data were consistent with a model intermediate between additive and multiplicative effects. However, the precise form of such a model is not clear. It should be emphasised that the choice of model for the interaction between smoking and radon should not have a very large effect on the estimate of radon-induced lung cancer risk in a general population of smokers and non-smokers. The main effect of the choice of model is on how the risk is distributed between smokers and non-smokers; the use of a sub-multiplicative rather than a multiplicative model would narrow the difference between the radon-induced risks for smokers and non-smokers, currently estimated to differ by an order of magnitude under the latter model.

3.6.1.3.3 Exposure rate. Darby and Doll (1990, 1991) have suggested that lower exposure rates may increase the risk of lung cancer associated with a given cumulative exposure. This was based in part on a comparison of risk coefficients obtained from studies in mines with differing mean exposure rates. In Table 3.6.2, which is an expanded version of a table given by Darby and Doll (1991), it can be seen that there is one mine (Port Radium, with a high exposure rate) for which the relative risk is particularly low, and there are three mines (Malmberget, Beaverlodge and Radium Hill, each with low exposure rates) for which the relative risk is particularly high. The risk coefficients for the other mines are fairly similar, taking into account the statistical uncertainties in these values, although the exposure rates in these mines covered almost as wide a range as those in mines cited above. However, one of the lowest risk coefficients in Table 3.6.2 arises for the French study (Tirmarche *et al.*, 1993), for which exposure rates were low. Darby and Doll attached particular importance to the comparison of the results for the Port Radium and Beaverlodge mines, since these mines were in the same broad geographical area and were operated by the same company. While a re-evaluation of exposures at Beaverlodge suggested that they have been underestimated by about 35% on average (SENES, 1991), this had little influence on the estimate of excess relative risk per unit exposure (Howe and Stager, 1996). The BEIR IV analysis used data for only some of the mines in Table 3.6.2, between which there was little evidence of heterogeneity in the risk coefficients. However, two of the mines included in the BEIR IV analysis, namely Malmberget and Beaverlodge, have among the highest risk coefficients in Table 3.6.2. Consequently, the overall risk coefficient derived by the BEIR IV Committee (ie, an excess relative risk of $2.5 \cdot 10^{-2}$ WLM^{-1} at ages 55-64 years following exposure 5-14 years earlier) is weighted by the data for these two mines and differs by only up to a factor of around 2 from the risk coefficients for the individual mines considered by BEIR IV.

Table 3.6.2: Relative risk coefficients in relation to exposure rate in mines (after Darby and Doll, 1991).

Mine(s)	Average exposure rate (WLM y^{-1})	Excess relative risk, 10^{-2} WLM^{-1}	
		Original (95% Confidence Interval)	BEIR IV[a]
Colorado Plateau, US	124	1.2	0.9[b]
Port Radium, Canada	109	0.27 (0.11, 0.43)	-
Newfoundland, Canada	67	0.66 (0.33, 0.99)	-
West Bohemia, Czech Republic	~28	1.1 (0.65, 1.8)	-
Yunnan Province, China	20	0.6 (0.4, 0.8)	-
New Mexico, US	~16	1.1	
Ontario, Canada	~10	1.3	1.8
Cornwall, UK	~10	0.9 (0.5, 1.4)	-
Beaverlodge, Canada[c]	8	3.3 (1.0, 9.6)	5.1
Radium Hill, Australia	7	~5	
France	~5	0.4 (-0.1, 0.9)	-
Malmberget, Sweden	5	3.6 (2.2, 5.0)	3.6

Notes
(a) Based on those aged 55-64 years with exposure 5-14 years previously.
(b) Based on cumulative exposures of less than 2000 WLM.
(c) Based on updated analysis by Howe and Stager (1996).

Within studies at a given mine, there was evidence of an increase in the risk per cumulative exposure with decreasing exposure rate in the Colorado Plateau study, after excluding the data on the highest cumulative exposures for which uncertainty exists in the exposure estimates (Hornung and Meinhardt, 1987). The BEIR IV Committee also noted a suggestion of such an effect within the Beaverlodge study but, because it was not apparent within the other cohorts that they examined (Ontario and Malmberget), it was not included in their final risk model. Re-evaluation of the Beaverlodge data based on revised exposure estimates does not now indicate an exposure rate effect (Howe and Stager, 1996). However, an exposure rate effect of the form described above has now been reported for the Czech and Chinese (Xuan *et al.*, 1993) studies. For example, in West Bohemia (Czech Republic), the risk per unit exposure among miners whose exposure rates never exceeded 10 WL (Working Levels) was about twice that for all miners – whose mean rate was $\sim$ 28 WLM y^{-1} in the cohort (Tomášek *et al.*, 1994). The reports of other studies have generally either not addressed this topic or have been unable to provide information owing to the lack of variation in exposure rates within certain mines (for example, as in Cornwall (Hodgson and Jones, 1990)). There are also indications from studies in rats of an increased effect at lower exposure rates based on the same total exposure (Cross, 1991), perhaps due to cell-killing at high exposures. However, the estimates of risk at the lower exposure rates in these experiments are comparable to the BEIR IV value.

Based on the combined analysis of 11 miner studies, Lubin *et al.* (1995b) concluded that there is evidence of lower radon risk coefficients at higher exposure rates. However, any exposure rate effect would disappear at sufficiently low exposure rates (Brenner, 1994). Indeed, Lubin *et al.* (1995b) noted that the inverse exposure-rate effect seemed to diminish below a total exposure of 50 WLM. It should also be emphasised that the BEIR IV (1988) analysis took account of the two cohorts with the highest reported risk coefficients. As a consequence, the BEIR IV risk coefficient differs from the highest value in any individual study by only about a factor of two, and so any under-estimation of risks from indoor radon is likely to be quite low. As with the topics discussed earlier, it may ultimately be possible to examine the question of exposure rate effects using information from the studies of indoor radon exposure.

3.6.1.4 Epidemiological studies of lung cancer and indoor radon exposure

While a number of correlation studies of lung cancer and mean radon concentrations have been carried out in various countries, these have generally been unsatisfactory, owing to the general lack of information on individual smoking habits in these studies and the fact that they tend to be based on fairly large geographical areas (Stidley and Samet, 1993). Consequently emphasis is being placed on studies that collect data on individual radon exposures, as well as smoking habits and other possible confounders. In particular, about 15 case-control studies have been or are currently being performed throughout the world, mainly in Europe and North America. Because of the need to collect detailed data, such studies can be time-consuming to undertake. There is also the problem of attempting to estimate exposures that were received many years previously, possibly in one or more previous dwellings. In particular, random errors in the assessment of indoor radon exposure would tend to obscure the relationship with the risk of lung cancer (Lubin *et al.*, 1990). Researchers are therefore discussing plans for pooling the individual studies when these are completed, so as to increase the statistical power of the test for trend in the lung cancer risk with radon exposure.

At present only a few of these studies have reported results. There were indications from early studies in Sweden (Edling *et al.*, 1984; Svensson *et al.*, 1987, 1989) of an increase in lung cancer risk in relation to indirect estimates of radon exposures, based on house type and geology. In a later study of 210 women in Stockholm county with lung cancer and 410 controls for which radon measurements were made in dwellings, Pershagen *et al.* (1992) showed an increasing trend in risk with cumulative radon exposure; for exposure in excess of 5000 Bq m^{-3} y compared with less than 1250 Bq m^{-3} y, the relative risk was 2.3 (95% confidence interval 1.1-4.5). The excess relative risk per unit exposure estimated from this study, namely $2.7 \cdot 10^{-2}$ WLM^{-1}, is within the range of values arising from studies of miners, as shown in Table 3.6.2. Schoenberg *et al.* (1990) conducted a study in New Jersey (USA) in which year-long alpha track detector measurements were made in dwellings that had been occupied for at least ten years by 433 female lung cancer cases and 402 controls. The study yielded an excess relative risk of $3.4 \cdot 10^{-2}$ WLM^{-1} (90% confidence interval 0.0-$8.0 \cdot 10^{-2}$ WLM^{-1}) which is within the range cited in Table 3.6.2 for the miner studies. The authors did caution that less than 50% of the potential subjects were included in this study, owing in part to the exclusion of the more mobile section of the population who did not meet the above residence criterion. A study in China (Blot *et al.*, 1990) of about 300 female lung cancer cases and a similar number of controls failed to show a correlation with indoor radon; radon data were available for a smaller proportion of the eligible cases (74%) than for the controls (90%), although the population was less mobile than in other studies. It should be noted that this study was performed in an area of China with high levels of other indoor air pollutants. In Finland, a study of 238 male lung cancer cases and 434 controls (mostly smokers in both groups) indicated an increasing risk with increasing radon exposure which, whilst not statistically significant, was consistent with the BEIR IV and ICRP risk estimates (Ruosteenoja, 1991). However, the controls tended to differ from the cases in terms of vital status, and measured radon levels were obtained for only 72% of the cases and 79% of the controls. Analysis of data from a study of 750 lung cancer cases and an equal number of controls in Winnipeg (Canada) found associations with radon ranging from weakly positive to weakly negative (Létourneau *et al.*, 1994). Similar results arose from a study of 538 non-smoking white women and 1,183 age-matched controls in Missouri, USA (Alavanja *et al.*, 1994), although there was some evidence of an increasing trend in the risk of adenocarcinoma with increasing exposure.

By far the largest case-control study of indoor radon and lung cancer to have reported to date is the national Swedish study of Pershagen *et al.* (1994). This study was based on 1,360 lung cancer cases (586 female, 774 male) diagnosed during 1980-84 and 2,847 controls (1,380 female, 1,467 male) selected from population registers. Radon measurements were made in nearly 9,000 dwellings occupied by the study subjects since 1947. The lung cancer risk was shown to increase to a statistically significant extent with increasing radon, expressed either as the time-weighted mean or as cumulative exposure. In particular, the excess relative risk was 1.3 (95% CI 1.1-1.6) for exposure over the average period considered in the study (ie, 32.5 years) at an average radon concentration in the range 140-400 Bq m^{-3}, and 1.8 (95% CI 1.1-2.9) for concentrations in excess of 400 Bq m^{-3}. These risks were approximately doubled after excluding those subjects who slept near an open window, and whose radon exposures may consequently have been poorly classified. The risk estimates are in the same range as those from the miner studies. There was statistically significant evidence that the joint effect of radon and smoking on the lung cancer risk exceeded additivity, and the data were consistent with a multiplicative effect.

Taken overall, Lubin (1994) has pointed out that the results from the studies published to date are consistent with various levels of risk, including that estimated by the BEIR IV Committee (1988). As stated earlier, it will be necessary to wait until results from further studies are published and then pooled before it will be possible to obtain an overall idea of the magnitude of the risks associated with indoor exposure.

3.6.2 Cancers Other Than Lung

3.6.2.1 Dosimetry

Doses to organs other than the lung from radon and progeny have been reviewed by Miles and Cliff (1992). Inhaled radon may be transported to various organs after dissolving in blood. The concentration of radon in organs and the consequent dose both depend on the fat content of the organs, since radon is about 16 times more soluble in fat than in blood. Miles and Cliff (1992) quoted an annual dose of 16 μSv to soft tissue containing 1% fat arising from exposure at 20 Bq m^{-3} in dwellings. This dose is about two orders of magnitude lower than that commonly arising from penetrating natural radiation per year.

Richardson *et al.* (1991) suggested that because of the high fat content in bone marrow, there may be a significant risk of leukaemia induction from indoor radon exposure. The red bone marrow contains 40% fat. Based on the assumption that this fat is distributed uniformly in marrow, the annual dose arising from exposure at 20 Bq m^{-3} can be calculated as 96 μSv. However, since the fat is not distributed uniformly, much of the dose will not be delivered to sensitive cells in the bone marrow (Miles and Cliff, 1992). As a consequence, the value of 96 μSv is likely to over- estimate the annual dose. Furthermore, this value is still about an order of magnitude less than the annual dose commonly arising from penetrating natural radiation.

In terms of effective dose (ICRP, 1991), the annual value arising from exposure at the action levels of 200-600 Bq m^{-3} suggested internationally by ICRP (1993) is likely not to exceed 0.1-0.3 mSv for irradiation of the red bone marrow, but to be approaching two orders of magnitude higher for lung irradiation. Thus lung cancer is the major health effect arising from radon exposure. While it is therefore appropriate to base action against radon on the limitation of lung doses and associated risks, it is worth noting that such programmes will also reduce doses to other organs.

3.6.2.2 Epidemiological studies

3.6.2.2.1· Studies of miners. Darby *et al.* (1995) have carried out a detailed collaborative analysis of cancers other than lung cancer in 11 cohorts of radon-exposed miners. This covered over 64,000 men, employed on average for 6.4 years and with a mean radon exposure of 155 WLM. For all non-lung cancers combined, the 1,179 deaths observed was virtually identical to that expected in the corresponding regions. Furthermore, mortality did not increase with increasing cumulative exposure. A total of 28 individual cancer categories were also examined. It should be noted that, in the absence of an underlying effect, some statistically significantly increased and decreased results would be observed simply by chance when a large number of endpoints is studied. In this instance, raised mortality was observed for stomach and liver cancers, whereas decreased rates were found for cancers of the tongue and mouth combined, pharynx and colon. In the case of leukaemia, mortality was raised in the period within 10 years of starting work but not subsequently. Furthermore, for none of the aforementioned cancer types was mortality related significantly to cumulative exposure. Only for pancreatic cancer was a statistically significant (increasing) trend with cumulative exposure found. However, the authors thought that this was likely to be a chance finding.

Darby *et al.* (1995) therefore concluded that their investigation of underground miners provides considerable evidence that high radon concentrations do not cause a material risk of mortality from cancers other than lung cancer. As a consequence, they recommended that protection standards for radon should continue to be based on considerations of lung cancer risks.

3.6.2.2.2 Indoor radon exposure. Henshaw *et al.* (1990) claimed that indoor radon exposure is associated with the risk of leukaemia and certain other cancers, such as melanoma and cancers of the kidney and prostate. These claims were based in large part on an international correlation study, in which national cancer incidence rates and mean indoor radon concentrations in different countries were compared. However, as Butland *et al.* (1990) pointed out, it is very difficult to place any interpretation on these results, owing to the wide variation in the quality of the radon data between countries, as well as some variation in the quality of the cancer data. In particular, Nordic countries, where cancer registries have been operating for a long period and the quality of the cancer data is high, tend to have the highest national radon levels. It is also difficult to interpret results presented by Henshaw *et al.* (1990) within certain countries (e.g., Canada) since they still relate to large geographical areas.

An analysis has been performed of the geographical distribution of childhood leukaemia and non-Hodgkin's lymphoma throughout Britain in relation to indoor radon concentrations and indoor and outdoor gamma dose rates (Muirhead *et al.*, 1991). The study had the advantage of being based on data collected in a uniform manner for a large number of fairly small areas, namely districts. Taken over all districts, there were no significant correlations between childhood leukaemia incidence and the levels of natural radiation. When the data were amalgamated into larger areas, namely counties, there were indications of a positive trend in leukaemia incidence with indoor radon level and of a negative trend with indoor gamma dose rate. However, when the district-level data adjusted for county were studied, these trends were reversed although they were not statistically significant. The differences between the results from the analyses based respectively on county-level data and on district-level data adjusted for county did not seem to be explainable on the basis of uncertainties in the estimates of natural radiation levels. Rather, Muirhead *et al.* (1991) concluded that geographical confounding factors are likely to have affected the former analysis (which was based on larger areas) and that, after allowing for county effects, the data do not support the claims of Henshaw *et al.* (1990). A later analysis that additionally took account of socio-economic variables confirmed this conclusion (paper in preparation).

The results obtained by Muirhead *et al.* (1991) reinforce the view that correlation studies based on large areas are often likely to be misleading because of confounding factors, and that it is better to look at data pertaining to small areas (as did Muirhead *et al.*, 1991) or to individuals, as in a case-control study. The latter approach is being taken in a large study of childhood cancer in England, Wales and Scotland that is being managed under the auspices of the UK Coordinating Committee on Cancer Research. This study, which will ultimately include about 3,000 childhood cancer cases and twice as many controls, involves the measurement of radon levels in the dwellings of study subjects. Measurements are also being made for a study in the UK by the Leukaemia Research Fund of adult leukaemias, and for a study in the USA by the US National Cancer Institute and the Children's Cancer Group of about 600 newly-diagnosed cases of childhood acute lymphoblastic leukaemia and 600 matched controls. These studies should have reasonably high statistical power to detect any associations, whilst avoiding the problems inherent in comparing data across large geographical areas.

3.6.3 Concluding Remarks

Epidemiological studies of over 60,000 radon-exposed miners, together with evidence from animal studies, demonstrate clearly that radon is causally related to the risk of lung cancer. Based on studies of miners, ICRP and the US BEIR IV Committee have assessed the lifetime risk of radon-induced lung cancer to be about 3-3.5$\cdot 10^{-4}$ WLM^{-1}. There are, however, several uncertainties associated with the estimation of risks from indoor radon exposures based on the miner studies. These include: possible differences in the lung dose per unit exposure between domestic and mining environments; lack of data for females and those exposed early in life; uncertainty about the joint effect of radon and smoking on lung cancer risks; and the possibility that the risk per unit cumulative exposure may be greater at lower rather than higher exposure rates.

In view of these uncertainties, direct information on the risk of lung cancer associated with indoor exposures is being sought from a series of case-control studies, mainly taking place in Europe and North America. Of the studies that have been published so far, most have been consistent with the miner studies, although the risk estimates had wide confidence intervals. The largest domestic study to date, conducted in Sweden and involving measurements in nearly 9,000 dwellings, found a statistically significant trend in risk with increasing exposure that was in line with the miner studies. Further information will be obtained during the next few years as results from other studies are published and subsequently pooled.

Doses to organs other than lung are considerably lower than lung doses, and it is clear that lung cancer is the major health effect associated with radon exposure. Results from studies of radon-exposed miners have not overall indicated excesses of cancers other than lung. Whilst it has been claimed, based on a comparison across countries, that indoor radon may be associated with leukaemia and certain other cancers, there were methodological problems with this study. A study based on small-area data throughout Great Britain, on the other hand, did not show an association between the geographical patterns of indoor radon and childhood leukaemia. Case-control investigations of childhood cancers and radon, which are based on data for individuals rather than geographical areas, are currently being conducted in the UK and USA. Investigators who studied cancer risks in miners have suggested that protection standards should continue to be based on considerations of the lung cancer risk alone.

References

Alavanja M. C. R., Brownson R. C., Lubin J. H., Brown C., Berger E. and Boice J. D. (1994) Residential radon exposure and lung cancer among nonsmoking women. *J. Natl. Cancer Inst.* **86**, 1829-1837.

BEIR IV (Committee on the Biological Effects of Ionising Radiation) (1988) *Health Risks of Radon and Other Internally Deposited Alpha-Emitters.* NAS/NRC, National Academy Press, Washington DC.

Blot W. J., Xu Z-Y., Boice J. D., Zhao D-Z., Stone B. J., Sun J., Jing L-B. and Fraumeni J. F. (1990) Indoor radon and lung cancer in China. *J. Natl. Cancer Inst.* **82**, 1025-1030.

Brenner D. J. (1994) The significance of dose rate in assessing the hazards of domestic radon exposure. *Health Phys.* **67**, 76-79.

Butland B. K., Muirhead C. R. and Draper G. J. (1990) Letter to the editor: Radon and leukaemia. *Lancet* **335**, 1338-1339.

Cross F. T. (1991) Experimental studies on lung carcinogenesis and their relationship to future research on radiation-induced lung cancer in humans. In *The Future of Human Radiation Research* (eds. G. B. Gerber *et al.*), British Institute of Radiology, London, BIR Report 22, pp. 27-35.

Darby S. C. and Doll R. (1990) Radiation and exposure rate. *Nature* **344**, 824.

Darby S. C. and Doll R. (1991) Radon in houses: How large is the risk? *Radiat. Prot. Australia* **8**, 83-88.

Darby S. C., Whitley E., Howe G. R., Hutchings S. J., Kusiak R. A., Lubin J. H. *et al.* (1995) Radon and cancers other lung cancer in underground miners: A collaborative analysis of 11 studies. *J. Natl. Cancer Inst.* **87**, 378-384.

Edling C., Kling H. and Axelson O. (1984) Radon in homes - a possible cause of lung cancer. *Scand. J. Work Environ. Hlth.* **10**, 25-34.

Henshaw D. L., Eatough J. P. and Richardson R. B. (1990) Radon as a causative factor in induction of myeloid leukaemia and other cancers. *Lancet* **355**, 1008-1012.

Hodgson J. T. and Jones R. D. (1990) Mortality of a cohort of tin miners, 1941-86. *Br. J. Indust. Med.* **47**, 665-676.

Hornung R. W. and Meinhardt T. J. (1987) Quantitative risk assessment of lung cancer in US uranium miners. *Health Phys.* **52**, 417-30.

Howe G. R., Nair R. C., Newcombe H. B., Miller A. B. and Abbatt J. D. (1986) Lung cancer mortality (1950-80) in relation to radon daughter exposure in a cohort of workers at the Eldorado Beaverlodge uranium mine. *J. Natl. Cancer Inst.* **77**, 357-362.

Howe G. R., Nair R. C., Newcombe H. B., Miller A. B., Burch J. D. and Abbatt J. D. (1987) Lung cancer mortality (1950-80) in relation to radon daughter exposure in a cohort of workers at the Eldorado Port Radium uranium mine: Possible modification of risk by exposure rate. *J. Natl. Cancer Inst.* **79**, 1255-1260.

Howe G. R. and Stager R. H. (1996) Risk of lung cancer mortality after exposure to radon decay products in the Beaverlodge cohort based on revised exposure estimates. *Radiat. Res.* **146**, 37-42.

IARC (1988) *IARC Monographs on the Evaluation of Carcinogenic Risks to Humans.* Vol. 43: Man-made mineral fibres and radon, International Agency for Research on Cancer, Lyon.

ICRP (1991) 1990 Recommendations of the International Commission on Radiological Protection. *ICRP Publication 60, Ann. ICRP* **21**, No. 1-3.

ICRP (1993) Protection against ^{222}Rn at home and at work. *ICRP Publication 65, Ann. ICRP* **23**, No. 2.

ICRP (1994) Human respiratory tract model for radiological protection. *ICRP Publication 66, Ann. ICRP* **24**, No. 1-4.

Kopecky K. J., Nakashima E., Yamamoto T. and Kato H. (1986) Lung cancer, radiation and smoking among A- bomb survivors, Hiroshima and Nagasaki. Radiation Effects Research Foundation, Hiroshima, RERF TR 13-86.

Kusiak R. A., Ritchie A. C., Muller J. and Springer J. (1993) Mortality from lung cancer in Ontario uranium miners. *Br. J. Ind. Med.* **50**, 920-928.

L'Abbé K. A., Howe G. R., Burch J. D., Miller A. B., Abbatt J., Band P., Choi W., Du J., Feather J., Gallagher R., Hill G. and Matthews V. (1991) Radon exposure, cigarette smoking, and other mining experience in the Beaverlodge uranium miners cohort. *Health Phys.* **60**, 489-495.

Létourneau E. G., Krewski D., Choi N. W., Goddard M. J., McGregor R. G., Zielinski J. M. and Du J. (1994) Case-control study of residential radon and lung cancer in Winnipeg, Manitoba, Canada. *Am. J. Epidemiol.* **140**, 310-322.

Lubin J. H. (1994) Invited commentary: Lung cancer and exposure to residential radon. *Am. J. Epidemiol.* **140**, 323-332.

Lubin J. H., Samet J. M. and Weinberg C. (1990) Design issues in epidemiologic studies of indoor exposure to Rn and the risk of lung cancer. *Health Phys.* **59**, 807-817.

Lubin J. H., Boice J. D., Edling C., Hornung R. W., Howe G., Kunz E. *et al.* (1994) *Radon and Lung Cancer Risk: A Joint Analysis of 11 Underground Miners Studies*, National Institutes of Health, Washington DC, NIH Publication No. 94-3644.

Lubin J. H., Boice J. D., Edling C., Hornung R. W., Howe G. R., Kunz E. *et al.* (1995a) Lung cancer risk in radon-exposed miners and estimation of risk from indoor exposure. *J. Natl. Cancer Inst.* **87**, 817-827.

Lubin J. H., Boice J. D., Edling C., Hornung R. W., Howe G., Kunz E. *et al.* (1995b) Radon-exposed underground miners and inverse dose-rate (protraction enhancement) effects. *Health Phys.* **69**, 494-500.

Miles J. C. H. and Cliff K. D. (1992) Dose to lung and other organs from radon and thoron as a function of age. *Radiat. Prot. Dosim.* **41**, 251-253.

Morrison H. I. and Villeneuve P. J. (1995) *Radon Progeny Exposure and Lung Cancer Risk: Analyses of a Cohort of Newfoundland Fluorspar Miners*, Atomic Energy Control Board, Ottawa, AECB INFO-0570.

Muirhead C. R., Butland B. K., Green B. M. R. and Draper G. J. (1991) Childhood leukaemia and natural radiation. *Lancet* **337**, 503-504.

Muller J., Wheeler W. C., Gentleman J. F., Suranyi G. and Kusiak R. (1984) Study of mortality of Ontario miners. *Int. Conf. on Occupational Radiation Safety in Mining*, 1984, Toronto.

NRC (1991) *Comparative Dosimetry of Radon in Mines and Homes.* National Academy Press, Washington DC.

NRPB (1990) Human exposure to radon in homes: Recommendation for the practical application of the Board's statement. *Doc. NRPB* **1**, 17-32.

Pershagen G., Liang Z-H., Hrubec Z., Svensson C. and Boice J. D. (1992) Residential radon exposure and lung cancer in Swedish women. *Health Phys.* **63**, 179-186.

Pershagen G., Åkerblom G., Axelson O., Clavensjö B., Damber L., Desai G. *et al.* (1994) Residential radon exposure and lung cancer in Sweden. *New Engl. J. Med.* **330**, 159-164.

Radford E. P. and St Clair Renard K. G. (1984) Lung cancer in Swedish iron ore miners exposed to low doses of radon daughters. *New Engl. J. Med.* **310**, 1485-1494.

Richardson R. B., Eatough J. P. and Henshaw D. L. (1991) Dose to red bone marrow from natural radon and thoron exposure. *Br. J. Radiol.* **64**, 608-624.

Roscoe R. J., Steenland K., Halperin W. E., Beaumont J. J. and Waxweiler R. J. (1989) Lung cancer mortality among non-smoking uranium miners exposed to radon daughters. *J. Amer. Med. Assoc.* **262**, 629-633.

Ruosteenoja E. (1991) Indoor radon and risk of lung cancer: an epidemiological study in Finland. Finnish Centre for Radiation and Nuclear Safety, Helsinki, STUK-A99.

Samet J. M., Pathak D. R., Morgan M. V., Key C. R., Valdevia A. A. and Lubin J. H. (1991) Lung cancer mortality and exposure to radon progeny in a cohort of New Mexico underground uranium miners. *Health Phys.* **61**, 745-752.

Schoenberg J. B., Klotz J. B., Wilcox H. B., Nicholls G. P., Gildel-Real M. J., Sternhagen A. and Mason T. J. (1990) Case-control study of residential radon and lung cancer among New Jersey women. *Cancer Research* **50**, 6520-6524.

SENES Consultants Limited (1991) Detailed reconstruction of radon daughter exposures of Eldorado Beaverlodge uranium mine employees. Report prepared for the Atomic Energy Control Board, Ottawa.

Ševc J., Kunz E., Tomášek L., Plaček V. and Horáček J. (1989) Cancer in man after exposure to Rn daughters. *Health Phys.* **54**, 27-46.

Stidley C. A. and Samet J. M. (1993) A review of ecologic studies of lung cancer and indoor radon. *Health Phys.* **65**, 234-251.

Svensson C., Eklund G. and Pershagen G. (1987) Indoor exposure to radon from the ground and bronchial cancer in women. *Int. Arch. Occup. Environ. Health* **59**, 123-131.

Svensson C., Pershagen G. and Klominck J. (1989) Lung cancer in women and type of dwelling in relation to radon exposure. *Cancer Research* **49**, 1861-1865.

Tirmarche M., Raphalen A., Allin F., Chameaud J. and Bredon P. (1993) Mortality of a cohort of French uranium miners exposed to relatively low radon concentrations. *Br. J. Cancer* **67**, 1090-1097.

Tomášek L., Darby S. C., Fearn T., Swerdlow A. J., Plaček V. and Kunz E. (1994) Patterns of lung cancer mortality among uranium miners in West Bohemia with varying rates of exposure to radon and its progeny. *Radiat. Res.* **137**, 251-261.

Woodward A., Roder D., McMichael A. J., Crouch P. and Mylvaganam A. (1991) Radon daughter exposures at the Radium Hill uranium mine and lung cancer rates among former workers, 1952-87. *Cancer Causes and Control* **2**, 213-220.

Xuan X. Z., Lubin J. H., Li J. Y., Yang L. F., Luo Q. S., Yang L., Wang J. Z. and Blot W. J. (1993) A cohort study in Southern China of tin miners exposed to radon and radon decay products. *Health Phys.* **64**, 120-131.

4.

APPLICATIONS IN EARTH SCIENCES

J. Hakl[1], I. Hunyadi[1] and A. Várhegyi[2]

4.1 Radon Monitoring in Caves

4.1.1 Introduction

Caves are commonly considered as static environment, where not only the continual darkness but the temperature and high relative humidity are stable. In most cases this belief is not justified, because air currents can cause measurable changes in the physical parameters. Caves occur everywhere on the earth, although mainly in karst areas - as they are formed mostly in limestone environments. It is obvious that the noble gas radon, which moves freely through the pores of permeable rocks, will easily penetrate into subsurface cavities and even huge caves.

In the last two decades the development and the widespread application of etched track techniques have found their place also in radon measurement projects aimed to study the cave environment. These measurements have ranged from sporadic observations up to regular long-term studies, and they were motivated by dosimetric and transport study approaches as well. In Hungary the first *in situ* field radon measurements in caves performed by etched track techniques started as early as 1977 and were initiated by Dr G. Somogyi, one of the pioneers on the etched track field. Since then, the number of studied caves has increased substantially and we have reached 10,000 observation data in 31 Hungarian caves.

After recognising the link between the secondary porosity, fracturization and formed radon levels the ever-pressing interest to understand underground radon transport in karstic environment has been increased. During the last several years active electronic devices, automatic multiparameter (radon - typically one hour integration time, pressure, temperature, humidity) monitors (Dataqua Ltd, Hungary, Kölcsey F. u. 1, 8220 Balatonalmádi), have been included into cave studies. These studies were partly financed by Hungarian funds (projects AKA 1-3-86-185, OTKA 2011, 3005 and T 016558, T 017560) during the course of the last ten years. The accomplishment of this section thanks is to this support, where of course international experiences are also appreciated.

4.1.2 Sources of Radon Isotopes in Cave Environment

The sources of radon in caves are the bedrock and deposits. Radon levels in caves are influenced primarily by the uranium content of the rock. Limestone and other sedimentary rocks are found to contain about 1.3-2.5 ppm ^{238}U (16-31 Bq kg^{-1}) on average.

[1]Institute of Nuclear Research, Debrecen, Hungary
[2]Mecsekurán Ore Mining Ltd, Pécs, Hungary

The relatively high values of radon found in caves are due to these minute quantities of parent substances that occur naturally on and within the interior surfaces of the caves. Increased ^{238}U concentrations can be associated either with fluorite mineralizations or hydrocarbons present in the surrounding limestone.

Uranium can be oxidised and mobilised by groundwater flow. Once reducing conditions are encountered, the uranium is readily precipitated from solution. This leaching fixation process leads to the enrichment of uranium in adjacent deposits. This secondary transport and enrichment process is important in cave environment, as fractures in rock can increase the surface area interacting with water. Experimental evidence of this effect in cave environment was found by Navrátil *et al.* (1993), who observed enrichment of uranium on cave walls.

The primary radon source is the ^{226}Ra content of the rock. Nazaroff *et al.* (1988) report mean ^{226}Ra content of carbonates to be 25 Bq kg^{-1} (range 0.4-233), which were found to be distributed lognormally. The reports of other authors (Wilkening, 1981): 96.2 Bq kg^{-1}; (Surbeck, 1990): 20 Bq kg^{-1}, well agree with this value. The results of ^{226}Ra determinations of bedrock and soil samples from the Hungarian caves examined fall in range 0.6-26 Bq kg^{-1} (Dezső, 1995). The relatively impermeable soils (deposits) such as clay do not have sufficient porosity to allow transfer of significant amounts of soil gas; therefore their contribution to radon budget is small (Michel, 1987). Accordingly, Burkett (1993) found radon emitted from the clay to be not sufficient to account for the radon concentration measured in the cave.

4.1.3 Review of Some Published Radon Data in Caves

Continuous radon measurements covering at least one-year long period were performed by the etched track technique in Hungary, Italy, Slovakia and Luxembourg. Seasonally one-week-long observations were performed in England, and measurements not covering a full year are reported from Mexico and the USA.

Based on long-term continuous radon observation the relative variations of radon between monitor locations indicate that the cave atmosphere is not a uniform radon environment over any given time period; so it is very hard to represent one cave with one radon value. Therefore, we decided to compile all the globally available radon data from 220 different caves (Ahlstrand, 1980; Amano *et al.*, 1985; Borau *et al.*, 1993; Burkett, 1993; Cappa *et al.*, 1995; Cigna and Clemente, 1981; Collar and Odgen, 1991; Cunningham and LaRock, 1991; Fernandez *et al.*, 1984; Géczy *et al.*, 1988; Gunn *et al.*, 1991; Hakl *et al.*, 1992; Hakl, 1993, 1994; Hyland and Gunn, 1994; Hyland *et al.*, 1994; Kobal *et al.*, 1986, 1987, 1988; Lénárt *et al.*, 1988; Kaufmann *et al.*, 1995; Lively and Krafthefer, 1995; Marx, 1996; Massen *et al.*, 1995; Miki and Ikeya, 1980; Navrátil *et al.*, 1994; Papastefanou *et al.*, 1986; Quinn, 1988, 1990; Roda *et al.*, 1994; Seymore *et al.*, 1980; Solomon *et al.*, 1992; Surbeck, 1990; Vibranov *et al.*, 1975; Vičanová *et al.*, 1994; Wilkening, 1979, 1980; Wilkening and Watkins, 1976; Yarborough, 1980, 1981) including our not yet published data from 12 Hungarian caves. In the distribution represented in Fig. 4.1.1, we have included either cave minimum and maximum data (given by authors) or average values. The distribution of the data is close to lognormal.

We note that, taking into account results obtained only by etched track methods, the "annual average" radon concentrations in karstic caves range from 0.1 to 20 kBq m^{-3} with a 2.8 kBq m^{-3} arithmetic average. The lower end of the scale is associated with big chamber volumes or high ventilation rates; the upper end is characterised by closed, badly ventilated places and uranium-rich sediments. The highest annual average was observed in 1992 in the Cserszegtomaj well-cave, Keszthely Mountains, Hungary. The maximum two-monthly average reached 33 kBq m^{-3} in the same cave. On the other hand, Hyland and Gunn (1994) report 46 kBq m^{-3} one-week average from the Peak District region, England. The maximum radon concentration is also reported from the same region, viz. from Giants Hole, Derbyshire Peak District, by Gunn *et al.* (1991). The result of the spot measurement was 155 kBq m^{-3}.

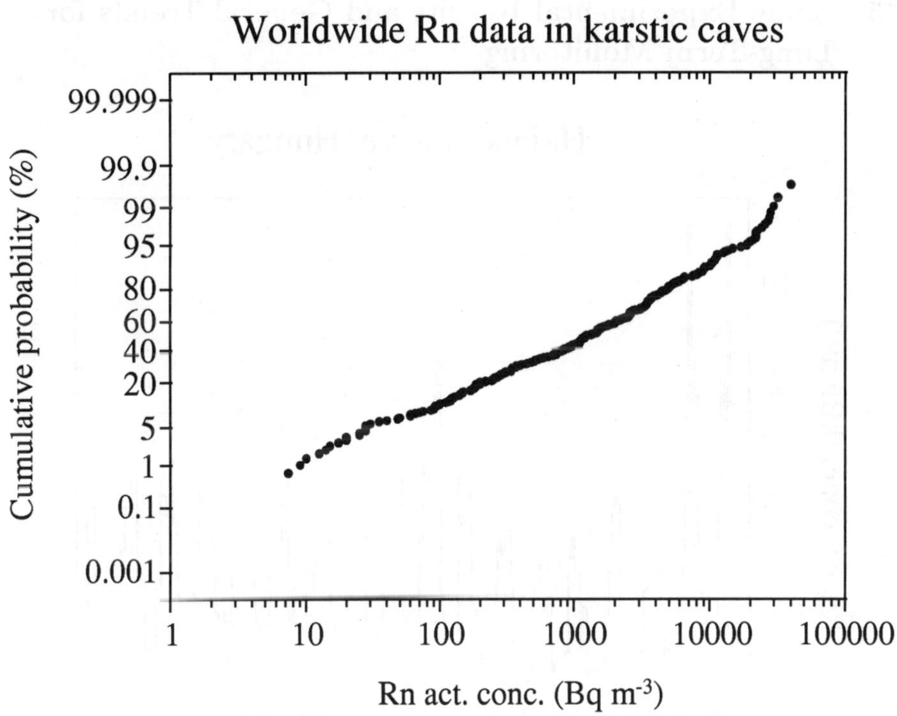

Figure 4.1.1: The distribution of radon activity concentrations reported from different caves worldwide.

4.1.4 Techniques of Radon Measurement in Caves

For the purpose of cave radon measurement, different types of opened and closed diffusion chambers are used, equipped with alpha sensitive polymer track detectors. Somogyi *et al.* (1983) described different types of multi-detector devices suitable for measurement in caves. The most significant improvement in the measuring technique was the thermally stabilised double-wall diffusion chamber (filled with water), which avoided the frequent problem of wetting of detector surfaces. Nowadays several commercially available radon monitors developed for indoor radon measurements are used in cave environment, which, owing to the high relative humidity, may not always be optimal. The typical exposure time ranges from one week to several months.

From the beginning of the 1990's, diffusion chambers equipped with Si based semiconductor detectors and connected to microprocessor controlled data storing units were also introduced into cave studies. Their time resolution is limited by diffusion and detection efficiency; typically it is one hour.

4.1.5 Some Experimental Results and General Trends for Long-Term Monitoring

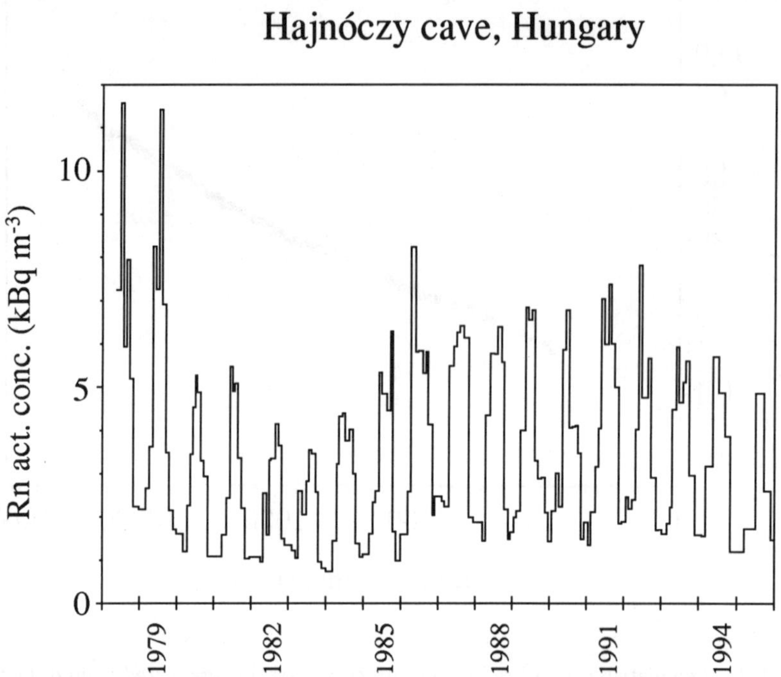

Figure 4.1.2: Radon time series observed by using etched track technique in the Great Hall of the Hajnóczy cave, Bükk Mountains, Bükk National Park, NE Hungary. Each observed value is the average of three simultaneous readings with one month integration time.

It is a general phenomenon that in each cave more or less periodical fluctuations of smaller or larger amplitude can be found around the mean value of the radon activity concentration. The frequency and amplitude distributions of the data are characteristic of the cave and its environment, of the uranium (radium) content of the enclosing rocks and stones and of the extension of that porous surrounding which is in connection with the cave air by the intrusion of atmospheric air and radon traced subsurface fluids.

The longest time series were obtained by us in the Hajnóczy cave, Bükk Mountains, Bükk National Park, NE Hunary (Fig. 4.1.2), which shows a very regular seasonal variation. It is clearly seen in the figure that the seasonal variation is superposed on a long-term change of the mean activity concentration. The same effect was found in a few other Hungarian caves; but it is also reported by Lively and Krafthefer (1995). In Fig. 4.1.3, the annual average radon data from three different types of caves are shown.

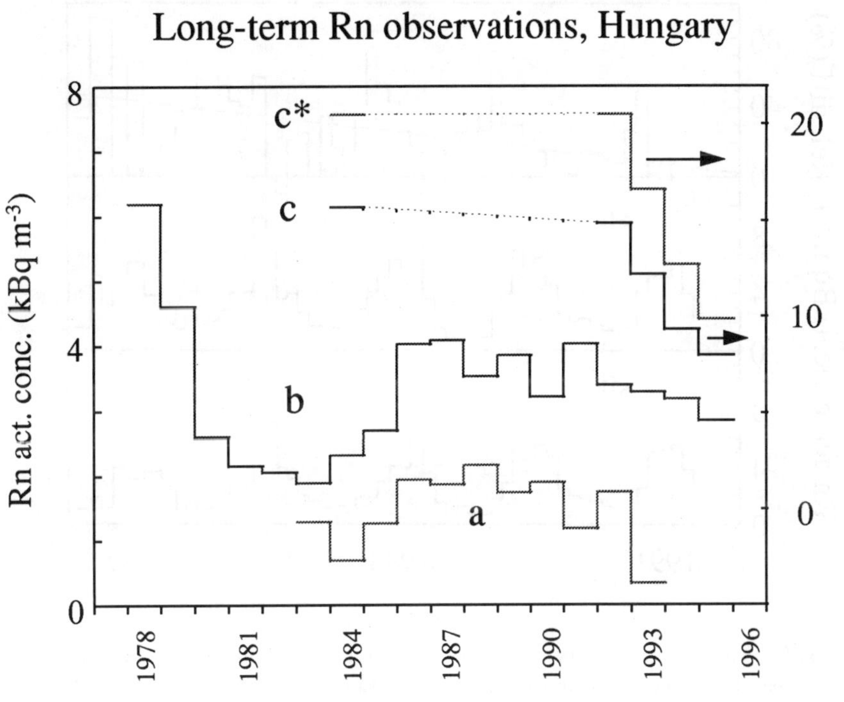

Figure 4.1.3: Long-term change observed in three Hungarian caves of different types. Hajnóczy cave, Bükk Mountains, Bükk National Park, NE Hunary, is dominated by chimney effect winds; Létrási-Vizes cave, Bükk Mountains, Bükk National Park, NE Hungary, is characterised by temperature difference driven convective air motions; and Cserszegtomaj well-cave, Keszthely Mountains, SW Hungary, is influenced mainly by atmospheric pressure changes. The curves represent the middle part of the Létrási-Vizes cave (a); Great Hall in the Hajnóczy cave (b); and cave average (c) and a remote part (c*) in the Cserszegtomaj well-cave.

The long-term change can be amplified selectively at different test sites of a given cave. This effect is markedly shown on time series taken at two different depths of a vertical cave (Sátorkő-puszta cave, Pilis Mountains, N Hungary). The relative difference of the two data series [(a-b)/((a+b)/2) x 100%] shows an increasing tendency with years (see Fig. 4.1.4).

The observed phenomenon displays the effect of slowly changing environmental parameters on radon transport processes. Such an external parameter may be the change in radon emanation power due to the slow change of water content of the cave surrounding, which may reflect the variation in annual precipitation and global meteorological situation influenced by sunspot cycles (Hunyadi *et al.*, 1988).

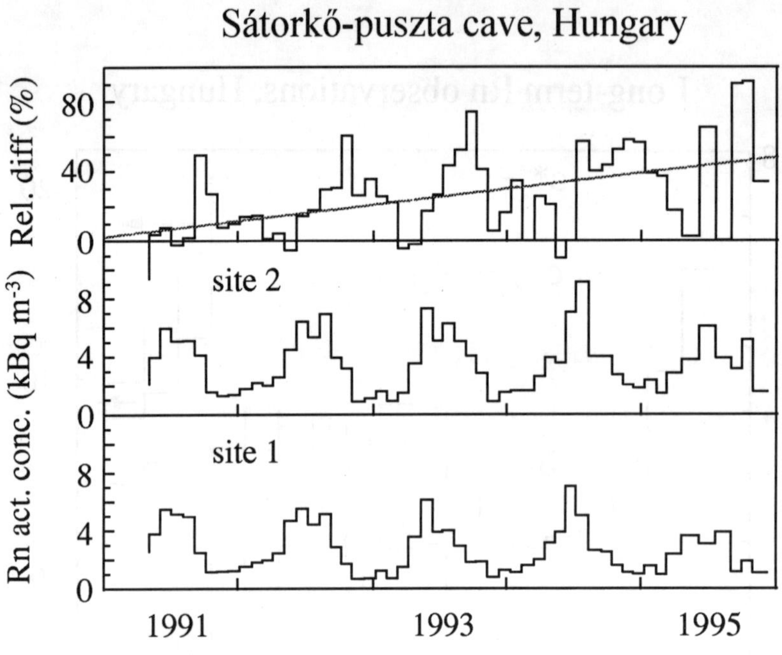

Figure 4.1.4: The radon activity concentrations at two measuring sites and the relative difference between them representing the Sátorkő-puszta cave, Pilis Mountains, N Hungary.

4.1.6 Radon as a Natural Radioactive Tracer of Subsurface Transport Processes

Subsuface natural fluids in the majority of cases carry small amounts of environmental isotopes. The behaviour of these elements, and the variation of their concentration in time and space, is the result of physical, chemical and biological interactions. These elements whose physical properties, concentrations, etc., provide information on flow and kinetics of the carrying substances are called natural tracers.

The application of radon as a natural tracer is not yet common and widespread. Among the natural tracers it would be considered on the one hand as ideal, since it is easily detectable even in small quantities, and does not modify the characteristics of the environment. On the other hand, unfortunately its sources appear everywhere and are spread over in a manner unknown *a priori*. Therefore the interpretation of the concentration data is not straightforward: it needs interdisciplinary expertise of hydrogeologists, geologists, physicists, radiogeochemists, etc. Joint efforts have given results in different fields. The observations of subsurface fluid motions traced by natural radon were followed by new ideas about the basic transport phenomena and, later, by new interdisciplinary applications: as, for example, mapping of active faults; investigations of volcanic and seismic activities; earthquake prediction; hydrogeological research, etc. (Fleischer, 1988).

In the speleology, similarly to the previously mentioned fields, these types of measurements have already found their applications, and they make important contributions to the better understanding of the natural regimes of caves. Cunningham and LaRock (1991) delineated six microclimate zones in Lechuguilla cave, Carlsbad caverns, National Park, New Mexico, using radon grab sampling in conjunction with observed air flow data. Atkinson *et al.* (1983) from a single set of etched track measurements in the Castleguard cave, Columbia Icefields, Alberta, Canada, identified the effect of tributary air currents from larger fissures.

The most common and most apparent phenomenon which was discovered in the majority of the investigated caves throughout the world was the annual change of radon activity concentration. Wilkening and Watkins (1976) identified temperature gradients favourable to vertical convective transport through relatively large openings. They identified as well transport of radon by air movement through cracks and fissures due to pressure gradients (Wilkening, 1980). As karstic caves are situated generally in highly fracturized rocks, such a configuration is favourable for the emergence of air circulation through this fracture system. The strength of such air motions is taken to be proportional, to a first approximation, to dT/f, where dT is the temperature difference between the cave and outside and f is a friction factor characterizing the flow resistance (Atkinson *et al.*, 1983; Quinn, 1988; Wilkening and Watkins, 1976). A typical radon time-series indicating the temperature control is shown in Fig. 4.1.2 (Hajnóczy cave, Bükk Mountains, Bükk National Park, NE Hungary). On approaching deeper parts of the karst, the radon levels are mostly stable, as the strength of air motions decreases due to their 1/f dependence; and owing to saturation effects the amplitude of the changes also decreases (Hakl *et al.*, 1992). Radon levels in deep, complex caves cannot be simply related to outside atmospheric parameters (Cunningham and LaRock, 1991).

The actual value of the radon concentration in the cave is influenced by subsurface fluid motions due to the periodically or randomly changing gradients in the environmental parameters (temperature, pressure, humidity, stresses, ...), and by the radon concentration saturated in the pore spaces of the surrounding rocks. Using long-term temporal and large-scale spatial variation measurements from the analysis of temporal radon variations, the complex interplay of the two traced substances, water and air, can be identified. Such measurements were performed in the Létrási-Vizes cave, Bükk Mountains, Bükk National

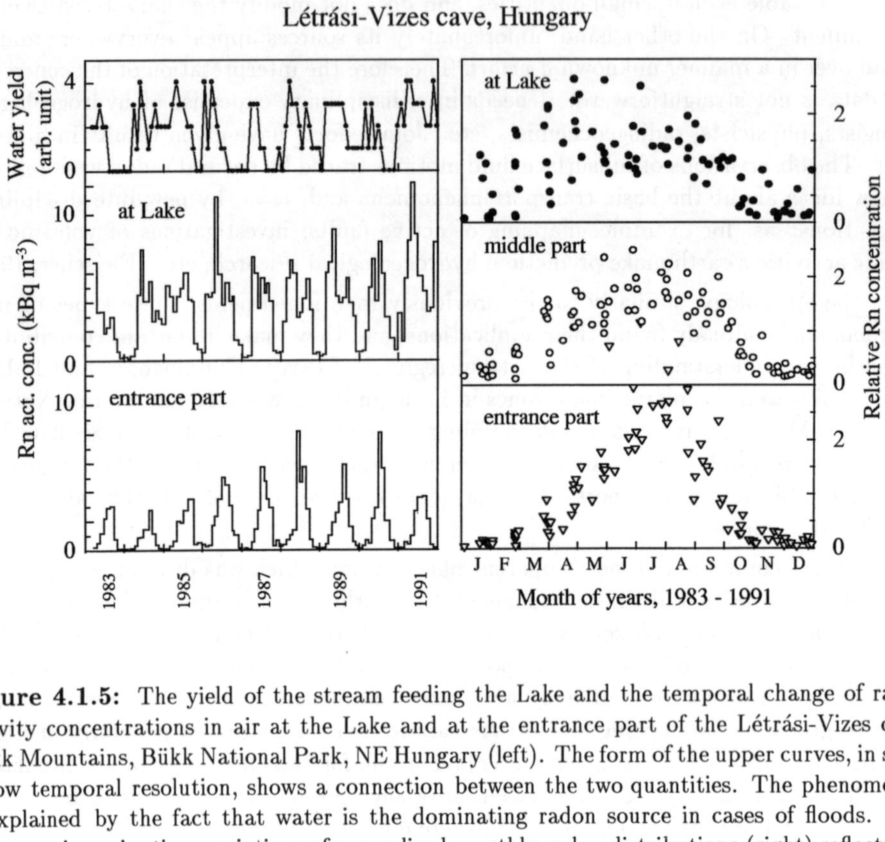

Figure 4.1.5: The yield of the stream feeding the Lake and the temporal change of radon activity concentrations in air at the Lake and at the entrance part of the Létrási-Vizes cave, Bükk Mountains, Bükk National Park, NE Hungary (left). The form of the upper curves, in spite of low temporal resolution, shows a connection between the two quantities. The phenomenon is explained by the fact that water is the dominating radon source in cases of floods. The difference in springtime variations of normalized monthly radon distributions (right) reflects the same phenomenon, simultaneously showing radon regimes of the three distinct microclimatic zones.

Park, NE Hungary (Lénárt *et al.*, 1988). Typical patterns of temporal changes are characterised by summer maxima and winter minima. Similar variations were found at almost all measuring sites, with the exception of that part where more or less continuous water inlets were present. According to the temporal variation, Lénárt *et al.* (1988) differentiated three parts in the cave. In the first part (entrance part) the mean radon activity concentration in air increased from 1 kBq m^{-3} to 2.2 kBq m^{-3}. In the second part it fluctuated around 2.2 kBq m^{-3}. We note that in these two parts there was a slight asymmetry in the way the radon level changes between summer and winter. In the springtime, radon levels in the cave increase gradually, whereas in the fall they decline rapidly. (On increasing the time resolution of the measurements, the same effect can be made much more pronounced, as was found by Lively and Krafthefer (1995) using active techniques.) The highest air radon values, around 4 kBq m^{-3}, were found in the third part, which was characterised by the presence of several streams and a Lake fed by them. In the third part the dominant role of the periodic stream is evident from the long-term measurement (see Fig. 4.1.5); while in the first two parts the change of the radon activity concentration showed close connection with the temperature difference between the cave and outside.

Radon levels in some cases show significant correlation with barometric pressure, but as the characteristic time of pressure changes is mostly less than the exposure time of track detectors, this type of phenomenon can be revealed only by active methods.

4.1.6.1 Modelling of observed features

The underground radon transport can be described by the following transport equation

$$\frac{\partial C}{\partial t} = D_{eff}\Delta C - \nabla(\vec{v}C) - \lambda C + \phi \tag{4.1.1}$$

where C [m^{-3}] is the radon concentration in pore space, D_{eff} [m^{-2} s^{-1}] is the effective diffusion coefficient of radon, $\vec{v}$ [m] is the velocity of the carrying substance, λ [s^{-1}] is the decay constant of radon and ϕ [m^{-3} s^{-1}] is the source term. In the equation, the first term describes diffusion, second term convection, third term decay and fourth term radon sources. For the solution of transport equation, first of all it is necessary to know the velocity field (e.g. Navier-Stokes equation, which by itself is a sufficiently complicated problem); then taking into account the source term, and the initial and boundary conditions C can be determined. The emerging phenomena are determined by the form, shape and structure of the underground void space. The above equations can be generally solved only numerically.

The realisation of radon transport processes sharply depends on the configuration of the interconnected underground cavities. In the case of blind-end systems, atmospheric pressure changes are the main control parameter (Ahlstrand, 1980; Hakl *et al.*, 1995; Wigley, 1967), which are superimposed by convective air exchange due to temperature differences in cave systems with vertical extension. In case of relatively large entrances the convective air exchange due to temperature differences can mostly account for the radon transport process taking place (Wilkening, 1979). In the case of two or more entrance systems, where the other 'entrances' can be complexes of smaller fissures and fractures, chimney effect winds may dominantly govern the radon transport; or in some cases atmospheric winds may do so.

The interpretation of the data can be affected by the presence of these unknown (unassumed) 'entrances'. Yarborough (1980), in a study of nine caves, identified two general types of physical cave configurations that affect airflow patterns and radon concentrations. Type 1 caves have most passages at or above the entrance elevation; Type 2 caves have most passages below the elevation. When the outside temperature exceeded the cave temperature, he found that Type 1 caves exhaled, while stagnation occurred in Type 2 caves; when the outside temperature fell below the cave temperature, both types inhaled. As Type 1 cave is a horizontal mirror of Type 2 cave, the seasonal phenomena should be opposites of each other, so stagnation should occur in Type 1 cave in winter time. Winter time stagnation was identyfied in Pisznyice cave, Gerecse Mountains, Hungary, resulting in high radon concentrations in the cave during the cold season (Hakl, 1994).

The above examples show the problems of interpretation and modelling. In a few special cases, however, the emerging processes are analytically surveyable. These are the cases of idealised two-entrance horizontal flow-through, and one narrow entrance vertical caves. In the case of the horizontal model cave, the second entrance may represent the set of vertical fractures which connect the main passage to the surface through the overburden.

First, let us consider a schematic horizontal cave (see Fig. 4.1.6, left). In this case, owing to the temperature dependence of air density the pressure exerted at the lower end of the cave by the outside air will differ from that inside. This pressure difference in this case is (Atkinson *et al.*, 1983)

$$\Delta p \approx -gh\rho_{in}\frac{\Delta T}{T_{out}} \tag{4.1.2}$$

where ρ_{in} is the density inside air, g is the gravitational acceleration, h is the height difference between the two entrances, ΔT is the temperature difference between the cave and the outside, and T_{out} is the outside temperature. According to this relation, the direction of air flow through the entrance depends on the season; in warm season air flows out of, and in cold season air flows into, the cave.

In the case of narrow-entrance vertical caves, the most effective processes in inducing air motions are the atmospheric pressure changes (Fig. 4.1.6, right). Falling ambient atmospheric pressure drains air from the cave; increasing atmospheric pressure presses air into the cave through the entrance. The volume of air passing the cave entrance

$$\Delta V \approx -\frac{V_{cave}}{p_{out}}\Delta p \tag{4.1.3}$$

where V_{cave} is the cave volume, p_{out} and Δp are the ambient atmospheric pressure and the change in the ambient atmospheric pressure, respectively.

We note that air flows induced by atmospheric pressure changes can be rather quick. At the entrances of giant caves their speed can reach several tens of km h^{-1}.

The above considerations may suggest that advection is the main physical process, which governs the radon transport. This implication depends, however, on the size and density of openings and fractures, as they play an important role in the forming of radon concentrations as well as in the dynamics of transport processes inside them (Hakl, 1993; Scherry and Siegel, 1986). The effect of geometrical parameters can be well investigated in a three-dimensional model of a cylindrical void embedded in the rock matrix. By solving the appropriate transport equation, the transport length (z) may be derived as the parameter which characterises the radon transport inside the cylinder

$$z = \frac{v}{\lambda} \frac{1}{1 + 2(z_d/r_0)R(r_0)} \tag{4.1.4}$$

where v is the air flow velocity, λ is the decay constant of radon, z_d is the effective radon diffusion length in rock matrix, r_0 is the radius of the cylinder, and R is the negative of the ratio of the derivative of the zeroth order modified Bessel function to the zeroth order Bessel function ($i.e. R = -K_0'/K_0$) evaluated at r_0/z_d.

According to this formula two extreme cases can be distinguished: (a) In the region where $z \ll z_d$, diffusion dominates as the radon concentration of the moving air is adapted to its bounding environment at very short distances; and (b) If $z \gg z_d$, advection dominates: radon is transported much more quickly than is characteristic for diffusion.

Physically, the above two conditions mean that the convective processes taking place along the underground openings, and the diffusive processes perpendicular to the walls of channels, compete with each other. The critical speed v_{crit}, of a flowing substance in a channel, therefore, can be defined in such a way, that the characteristic transport lengths in porous matrix and in the channel equals with each other, i.e.

$$\sqrt{\frac{\lambda}{D_{eff}}} = \frac{\lambda}{v_{crit}} \left(1 + 2\frac{z_d}{r_0} R(r_0)\right) \tag{4.1.5}$$

Substituting typical values for diffusion length for limestone (0.2-0.4 m) yields $v_{crit} = 0.2$-20 mm h^{-1}, depending on the size of the channel (0.002-2 m).

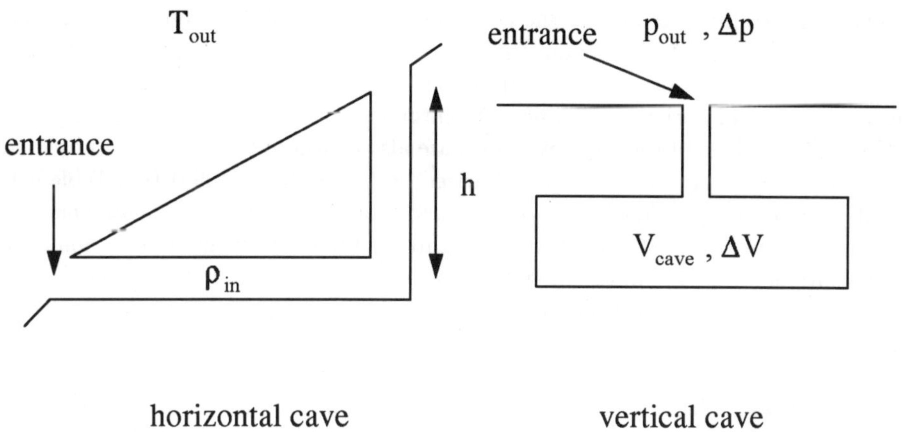

Figure 4.1.6: Schematic horizontal and vertical caves; h is the elevation difference between the upper and lower entrances (horizontal cave), ρ_{in} is the air density of inside air, T_{out} is the outside temperature, p_{out} is the ambient outside pressure, V_{cave} is the cave volume and ΔV is the volume of the air passing the vertical cave entrance in case of Δp change in the ambient outside pressure (not to scale).

A more realistic model of fracturized karstic overburden above a horizontal cave is the model of parallel voids connecting the main passage to the surface (Géczy *et al.*, 1989; Scheidegger and Liao, 1972). This model is a self-evident synthesis of the idealised two entrance horizontal cave model and cylindrical void embedded in the rock matrix model. In the warm season, according to the Δp pressure difference, air will flow into the cave from the direction of the fracture system (and will flow out of the cave through the entrance at lower altitude). If the velocity of the flow is 2-3 times higher than the critical value, the flow will be able directly to wash out radon from the fractures. If the velocity is less than the critical velocity, then the diffusion profiles in the rock matrix are shifted in the first step, and therefore increased radon flux can reach the cave passage only with delay. The level and temporal variation of radon concentration in the cave gallery therefore depend on the density and size of the fractures. In the cold season, the direction of the flow is reversed. From the direction of the lower entrance fresh outside air flows into the cave forming considerably lower radon concentration levels.

By the application of etched track technique alone, the above-mentioned two competitive processes can hardly be discriminated between, owing to the limited time-resolution of etched track technique. On applying the combination of active electronic and passive etched track techniques, however, most of the hidden phenomena can be discovered (see subsubsection 4.1.8.1).

4.1.7 Radon Risk Inside and Above the Caves

It is known that high radon exposure results in excess lung cancer mortality rate among miners. A majority of experts believe that elevated radon levels in homes induce excess lung cancer mortality rate among the general public, too. High radon levels that occur in caves may raise the question whether any people are at risk or not. Most concerned groups are the cavers and tour guides. Moreover, some cold karstic caves are used for therapeutic treatment of patients suffering from chronic respiratory diseases. These patients, and the staff members of the therapy teams, are also concerned about being subjected to higher risks. Hungarian data were summarised by Csige *et al.* (1996) (see Table 4.1.1). Currently, in Hungary radon is continuously monitored in all the five caves where therapeutic treatment of bronchial asthma and chronic bronchitis is going on. Some caves open to tourists are also monitored. In the case of the best known of these (Baradla cave, Aggtelek Karst, Aggtelek National Park, Hungary), the radon doses of tour guides were estimated for the years 1990-1994.

According to Navrátil *et al.* (1994) the range of patients' doses from nine different therapeutic caves in Europe is 0.07-1.32 mSv, which well fits with the above table. Based on internal working time reports, annual effective doses for therapy staff members are about 0.4 mSv in Abaliget cave, 0.12 mSv in Szent István cave and 6 mSv in Béke cave.

Annual personnel exposures also cover a wide range. Tour guides in Abaliget cave receive about 12 mSv annually, Cigna and Clemente (1981) cite Yarborough, who has found 0.005-19.9 mSv per year for seven different US caves; whereas Nikodemová (1995) reports effective dose rates 0.17-4.05 mSv per month for personnel in Slovakian caves.

Table 4.1.1: Summary of radon doses received on the course of speleotherapy in Hungary (Csige *et al.*, 1996).

Cave name	Period	Sample size	Annual effective dose (mSv)	
			Mean, (Range)	Geometric mean, (Geometric STD)
Abaliget[1]	1994	127 patients	0.54, (0.03-1.26)	0.40, (0.77)
Béke[2]	1994 summer	56 patients	1.91, (1.86-1.97)	1.91, (0.02)
Szemlőhegy[*,3]	1990-1992	229 patients	0.85, (0.10-5.00)	0.62, (0.76)
Szent István[2]	1994	360 patients	0.06, (0.01-0.17)	0.04, (0.88)
Tapolca[4]	1994	481 patients	0.87, (0.04-2.19)	0.45, (1.32)
Baradla[2]	1990-1994	12 tour guides	2.66, (0.12-5.55)	2.13, (0.80)

[1] - Mecsek Mountains, S Hungary;
[2] - Aggtelek Karst, Aggtelek National Park, NE Hungary;
[3] - Buda Mountains (at Budapest);
[4] - Balaton Highland;
* - In the case of Szemlőhegy cave cumulative effective doses are given over 1990-1992, as not all the patients attended the therapeutic courses each year.

Most active cavers may be exposed to even higher doses; so personal radon dosimetry is highly recommended for those people. One of the authors (J. Hakl) has measured his own annual effective dose due to radon inhalation while working in different caves, and found that it was higher than 30 mSv in 1992. Even higher personal doses can be received in caves with unusually high radon concentrations. Hyland and Gunn (1994) estimated that it took 33 hours to reach 15 mSv limit in some caves of North Pennines. Two important radiological consequences follow from the periodical behaviour of subsurface air circulation in karst. First is the seasonally varying underground radon level which results in a large difference between summer and winter values, and hence also in the radon exposures received by cave visiting people.

The other consequence of the seasonally directed transport phenomenon is that variation in radon exhalation can also be expected on karstic terrains seasonally. A very convincing experimental result of the phenomenon is represented in Fig. 4.1.7, where radon time-series measured inside the Hajnóczy cave, Hungary (with minima in the cold winter season), and in a slit above the cave (with winter maxima) are shown. Similar winter maxima were found on several karstic terrains of Hungary (Hakl *et al.*, 1992). Elevated radon levels already observed in houses in summertime (Wilson *et al.*, 1991) and wintertime (Gammage *et al.*, 1992) are attributed to this phenomenon, showing that the season of maxima depends on the position of the house with respect to the underground air circulatory system. The inverse correlation of radon levels inside a cave and in a house above the cave was found directly by Lively and Krafhefer (1995).

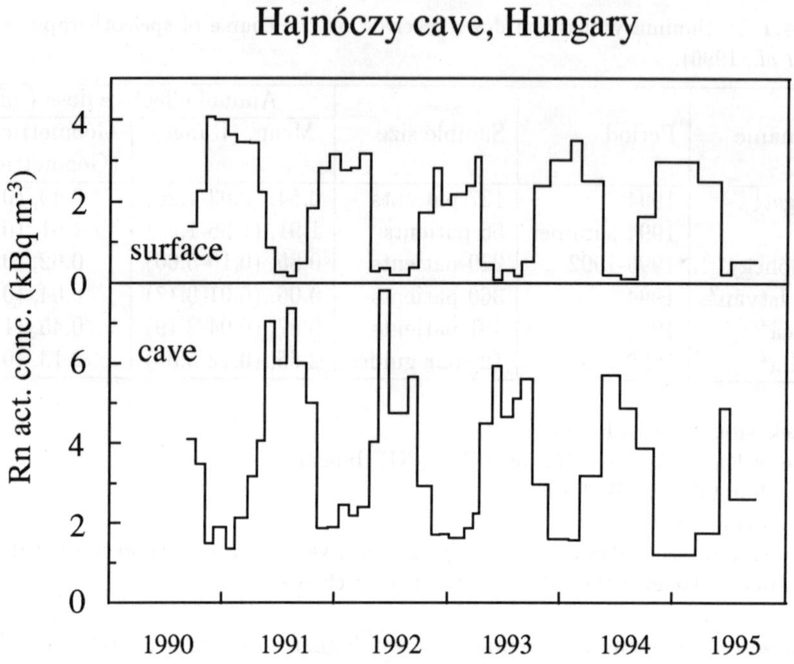

Figure 4.1.7: Push-pull type change of seasonal radon activity concentrations observed in the Great Hall of the Hajnóczy cave, Bükk Mountains, Bükk National Park, NE Hungary, and in a surface slit above the cave.

4.1.8 Future Developments

4.1.8.1 Combination of etched track and real-time detection techniques

Combining etched track and real-time detection techniques, spatial and detailed temporal variation of radon concentration can be obtained. The information content of data can be increased substantially, but the choice of proper measuring places is also desirable. By choosing proper and right number of test sites, the functioning of the subsurface system can be understood in greater detail. Such type of combination of continuous and integrating radon measurements was performed in the Vass Imre cave situated in the Aggtelek Karst, Aggtelek National Park, NE Hungary. This cave is very similar to the model cave surmised by the air circulation model. It has one entrance, and is situated practically horizontally with no visible vertical connection to the surface. In the middle of the cave there is a syphon, which was closed by water several times during the observation period (1987-1993). When the syphon is open, there is continuous air flow either in or out from the cave through the entrance (winter and summer, respectively). When the syphon is closed, there is no measurable air flow through the entrance.

Table 4.1.2: Mean radon activity concentrations of air (kBq m^{-3}) in the Vass Imre cave, Aggtelek Karst, Aggtelek National Park, NE Hungary (1987-93).

	Before syphon		After syphon		Annual mean
	Winter	Summer	Winter	Summer	
Syphon opened (air flow exists)	1.2	8.5	1.2	8.5	4.8
Syphon closed ('no' air flow)	2.5	4.0	3.5	3.1	3.3

The radon concentration is being measured with etched track detectors since 1987 at 12 sites located at equal intervals along the cave passage, and in 1991 three more continuous real-time radon monitors were installed at characteristic points of the cave.

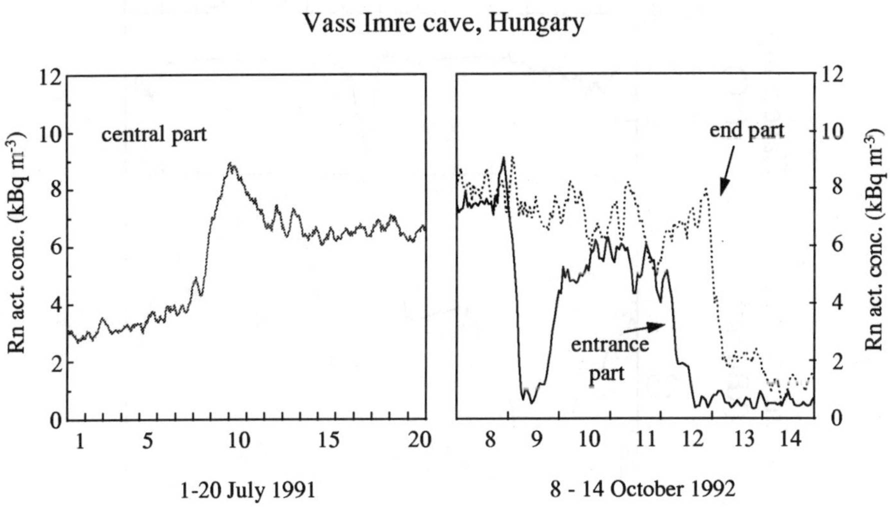

Figure 4.1.8: The real time records of radon time-series from Vass Imre cave, Aggtelek Karst, Aggtelek National Park, NE Hungary, at the times of syphon opening (left) and arrival of early cold fronts during fall (right).

Etched track detectors changed monthly showed elevated levels in the summer and lower levels in the winter: the temporal change is of a smooth sinusoidal type. The horizontal radon distributions along the main cave passage were uniform. The characteristic radon values found are summarised in Table 4.1.2.

The penetrability of the syphon affects markedly the mass of infiltrated air through the cave system by controlling the strength of the emerging air flows. When the syphon is closed, the restriction of the air flows has a different effect on the forming of radon levels, which depends on the season. In summertime the restricted air flow decreases, but in wintertime it increases, the mean radon levels in the cave, resulting in an overall drop of 30% in the annual mean radon concentration (see Table 4.1.2). This asymmetric effect can be explained on the basis of the air circulation model (Géczy *et al.*, 1988) and is in agreement with the results of numerical calculations (Holford *et al.*, 1993) showing the increase of mean radon levels due to periodically changing flow conditions.

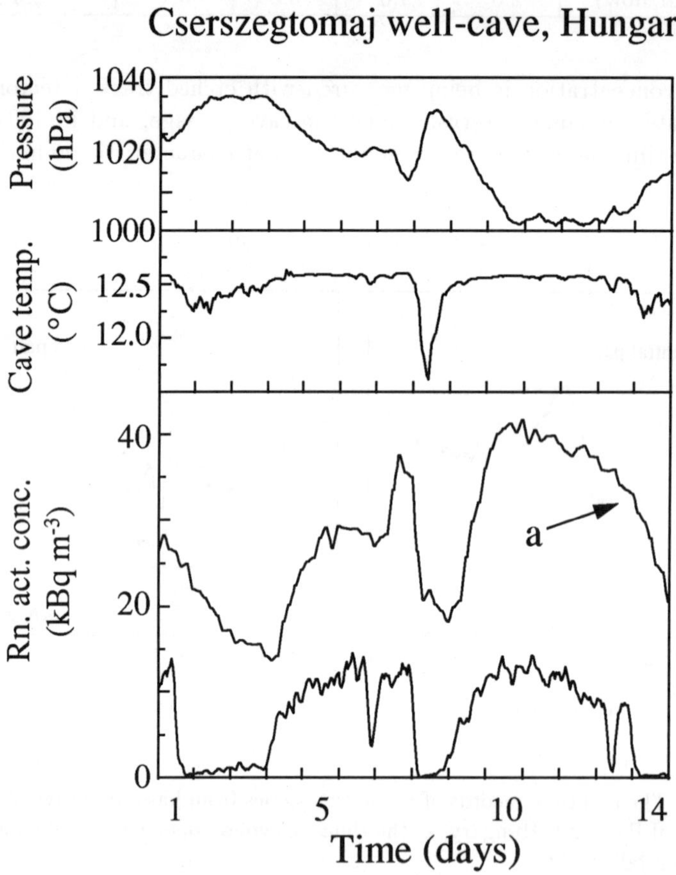

Figure 4.1.9: Temporal variation of temperature, pressure and radon activity concentration at the entrance part of the Cserszegtomaj well cave, Keszthely Mountains, SW Hungary. On the figure we also displayed the change of radon level measured deeper inside the cave (a).

On analysing temporal changes in radon time-series, we found advection as the dominant transport process. This is also shown on the transient part of the radon record from the Vass Imre cave, Aggtelek National Park, NE Hungary, which corresponds to a summertime syphon opening (see Fig. 4.1.8 left). The shape of the curve, on comparing it to the results of numerical calculations on radon exhalation from porous media (Janssens *et al.*, 1984), corresponds to the case when a sudden pressure drop results in instantaneous air-flow development. The pressure drop effect was also observed experimentally during a planned syphon closing and opening (Kérdő, 1994). On the other hand, the absence of strong daily changes in the radon record, which would correspond to the observed daily air flow fluctuations (Holl, 1993), shows the strong smoothing effect of diffusion, which can be due to relatively undeveloped fracture system of this cave. (We note that in caves located in more karstified environment, more or less pronounced daily fluctuations in radon records can be observed.)

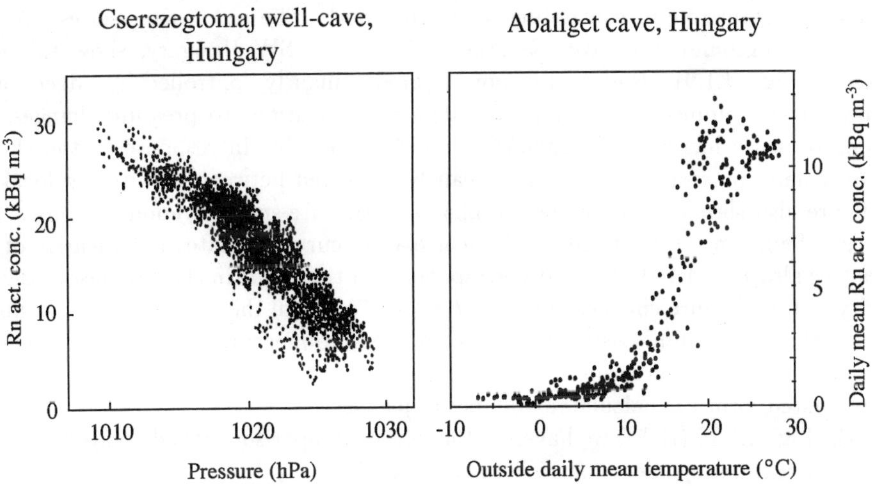

Figure 4.1.10: Radon activity concentration vs atmospheric pressure observed at one of the five continuous observation stations in the Cserszegtomaj well-cave, Keszthely Mountains, SW Hungary (left). The 'data cloud' corresponds to the 2 month section of the continuous data record. The inverse pressure control on radon levels is well seen in the figure. On the right-hand part of the figure we have displayed radon levels vs outside temperature observed in the Abaliget cave, Mecsek Mountains, S Hungary. The S-type dependence of radon concentration on the outside temperature is typical of caves dominated by chimney-effect winds.

Similar conclusion can be drawn by analysing passive etched track radon measurements. Because the radon concentration after the closed syphon does not depend on the season, the radon diffusion length in the rock matrix can be estimated, taking into account the geometry of the cave passage. Knowing the diffusion length, the air-flow velocity through the fracture system can be further appraised, using the summer data. The resulting speed is in the range (0.5-1)$\cdot 10^{-4}$ cm s^{-1}. This speed is less then the critical speed, which points to the diffusion control, as was deduced from analysing the real-time data.

In contrast to the above phenomena, the cave radon concentration 'shut down' is very quick in autumn, and is controlled purely by advection. An early cold front of short duration caused a fall in radon records at the first and second measuring sites, but it did not affect the end of the cave. High radon levels were recovered for two days by reversing the air-flow. Finally, permanent low radon values were formed in the cave air after the arrival of another, stronger, cold front. The time-difference between the radon falls corresponded to an air transport velocity of about 50 m h^{-1} along the main cave passage (see Fig. 4.1.8, right).

In some case, in caves with small entrances the changes in outside temperature have only a small effect on radon concentration inside the cave, and only on an annual scale. In contrast, the atmospheric pressure has strong influence on radon level: pressure changes play an important role in controlling radon transport. The radon time-series, observed in the Cserszegtomaj well-cave, Keszthely Mountains, SW Hungary, show this effect markedly (Fig. 4.1.9). Radon transport is predominantly controlled by advection, as is shown by the immediate response of radon concentrations to pressure changes. Decreasing pressure increases the radon level and conversely. In response to the changes in atmospheric pressure, radon 'pulses' can be observed between monitoring locations. The figure also shows the difference in phases of the site-specific radon responses. The hysteresis found in the pressure - radon correlation curve indicates nonlinearity of processes (see also Fig. 4.1.10, left). It corresponds to a time shift in the response of air-flow velocity to the pressure changes (Wigley *et al.*, 1967). From the data it is also possible to estimate the cave volume, using simultaneous pressure measurements (and utilising the known volume of the vertical entrance well). We found it to be 8-10 times larger than was estimated form the calculated volumes of passages.

In the Fig. 4.1.10 (right) we have, for the sake of comparison, also displayed the radon response of a chimney-effect controlled cave system.

4.1.8.2 Radon monitoring network

From the previous subsections it is evident that, from radon records, more information can be deduced unambiguously, taking account of other environmental parameters. This type of multiparameter approach is an emerging indispensable task in environmental transport studies.

As the role of the geological environment and the necessity of multiparameter studies has recently been recognised, there is an increasing need for instruments suitable for long-term field measurements. Owing to the complexity of the radon transport problems, these above-mentioned studies require numerous, continuously operating stations. From the scientific and economic point of view, an independent working ability is more and more frequently required (the usual on-line measurement and data transmission in real time are prohibited by cost and insufficient infrastructure).

These efforts initiated, in Hungary, the development of the prototype of the Dataqua field multiparameter radon monitor in the beginning of the 1990's, which opened new dimensions in radon research. In many places in Hungary, such complex radon monitoring stations have now been installed. At the test sites, radon concentrations together with other environmental parameters are continuously recorded with a sampling frequency of one measurement per hour. The main aims in operating these monitoring stations are:

- to indicate whether or not there are extremely high radon concentrations in places where people live or work (dwellings, cellars, caves, coal mines, uranium mines),
- to obtain radon output activity of some objects of interest (uranium mine, mill and tailings, coal mine and power plant, coal ash tailings, etc.) into the environment,
- to study the relationship between the variation of radon concentration and other environmental parameters (air temperature, barometric pressure, water level variation, etc.),
- to reveal the connection - if any - between radon exhalation and geodynamical events (earthquakes, fault movements, rock explosions in the mine, etc.).

The first field results proved the usefulness of the radon monitoring system. The advantage of this system, in comparison with the traditional solid state nuclear track detector technique alone, is obvious: the time resolution of observations has revealed previously unknown details of temporal variations, and by interpreting them one can define new approaches of physical models describing the objects of interest. Some of the results have been highlighted in foregoing sections. We note, however, that enormous efforts (including software and hardware) are required to analyse and interpret the great amount of time-series recorded (some 10,000 data per year per monitor).

4.1.9 Concluding Remarks

Research activity has been developed in the field of geologically related radon transport studies. As radon is an easily measurable radioactive element, it is a very convenient natural tracer for the transport processes taking place at the interfaces of the atmosphere, hydrosphere and lithosphere. It brings information from the depths of the earth, which may be helpful in the prediction of earthquakes, lava eruptions and other geodynamical events as mine outburst. Because of the improved time resolution, the proportion of active automatic radon measuring methods is continuously increasing in this field.

By interpreting the observed long-term radon time series, new results will be obtained from intercomparison of the dynamical behaviour of radon concentrations at different geographical and geological sites. In this way, general understanding of environmental (radon) transport processes will be improved.

The measurements in karstic caves will rely on examining the natural ventilation regimes. Because the interrelations between surface and cave environments are affected by flowing air masses, the exploration of transport processes is a precondition to gathering knowledge about the development and changes in the cave microclimate and about the propagation mechanism of external effects. In this way, these results should also find application in the protection of the karst environment.

The radon values should be compared with environmental parameters, e.g. microclimate, temperature, humidity, ventilation rates etc. The spin off would be to understand emanation control in these environments, and as a result this would help in controlling the radiation hazard.

References

Ahlstrand G.M. (1980) Alpha radiation levels in two caves related to external air temperature and atmospheric pressure. *NSS Bulletin* **42**, 39-41.

Amano H, Atsushi K. and Takeshi M. (1985) Simultaneous measurement of Rn and its progeny in cave air by liquid scintillation techniques and α-spectrometry. *Health Phys.* **49**, 3, 509-511.

Atkinson T.C., Smart P.L. and Wigley T.M.L. (1983) Climate and natural radon levels in Castleguard cave, Columbia icefields, Alberta, Canada. *Arctic and Alpine Research* **15**, 487-502.

Borau J., Gonzalez A., Espinosa G. and Golzarri J.I. (1993) Measurements of radon levels inside Mexican caves. *Nucl. Tracks Radiat. Meas.* **22**, 287-288.

Burkett C. (1993) Radon levels in limestone caves in central Pennsylvania. B.S. Thesis, University Park, Pennsylvania State University, Pa.

Cappa G., Cigna A.A., Tomassino L. and Torri G. (1995) Review of the radon concentration measurements in Italian caves and some aspects of radiation protection. In *Int. Symp. Show Caves and Environmental Monitoring* (ed. A. A. Cigna), Cuneo, Tipografia Saviglianse, pp. 149-161.

Cigna A.A. and Clemente G.F. (1981) Radiation hazards in natural caves. In *Proc. 8th Int. Cong. Speleology*, Bowling Green, Kentucky, Vol. **2**, pp. 420-423.

Csige I., Hunyadi I., Hakl J., Géczy G., Várhegyi A., Lénárt L. and Törőcsik, I. (1996) Radon exposures in caves in Hungary. In *Radiation Protection in Neighbouring Countries in Central Europe* (ed. D. Glavič-Cindro), J. Stefan Institute, Ljubljana, pp. 92-95.

Collar P.D. and Odgen A.E. (1991) Radon in soils, caves, and homes of north central Tennessee. In *The 1990 Int. Symp. on Radon and Radon Reduction Technology*, Vol. **3**, paper No. C-VI-3, pp. 6.21-6.33.

Cunningham K.I. and LaRock E.J. (1991) Recognition of microclimate zones through radon mapping, Lechuguilla cave, Carlsbad caverns, National park, New Mexico. *Health Phys.* **61**, 493-500.

Dezső Z. (1995) unpublished data.

Fernandez P.L., Quindos L.S., Soto J. and Villar E. (1984) Radiation exposure levels in Altamira cave. *Health Phys.* **46**, 2, 445-447.

Fleischer R.L. (1988) Radon in the environment - opportunities and hazards. *Nucl. Tracks Radiat. Meas.* **14**, 421-435.

Gammage R.B., Dudney C.S., Wilson D.L., Saultz R.J. and Bauer B.C. (1992) Subterranean transport of radon and elevated indoor radon in hilly karst terrains. *Atmospheric Environment*, **26A**, 12, 2237-2246.

Géczy G., Csige I. and Somogyi G. (1988) Air circulation in caves traced by natural radon. In *Proc. of the 10th Int. congress of Speleology* (ed. A. Kósa), Hungarian Speleological Society, Budapest, Vol. **2**, pp. 615-617.

Gunn J., Fletcher S. and Prime D. (1991) Research on radon in British limestone caves and mines, 1970-1990. *Trans. British Cave Res. Assoc.* **18**, 85-89.

Hakl J., Hunyadi I., Csige I., Géczy G., Lénárt L. and Tőrocsik I. (1992) Outline of natural radon occurrences in Hungary. *Radiat. Prot. Dosim.* **45**, 183-186.

Hakl J. (1993) Radon transport studies. PhD Thesis, Debrecen (In Hungarian).

Hakl J. (1994) Radon Transport Studies in Caves of Gerecse Mountains. *Limes* **94/2**, 89-92 (in Hungarian).

Hakl J., Hunyadi I. and Várhegyi A. (1995) The study of subsurface transport dynamics based on monitoring in caves. In *Gas Geochemistry* (ed. C. Dubois), Science Reviews, Nortwood, pp. 391-398.

Holford D.J., Schery S.D., Wilson J.L. and Phillips F.M. (1993) Modelling of Radon Transport in Dry, Cracked Soil. *J. Geophys. Res.* **98**, B1, 567-580.

Holl B. (1993) personal communication.

Hunyadi I., Hakl J., Lénárt L., Géczy G. and Csige I. (1991) Regular subsurface radon measurements in Hungarian karstic regions. *Nucl. Tracks Radiat. Meas.* **19**, 321-326.

Hyland R. and Gunn J. (1994) International comparison of cave radon concentrations identifying the potential alpha radiation risks to British cave users. *Health Phys.* **67**, 2, 176-179.

Hyland R., Gunn J. and Fletcher S. (1994) Preliminary results from the national cave radon survey conducted in Great Britain 1991-1992. In *Beiträge zu Speläeotherapie und Höhlenklima*, II (eds. D. Řičný, B. Sandri, H. Trimmel), Verband österreichescher Höhlenforscher, Wien, pp. 33-38.

Janssens A., Raes F. and Poffijn A. (1984) Transients in the exhalation of radon caused by changes in ventilation and atmospheric pressure. *Radiat. Prot. Dosim.* **7**, 81-86.

Kérdő P. (1994) unpublished data.

Kobal I., Smodiš B. and Škofljanec M. (1986) Radon-222 air concentrations in the Slovenia karst caves of Yugoslavia. *Health Phys.* **50**, 6, 830-834.

Kobal I., Smodiš B., Burger J. and Škofljanec M. (1987) Atmospheric ^{222}Rn in tourist caves of Slovenia, Yugoslavia. *Health Phys.* **52**, 4, 473-479.

Kobal I., Ančik M. and Škofljanec M. (1988) Variations of ^{222}Rn air concentration in Postojna cave. *Radiat. Prot. Dosim.* **25**, 3, 207-211.

Kaufmann O., Charlet J-M. and Quinif Y. (1995) Distribution des concentrations en radon les cavités karstiques Belges. In *Gas Geochemistry* (ed. C. Dubois), Science Reviews, Nortwood, pp. 49-62.

Lénárt L., Somogyi G., Hakl J. and Hunyadi I. (1988) Radon mapping in caves of eastern

Bükk region. In *Proc. 10th Int. Congress of Speleology* (ed. A. Kósa), Hungarian Speleological Society, Budapest, Vol. **2**, pp. 620-623.

Lively R.S. and Krafthefer B.C. (1995) ^{222}Rn variations in Mystery cave, Minnesota. *Health Phys.* **68**, 4, 590-594.

Massen F., Kies A. and Schintgen G. (1995) Problems and results of long-time climatological measurements in confined maze cave (Moestroff cave, Luxembourg). In *Int. Symp. Show Caves and Environmental Monitoring* (ed. A. A. Cigna), Cuneo, Tipografia Saviglianse, pp. 85-100.

Michel J. (1987) Sources. In *Environmental radon* (eds. C. R. Cothern and J. E. Jr. Smith), Plenum Press, New York, pp. 98-108.

Marx G. (1996) Shelf tectonics and radioactivity. *Fizikai Szemle* **1996/3**, 77-84 (In Hungarian).

Miki T. and Ikeya M. (1980) Accumulation of atmospheric radon in calcite caves. *Health Phys.* **39**, 2, 351-354.

Navrátil O., Surý J., Štelcl J. and Sas D. (1993) The results of one year long radiation monitoring at speleotherapeutical clinic at Ostrov u Macochy. *Chemické listy* **87**, 9A, 186-187 (In Czech).

Navrátil O., in D., Sandri B., Slavik P., Sur J. and Štelcl J. (1994) Die Radioaktivität in der Höhlenatmosphre. In *Beiträge zu Spelleotherapie und Höhlenklima*, II (eds. D. R̂ičný *et al.*), Verband österreichescher Höhlenforscher, Wien, pp. 47-56.

Nazaroff W.W., Moed A.M. and Nero A.V.Jr. (1988) Soil as a source of indoor radon: generation and entry. In *Radon and its decay products in indoor air* (eds. W. W. Nazaroff, A.V.Jr. Nero), John Wigley & Sons, New York, pp. 69-73.

Nikodemová D. (1995) Personal communication.

Papastefanou C., Monolopoulou M., Savvides E. and Charalamous S. (1986) Natural radiation dose in Petrolana cave. *Health Phys.* **50**, 2, 281-286.

Quinn J.A. (1988) Relationship between temperatures and radon levels in Lehman caves, Nevada. *NSS Bulletin* **50**, 59-63.

Quinn J.A. (1990) Factors that may affect radon daughter concentrations in Whiple cave and Goshute cave, Nevada. *NSS Bulletin* **52**, 104-109.

Roda S., Ščuka J. and Roda S. jun. (1994) Untersuchungen über emanationsercheinungen und sekundäre aureolen natürlicher Radioisotope in Höhlen der Slowakei. In *Beiträge zu Speläleotherapie und Höhlenklima* II (eds. D. Řičný *et al.*), Verband österreichescher Höhlenforscher, Wien, pp. 39-46.

Scheideggger A.E. and Liao K.H. (1972) Thermodynamic analogy of mass transport processes in porous media. In *Fundamentals of Transport in Porous Media*, IAHR, Elsevier, Amsterdam, pp. 3-13.

Scherry S.D. and Siegel D. (1986) The role of channels in the transport of radon from the soil. *J. Geophys. Res.* **91**, B12, 12366-12374.

Seymore F.W., Ryan R.M. and Corelly J.C. (1980) Radon and radon daughters in Howe caverns. *Health Phys.* **38**, 5, 858-859.

Solomon S.B., Cooper M.B., O'Brien R.S. and Wilkinson L. (1992) Radon exposure in a limestone cave. *Radiat. Prot. Dosim.* **45**, 171-174.

Somogyi G., Paripás B. and Varga Zs. (1983) Measurement of radon, radon daughters and thoron concentrations by multi-detector devices. *The Nucleus* **20**, 3-4, 51-55.

Surbeck H. (1990) Rn-222 transport from soil to karst caves by percolating water. In *Proc. of the 22nd Congress of IAH* (ed. A. Parriaux), International Association of Hydrogeologists, pp. 349-355.

Várhegyi A. and Hakl J. (1984) A silicon sensor based radon monitoring device and its use in environmental geophysics. *Geophysical Transactions* **39**, 4, 289-302.

Vibranov P. and Velikov V., Vasiliev G. (1975) Study on the radioactivity of caves in Bulgaria. *Higenia i zdraveopazvanie* **18**, 5, 491-495 (in Bulgarian).

Vičanová M., Ďurčik M., Havlik F. and Nikodemová D. (1994) Annual variations of equilibrium equivalent concentrations of radon air in Slovak caves. In *Proc. of Abstracts and Papers from XVIII. Radiohygienic Days* (ed. J. Thomas), Center of Medical Education, Prague, pp. 132-134.

Wilkening M.H. and Watkins D.E. (1976) Air exchange and ^{222}Rn concentrations in the Carlsbad caverns. *Health Phys.* **31**, 2, 139-145.

Wilkening M.H. (1979) Radon 222 and air exchange in the Carlsbad caverns. In *Proc. First Conference on Scientific Research in the National Parks* (ed. R. M. Linn), US Department of the Interior, National Park Service Transactions and Proceedings Series, No. 5, Vol. II, pp. 687-690.

Wilkening M.H. (1980) Radon transport process below the earth's surface. In *Proc. Natural Radiation Environment III* (eds. T. F. Gessell and W. M. Lowder), Technical Information Center, United States Department of Energy, Springfield, VA, Vol. **1**, pp. 90-104.

Wigley T.M.L. (1967) Non-steady flow through porous medium and cave breathing. *J. Geophys. Res.* **72**, 12, 3199-3205.

Wilson D.L., Gammage R.B., Dudney C.S. and Saultz R.J. (1991) Summertime elevation of ^{222}Rn levels in Huntsville, Alabama. *Health Phys.* **60**, 2, 189-197.

Yarborough K.A. (1980) Radon- and thoron-produced radiation in National Park Service caves. In *Natural Radiation Environment III.*, Vol. **2** (eds. T. F. Gessel and M. L. Wayne), Springfield, Va., NTIS, U.S. Dept. Energy Rept., pp. 1371-1395.

Yarborough K.A. (1981) Alpha radiation in natural caves. In *Int. Conf Radiation Hazards in Mining: Control, Measurement and Medical Aspects* (ed. M. Gomez), Am. Inst. Mining, Metallurgical and Petroleum Engineers, Soc. Mining Engineers, New York, pp. 642-653.

R. L. Fleischer[1]

4.2 Radon and Earthquake Prediction

4.2.1 Introduction

Correlations between changes of radon in the ground and earthquakes have raised expectations that radon can be used to help predict earthquakes. The needs of proper prediction are knowledge of the size of an event to come, and of where and when it will occur. Presently routes to earthquake prediction are at most in their infancy; the use of radon shares in being at an early stage of development.

Many diverse measurements have been interpreted as providing precursors to earthquakes. Rikitake (1976) has listed 15 such parameters - including strain, tilt, foreshocks and foreshock sequences, seismic wave velocity changes, fault creep, underground water and oil flow, earth resistivity, earth currents, geomagnetic effects, and radon variations. Major encouragement for the usefulness of radon was provided by Ulomov *et al.* (1967) and Ulomov and Marashev (1968) in finding increasingly elevated radon in ground water from a deep well in Tashkent beginning six years prior to the magnitude (M) 5.2 earthquake of April 15, 1966. The nature of the radon variations was particularly promising for predictions: The radon concentration increased at an accelerating rate, leveled off a few months before the event, and dropped sharply after it occurred. The radon was from a well that penetrated 1 to 2 km in the epicentral region of the event, which nucleated at 8 km depth. The discharge rate from the well is not known to this author.

Fig. 4.2.1 gives an example of a more recent similar observation made in Alaska using track etching. Here the details are more completely accessible in English (Fleischer and Mogro-Campero, 1985). An increase in radon concentration by a factor of 40 was recorded prior to an M = 6.3 event that occurred 178 km from the measuring site, and then the radon returned to nearly its original value.

Certain inferences about possible and impossible mechanisms of such results are fairly straightforward. First, it is not possible for excess radon to have flowed 180 km from the earthquake focus - given the permeabilities and the finite stresses and stress gradients within rocks. Second, stresses are far too low from pre-earthquake strains to have caused rock fractures to release radon at a distance of 180 km. The only plausible mechanism that has been suggested is that pre-earthquake stress gradients move fluids in the pore spaces at the site of radon observations (Dobrovolsky *et al.*, 1979; Fleischer 1981). More detailed discussion of mechanisms is deferred to section 4.2.4. These points are summarized here because they bear directly on decisions as to where and how best to measure radon to detect local fluid flow in the earth.

[1]Rensselaer Polytechnic Institute, Troy, New York, USA

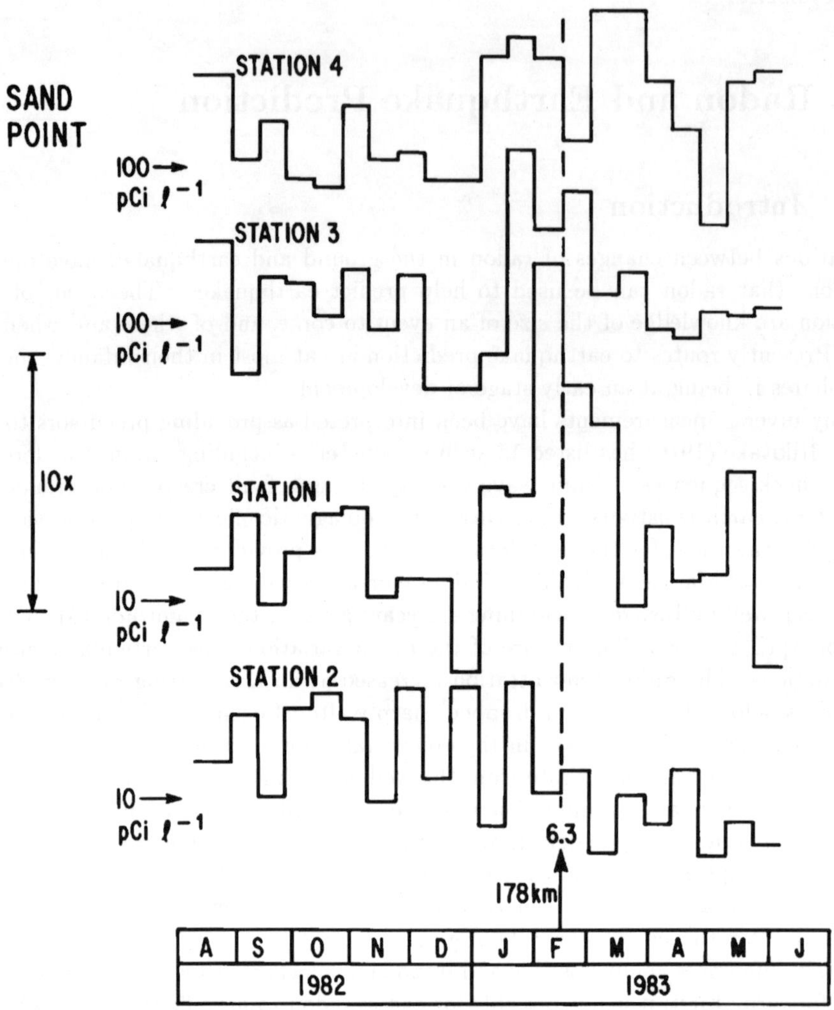

Figure 4.2.1: Variations of radon at Sand Point, Alaska, near 14 Feb. 1983, the time of an M = 6.3 earthquake at 178 km distance. ^{222}Rn was measured at a depth of 60 cm.

4.2.2 Techniques

Given that the objective is to use radon measurements to reveal the build-up of earth stresses, there are two major issues to be faced. One is where on the map to locate radon-monitoring emplacements. The other is what those emplacements should consist of in order that ^{222}Rn and only ^{222}Rn is measured and that extraneous factors such as moisture, temperature, and atmospheric pressure fluctuations do not obscure the wanted signals. Years of experimentation have allowed the latter issue to be resolved. The former, as the reader will quickly perceive, is still a matter of judgment and considerable uncertainty. Much of the following discussion is adapted from Fleischer and Mogro-Campero (1985).

4.2.2.1 Site selection

In general, site selection consists of two components, vertical and horizontal. The justification of the shallow depth that is recommended here (60 cm) is well-founded; that of horizontal locations is in an early exploratory stage and is in need of a firmer basis. One practice is to pick localities of seismic interest, emplace detectors at a number of locations, and monitor over a long enough period of time that the variability can be established and the "unresponsive" sites abandoned. Sites where radon varies considerably are expected to be ones where small changes in conditions are likely to have the largest effect on radon. In general the detailed, local hydrological conditions in the earth are unknown; yet what is wanted as a site is an analogue to a glass-thermometer where radon, as a measure of fluid flow, is observed in the calibrated stem and the bulb is strained by earthquake-related stresses.

As an example, at Blue Mountain Lake, New York, out of approximately three dozen sites, the group at General Electric selected four for continuing monitoring (Mogro-Campero *et al.*, 1980). Of these, one is clearly the most sensitive, and on one occasion gave readings that changed by a factor of 500 in radon concentration, systematically, over a period of four weeks (Fleischer and Mogro-Campero, 1980). Since in many cases radon variations of less than 50% are thought to be significant, this factor represents truly unusual sensitivity.

Choosing even the first site is not a totally random process. Locations that do not provide fluid connection with the underlying earth are unsuitable. For example, a near-surface layer of impervious clay was observed to shield the surface soil from significant radon changes at several Schenectady, NY, test sites, in an unpublished study.

The selection of the depth of measurement of about 60 cm is on a far firmer basis. This shallow depth is chosen to be close to a discontinuity in the radon emanation of the medium - in this case the air-ground interface. As described earlier (see section 1.1.3.3.2, Fig. 1.1.4), a location in the steep gradient between the high radon concentration deeper in the ground and the negligible concentration in the atmosphere is sensitively affected by vertical fluid flow in the ground - the concentration being raised by upflows and depressed by downflows. Systematic flows observed by arrays of detectors have been recognized by Fleischer and Mogro-Campero (1979a) and Fleischer *et al.* (1980a). Flows that might be caused by earthquake-related dilations or compressions of the near-surface interstitial spaces should be really recognized if they produce flow velocities of $\geq 3{\cdot}10^{-4}$ cm s^{-1} or, additionally, if they transport radon from nearby regions of significantly elevated or depleted concentration, such as are common in the earth.

Fig. 4.2.2 gives an example of radon-versus-depth measurements over areas with no flow, with upflow, and with downflow. Although the data are not extensive enough to confirm the analytic form of the theoretical curves where flow is present, they are consistent with the calculated relations. They also make clear that the depth range between 0.4 m and 0.7 m is optimal for detecting the effects of flow. The data are unpublished results from 1978 of R. L. Fleischer, M. D. Fiske, H. R. Hart, Jr., R. S. Likes, and A. Mogro-Campero.

The question as to the number density (or spacing) of measurement sites depends on the smallest earthquake that one wants to predict. This consideration will be discussed in a later subsection.

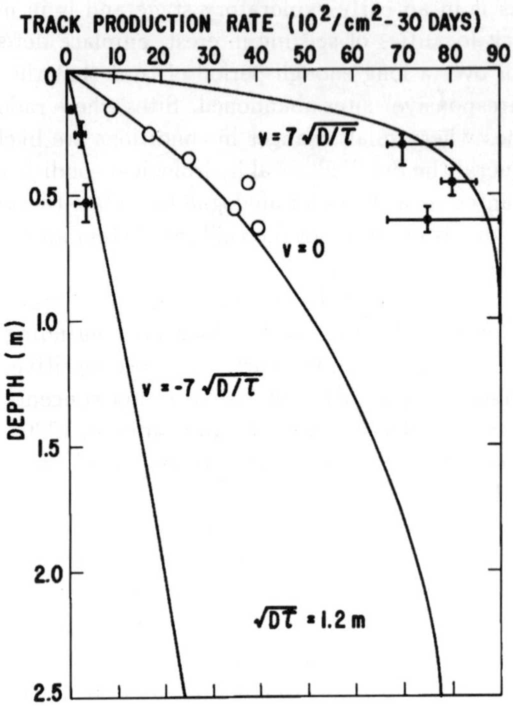

Figure 4.2.2: Radon measurements as a function of depth of different times reflect the presense of upflow, downflow, and the absence of flow. The data were collected near Thoreau, New Mexico, USA. The velocity v of flow is a function of the diffusion coefficient D and the mean life τ of radon.

4.2.2.2 In-ground measurement techniques

Measurements need to have reasonable reproducibility under unchanging radon environments, unaffected by weather, soil humidity, and unwanted background from thoron and other natural alpha emitters. We begin by describing best practice for in-ground measurements, show what the reproducibility is, and note the problems from using less complete techniques. Note that the state-of-the-art of radon monitoring devices, etching procedure, counting methods, track detectors, calibration and standardization are described in the present book in Chapter 2 (Etched Particle Track Methods) by Durrani; Ilić and Šutej; Tommasino; and Miles.

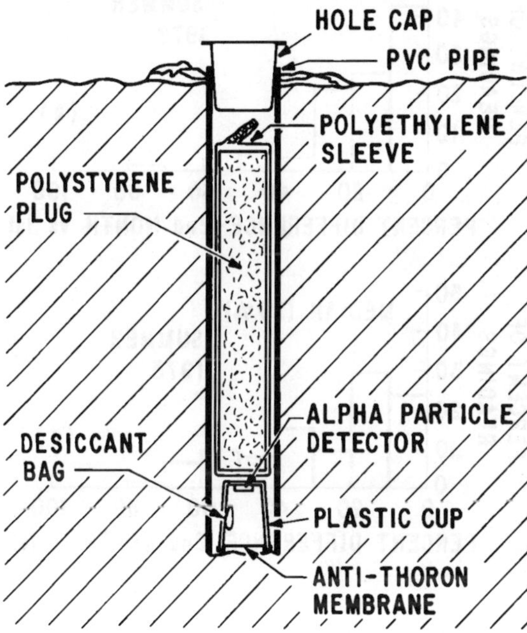

Figure 4.2.3: Sketch of subsurface radon-monitoring. A cup that contains a thermally insensitive alpha-particle detector and a desiccant to prevent condensation of moisture is closed at the bottom by a membrane that excludes ^{220}Rn and other alpha emitters except ^{222}Rn. Insulation between the cup and the surface of the ground suppresses loss of radon to the atmosphere. The depth of the hole is about 60 cm.

The method of installation in the ground is sketched in Fig. 4.2.3. The detector cup lies on soil at 60 cm depth in a hole that is prevented from collapsing by a plastic pipe (in this case PVC). A plastic-wrapped plastic-foam plug prevents atmospheric air from diluting the radon to be measured. A tapered hole cap further aids in this process and also minimizes accidents from people or animals stepping into the hole.

The detector cup, as sketched, uses the membrane and desiccant procedures described by Fleischer and Mogro-Campero (1978) and Likes *et al.* (1979). A plastic cup closed by a selective membrane (Ward *et al.*, 1977) admits ^{222}Rn and excludes all other natural alpha emitters. The alpha particles from the ^{222}Rn and its two alpha emitting daughters ^{218}Po and ^{214}Po are recorded by a CR-39 solid track detector located within the enclosure; and the integrated radon concentration is derived from the track count, calibrations having been performed in chambers with known ^{222}Rn concentrations (Fleischer *et al.*, 1980b).

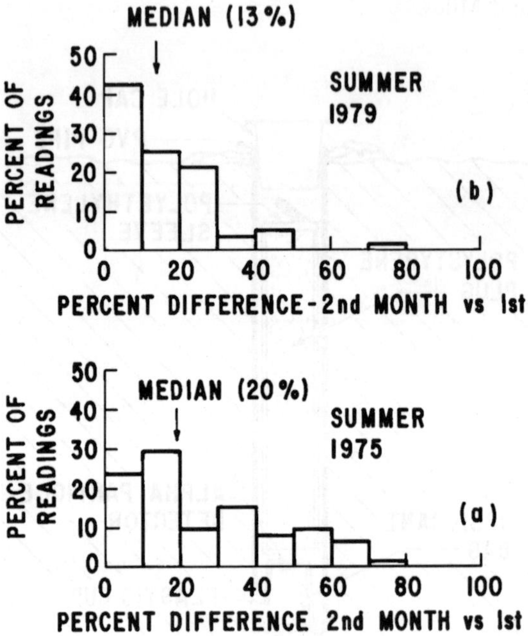

Figure 4.2.4: Month-to-month variability of readings, with the (a) original open-cup radon technology (1975), and (b) with the improved methods that are insensitive to mositure, thoron, and heat (1979), near Thoreau, New Mexico, USA.

The membrane serves three purposes - the exclusion of ^{220}Rn, rejection of soil or rock particles which carry other alpha activity, and slowing the entry of water vapor into the enclosed air space. Supressing the ^{220}Rn signal is important since, although typically having an activity that is similar to that of ^{222}Rn in the ground, it can vary widely depending on the Th/U ratio and the ease of escape from the minerals in which Th and U are located. Values at sites in New Mexico and Colorado in the USA for ($^{220}Rn/^{222}Rn$) vary by a factor of 15 (Fleischer and Mogro-Campero, 1979b).

The desiccant inside the detector cup in Fig. 4.2.3 prevents high water vapor contents from occurring inside the cup. This control prevents the following problem: If the humidity neared 100% during the warmer portion of the diurnal temperature cycle, then cooling at night would cause water to condense; if droplets form on the detector, fewer than normal alpha particles would reach it to produce tracks - thereby producing spuriously low readings.

The final critical component of proper measurements in the ground is the use of plastic detector materials such as CR-39 (Cartwright *et al.*, 1978) that is neither moisture sensitive nor sensitive to typical ambient temperatures.

Older systems used cellulose nitrate detectors that suffered from both these defects (Fleischer and Mogro-Campero, 1980; King, 1978), as shown in the more recent paper. Unfortunately the U.S. Geological Survey maintained for some years a policy of continuing the defective methods (for better intercomparison purposes), thus allowing them to compare compromised data with other compromised data.

In contrast, as indications of the reproducibility of the method as originally practiced and as now possible, Fig. 4.2.4 shows the month-to-month variability in (a) the summer of 1975 (old techniques), and (b) the summer of 1979 (newer techniques). The figure gives the difference between two consecutive month-long readings of 89 sites in 1975 and 55 in 1979, all in the same 0.08 km^2 area near Thoreau, New Mexico. For the latter period meteorological records show that there were 5.5 cm of precipitation in the first month and 1.0 in the second; thus the good reproducibility occurred in spite of a 5.5 times difference in rainfall.

In laboratory measurements the present system is reproducible within the counting statistics (Fleischer *et al.*, 1980b); and field measurements, which experience various meteorological and other natural causes of variability, are normally observed to reproduce with a $\pm$ 20% to 25% standard deviation in extensive measurements in different areas of low seismic activity, Fig. 4.2.4 being one example.

From the preferred methods for monitoring radon in the ground and the reason for them, the problems of less complete techniques are obvious. If no membrane and desiccant are used, extra tracks from ^{220}Rn background become possible, as does moisture condensation, which has the counter effect of lowering the track density. Detector depths of less than 0.3 m or more than 1 m will diminish sensitivity to genuine changes of ^{222}Rn in the ground. None of these lesser techniques guarantees that the real signal being sought will be lost, but by introducing noise they diminishe the number of true signals that are large enough to be recognized.

Another specific thoron-and-moisture-insensitive detector configuration was described by Humar *et al.* (1982), using polyvinyl chloride membranes to reject ^{220}Rn and reduce moisture. PVC has a 15 % lower water transmission rate than polyethylene of the same thickness.

4.2.3 Observations

Radon observations have been made at a variety of types of sites and by an assortment of methods - addressed at either short-term predictions (days to a week) or long-term ones (weeks to years). For long-term effects, track etching is a convenient method since periods of one or two weeks for individual measurements are readily possible. For short-term effects instruments with prompt readouts are used, normally employing some sort of electronic counter.

4.2.3.1 Types of sites

As noted earlier, the changes in radon that are sought are expected to result from fluid flow in the ground - motion that can deliver a different radon concentration to the measuring site than would exist in the absence of flow. One way to assure the nearby presence of different radon concentrations and consequent sensitivity to the presence of flow is to measure radon in the gradient near the earth's surface where the radon concentrations change from the soil values to much lower atmospheric ones. In this case readings are taken in relatively shallow ($\sim$ 0.5 m) holes in the ground.

One alternative is to drill much deeper holes and depend on discontinuities in radium content and/or fracturing in different strata to provide useful radon gradients. Another option is to sample well-water or springs and hope that liquid transport crosses gradients that are close enough to the measuring point that radon changes can be detected when the extent of transport alters.

The near-surface measurements have the virtue that the necessary radon gradient is reliably present. Caution is necessary, however, that no impervious layer shields the surface region from flow. The possible advantage of deeper readings lies in being below any shallow impervious layers; but their problem is that it is generally unknown when choosing a site whether significant radon gradients happen to be crossed by fluid flow.

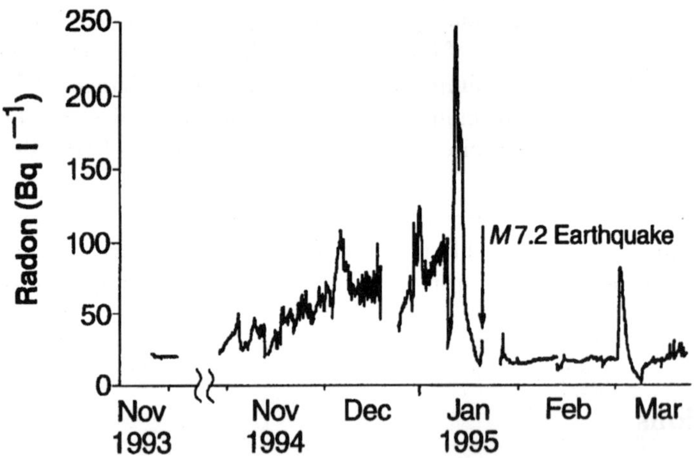

Figure 4.2.5: Radon data from a 17 m-deep well, 30 km from the epicenters of the January 17, 1995, Kobe M = 7.2 earthquake (Igarashi *et al.*, 1995).

4.2.3.2 Types of signals

The radon behavior before earthquakes has two sorts of premonitory characteristics. Although usually only one or the other type is seen, the record in Fig. 4.2.5 shows both (Igarashi *et al.*, 1995). For long-term prediction long-term effects are needed, and the figure shows an increase in radon concentration in deep well-water from 20 Bq l^{-1} to 60 Bq l^{-1} over a 2-month period prior to an M = 7.2 earthquake at Kobe, Japan, on January 17, 1995. The other type of feature - a short abrupt change - appeared on January 7, just 10 days before the event. Fig. 4.2.5 also illustrates why a single record is not enough for reliable forecasting. There is a second abrupt increase about 45 days later that has no known physical relation to an earthquake.

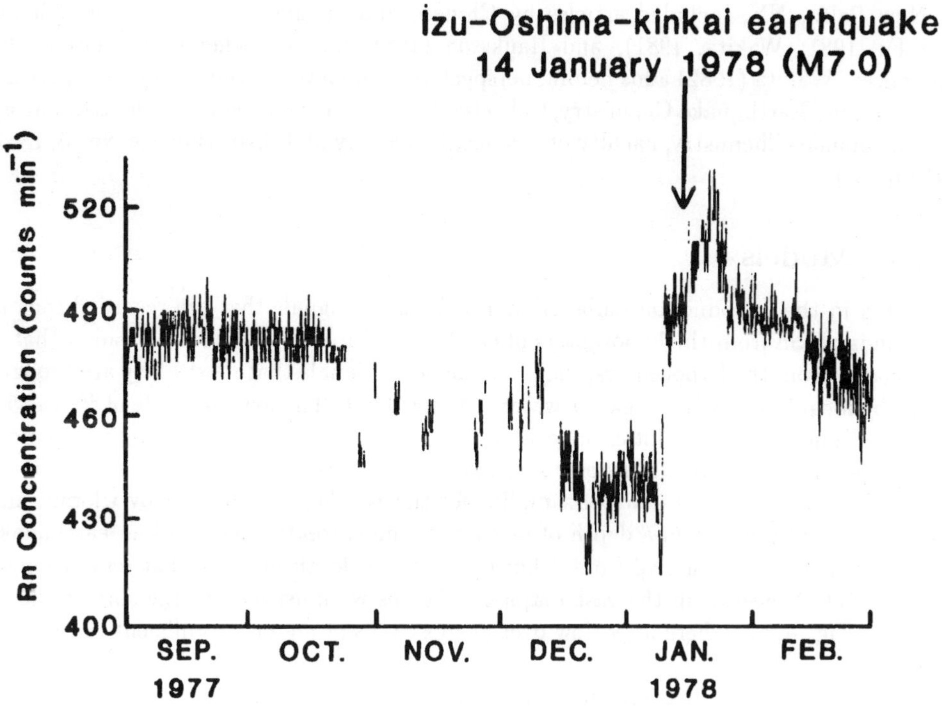

Figure 4.2.6: Radon concentration in groundwater 25 km from the 14 January 1978 (M = 7.0) earthquake.

Fig. 4.2.6 illustrates another point of note: Radon signals can be negative as well as positive. Prior to an M = 7.0 earthquake in Japan in January 1978, radon levels declined for 3 months (Wakita, 1981) until about a week before the event. The existence of both decreases and increases prior to earthquakes may well be associated with the fact that pre-seismic stresses can be either positive or negative, depending on what sector of the elastic strain field measurements are taken on. Fluid flows in different directions will give different results in near-surface sites and are likely to do so elsewhere.

4.2.3.3 Data compilations

It is beyond the available space to attempt a comprehensive listing of individual cases of associations between earthquakes and changes in radon in the ground. Having given some examples in Figs. 4.2.1, 4.2.5 and 4.2.6, this section merely gives sources where the interested reader can consult original references. American Geophysical Union meetings have given rise to three proceedings. See the *Journal of Geophysical Research* (1980) **85**, 3051-3121 for six articles; *Geophysical Research Letters* (1981) **8**, 421-484 for nine; and *Pure and Applied Geophysics* (1984/85) **122**, 141-618 for another ten. A conference in New Paltz, NY, included articles by Chinese and Japanese radon workers (Zhang and Fu, 1981; Wakita, 1981); and Hauksson (1981) and Fleischer (1981) cite useful references. Wakita (1982) adds helpful perspective; and eleven of the Japanese papers are published in "Earthquake Chemistry, Collected Papers", a 1988 volume of the Laboratory for Earthquake Chemistry, Faculty of Science, University of Tokyo (Bulletin No. 5, ISSN 0910-688X).

4.2.4 Models

As noted in the introduction (subsection 4.2.1), radon signals that are recorded tens to thousands of km from the hypocenters of earthquakes are not composed of radon that is propagated from the hypocenters; they must be local signals that most likely are brought about by local flows of either gas or water in the ground. The low mean life of ^{222}Rn (5.5 days) and known permeabilities of rock and soil are what justify this definite statement that long-distance radon transport is usually not involved. There may be occasional exceptions: for example, the water sampling for the Tashkent earthquake by Ulomov and Marashev (1968) was from a depth of $\sim$ 2 km in the epicentral region of an earthquake whose fault plane was active from 8 km up to 3 km depth or somewhat less (Ulomov *et al.*, 1967). However, in the vast majority of cases we must look to the sort of effects shown in Fig. 4.2.2, where local flow over at most tens of meters is sufficient to produce radon shifts.

The matters to be considered here are: what seismically related precursor can produce such distant flows, and what local conditions and processes may enhance radon non-uniformities.

4.2.4.1 Long-range forces

Earthquakes are generally the result of stresses that build within the earth and ultimately are relieved - at least in part - by shear deformation. Stress centers in elastic bodies are not confined; their strain fields extend to the boundaries of the solid. Thus, pre-seismic strains of diminishing magnitude exist far from the future slip region of an earthquake.

The long-range strain gradients are the proposed agent of fluid flow in both of the physical models that have been suggested as causes for long-range earthquake-related radon changes (Dobrovolsky *et al.*, 1979; Fleischer, 1981).

In the first of these descriptions - a volume-strain model - Dobrovolsky *et al.* (1979) described what they called an "earthquake preparation zone" by an ellipsoidal volume that, as rock is fractured, acquires a lower elastic modulus than the matrix, and hence produces a long-range strain field by relaxing in the presence of earth stresses. In the other model - a dislocation process - Fleischer (1981) describes the stress center as an asperity (i.e. an unslipped area) on a fault plane, the ultimate slipping of which will cause the earthquake.

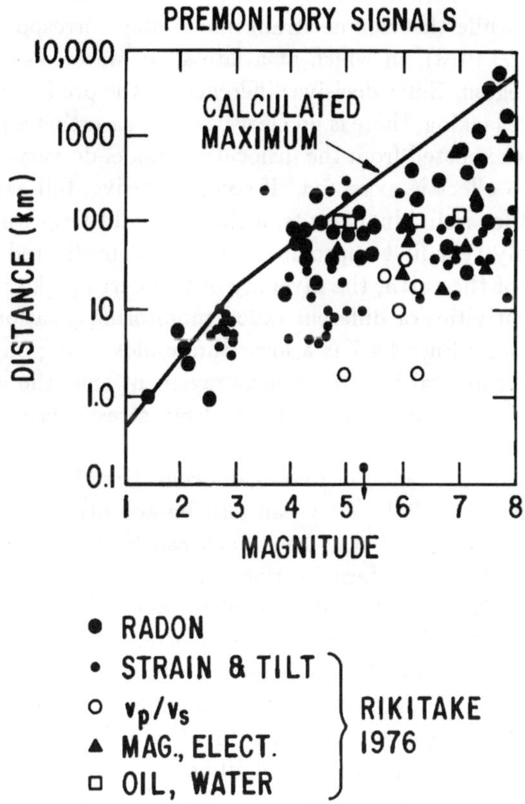

Figure 4.2.7: Premonitory signals on a magnitude-distance diagram. Data from surveys by Rikitake (1976) and Fleischer (1981), and a study by Fleischer and Mogro-Campero (1985). Strain and tilt, and radon measurements, cover similar fields in the diagram and most are consistent with cause-and-effect correlations. The solid line is the proposed upper limit.

In each case the long-range strains drop off as the inverse cube of the distance, and the distance for a given strain can be estimated from the elastic theory of solids and related to the magnitudes of earthquakes through established empirical relations between magnitude M and energy. The models lead to radii (in km) for the region that is strained to an empirical value that varies as $10^{0.43M}$ (Dobrovolsky *et al.*, 1979) or $10^{0.48M}$ (Fleischer, 1981). While the noise in existing data prevents choosing between the two exponents, Fig. 4.2.7 shows that existing radon data (and other precursors listed by Rikitake, 1976) are consistent with the proposed variation of the detection limit. It is critical to note about Fig. 4.2.7 that only one point is significantly above the line that specifies the maximum distance of detection – points below the line are permitted since the line is an upper limit.

Is there any good reason to prefer one of the two models as a physical description? The dislocation model leads more directly to shear deformation, which describes normal earthquake behavior; while the volume-strain model may correspond to a dilatancy description (Scholz *et al.*, 1973), in which at a late stage water is drawn into the highly stressed hypocentral region. Since decisive differences in the predictions of present-day results by the models are lacking, there is no clear basis for asserting a preference. However, the signs of the effects expected from the dislocation model do vary in a predictable way, with nodes where zero effect is expected. If comprehensive, full-azimuth, multi-station data are accumulated for individual events, a choice should be possible.

The models are only very first approximations to a complicated reality. They ignore the elastic variability of the earth, the diversity of real earthquake mechanisms, and the different inherent sensitivities of different radon monitoring locations and installations. Nevertheless, even though Fig. 4.2.7 is a logarithmic plot, the grouping of observations near and below the theoretical line gives encouragement that the gross features of the models are correct, i.e., that long-range strains from stress centers are responsible for radon redistribution.

On a more local scale the radon changes can occur directly from stress-induced fluid motion as sketched in Fig. 4.2.2. They can also be accentuated in cases where water is caused to enter an aquifer that was dry, which can lead to a surge in the release of radon from radiation damage at internal surfaces in rocks or soils (Fleischer, 1982, 1983). The inferred strains at the most distant sites where radon changes precede earthquakes can be as low as 10^{-8}, as derived from the models using linear elasticity. In reality, the elastic moduli of many rocks and of all soils increase with pressure, sometimes strikingly. As examples, for different Apollo 12 basalts, Kanamori *et al.* (1971) found the bulk modulus to increase by a factor of 6 between 0.1 MPa and 1 GPa (1 bar and 10 kbar), and Stephens and Lilley (1971) by a factor of 10 between 15 and 500 MPa (150 bar and 5 kbar). These effects provide higher strains at low stress - i.e. far from the epicenters and near the surface of the earth.

4.2.5 Prospects and Conclusions

In earthquake prediction the major use of radon monitoring has been as one more indicator amongst the many that were listed in section 4.2.1 (Rikitake, 1976). When several such indicators give signals, an earthquake alert may be appropriate. In such multi-technique systems usually only one or two radon stations would be involved.

For prediction using radon alone, an array of radon stations would be needed that gives overlapping, partially redundant coverage (Fleischer and Mogro-Campero, 1985). The relation shown in Fig. 4.2.7 allows areas of coverage to be defined for a given size earthquake. From the spacing of stations and the number that display pre-earthquake behavior, the size and location of an event should be derivable. It is convenient that for the larger earthquakes larger regions are expected to be affected and therefore fewer stations are needed. For example, from Fig. 4.2.7 one expects that a station spacing of <300 km would be required for $M = 6$ events, but <1500 km gives equivalent overlap for $M = 7$. Better localization is available, however, with more numerous radon sites.

A further localization possibility with an array of stations arises if the dislocation model is the correct description of most pre-earthquake strain fields. Since positive and negative strains are expected to alternate in circuits around the hypocenter, the nodal lines implied by changes of sign of the radon-changes should point to the position of earthquake threat.

By way of summary: reproducible methods of radon monitoring in the ground have been developed. At this time, numerous radon signals have been observed that the observers correlated with the impending earthquakes, and many of these correlations appear to have a high degree of reliability. Not all earthquakes give such signals, and presumably some of the correlations are by chance. More meaningful prediction would be possible in a region with a network of radon stations that would allow overlapping coverage for earthquakes above a threshold magnitude that is set by the spacing of the stations. The long range of radon effects implies that the cause is stress-induced flow of fluids within the earth; and two models of the strain fields are consistent with the observed distance-of-effect vs earthquake-magnitude data.

Deployment and testing of an adequate array requires money and persistent effort. The funding is a major obstacle. If it is to be overcome, progress may occur either in a country with low labor costs or by an international consortium whose purpose is to monitor radon thoroughly in a designated test area.

References

Cartwright B. G., Shirk E. K. and Price P. B. (1978) CR-39: A nuclear-track-recording polymer of unique sensitivity and resolution. *Nucl. Instr. Meth.* **153**, 457-460.

Dobrovolsky I. P., Zubkov S. I. and Miachkin V. I. (1979) Estimation of the size of earthquake preparation zones. *Pure Appl. Geophys.* **117**, 1025-1044.

Fleischer R. L. (1981) Dislocation model for radon response to distant earthquakes. *Geophys. Res. Lett.* **8**, 477-480.

Fleischer R. L. (1982) Alpha-recoil damage and solution effects in minerals: uranium isotopic disequilibrium and radon release. *Geochim. Cosmochim. Acta* **46**, 2191-2201.

Fleischer R. L. (1983) Theory of alpha recoil effects on radon release and isotopic disequilibration. *Geochim. Cosmochim. Acta* **47**, 779-784.

Fleischer R. L. and Mogro-Campero A (1978) Mapping of integrated radon emanation for detection of long-distance migration of gases within the Earth: Techniques and principles. *J. Geophys. Res.* **83**, 3539-3549.

Fleischer R. L. and Mogro-Campero A (1979a) Radon enhancements in the earth: Evidence for intermittent upflows? *Geophys. Res. Lett.* **6**, 361-364.

Fleischer R. L. and Mogro-Campero A (1979b) Integrated radon mapping in the earth – assessment of the ^{220}Rn signal and its exclusion. *Geophysics* **44**, 1541-1548.

Fleischer R. L. and Mogro-Campero A. (1980) Techniques and principles for mapping of integrated radon emanation within the earth. In *Natural Radiation Environment III* (Proc. Symposium, Houston 23-28 April, 1978) (eds. T. F. Gesell and W. M. Lowder) Tech. Info Center. U.S. Dept. of Energy, Springfield, VA, Conference 780422, **Vol. 1**, pp. 57-70.

Fleischer R. L. and Mogro-Campero A. (1985) Association of subsurface radon changes in Alaska and the northeastern United States with earthquakes. *Geochim. Cosmochim. Acta* **49**, 1061-1071.

Fleischer R. L., Hart H. R., Jr., and Mogro-Campero A (1980a) Radon emanation over an ore body: Search for long-distance transport of radon. *Nucl. Instr. Meth.* **173**, 169-181.

Fleischer R. L., Giard W. R., Mogro-Campero A., Turner L. G., Alter H. W., and Gingrich J. E. (1980b) Dosimetry of environmental radon: methods and theory for low-dose integrated measurements. *Health Phys.* **39**, 957-962.

Hauksson E. (1981) Radon content of groundwater as an earthquake precursor: evaluation of worldwide data and physical basis. *J. Geophys. Res.* **86**, 9397-9410.

Humar M., Najžer M. and Ilić R. (1982) Moisture-insensitive probes for separate measurements of radon isotopes. In *Proc. 11th Int. Conf. on Solid State Nuclear Track Detectors* (eds. P. H. Fowler and V. M. Clapham), Bristol, 1981, Pergamon Press, Oxford, pp. 549-552.

Igarashi G., Saeki S., Takahata N., Sumikawa K., Tasaka S., Sasaki Y., Takahashi M. and Sano Y. (1995) Groundwater radon anomaly before the Kobe earthquake in Japan. *Science* **269**, 60-61.

Kanamori H., Mizutani H. and Hamano Y. (1971) Elastic wave velocities of Apollo 12 rocks at high pressure. In *Proc. Second Lunar Science Conf.* **3**, MIT Press, Cambridge, pp. 2323-2326.

King C-Y. (1978) Radon emanaton on San Andreas Fault. *Nature* **271**, 515-519.

Likes R. S., Mogro-Campero A. and Fleischer R.L. (1979) Moisture-insensitive monitoring of radon. *Nucl. Instr. Meth.* **159**, 395-400.

Mogro-Campero A., Fleischer R. L. and Likes R. S. (1980) Changes in subsurface radon concentration associated with earthquakes. *J. Geophys. Res.* **85**, 3053-3057.

Rikitake T. (1976) *Earthquake Prediction. Developments in Solid Earth Geophysics* **9**, Elsevier, Amsterdam.

Scholz C. H., Sykes L. R. and Aggarwal Y. P. (1973) Earthquake prediction: A physical basis. *Science* **181**, 803-810.

Stephens D. R. and Lilley E. M. (1971) Pressure-volume properties of two Apollo 12 basalts. In *Proc. Second Lunar Sci. Conf.* **3**, MIT Press, Cambridge, pp. 2165-2172.

Ulomov V. I., Zakharova A. I. and Ulomova N. V. (1967) Tashkent earthquake of April 26, 1966, and its aftershocks. *Akad. Nauk. SSSR, Geophysics* **177**, 567-570.

Ulomov V. I. and Marashev B. Z. (1968) A precursor of a strong tectonic earthquake. *Dokl. Akad. Sci. USSR, Earth Sci. Sect.* **176**, (#1-6): 9-11, pp. 319-321 (In Russian).

Wakita H. (1981) Precursory changes in groundwater prior to the 1978 Izu-Oshima-Kinkai earthquake. In *Earthquake Prediction* (eds. D. W. Simpson and P. G. Richards), *Amer. Geophys. Un.*, Washington, D.C., pp. 527-532.

Wakita H. (1982) Changes in groundwater level and chemical composition. In *Earthquake Prediction Techniques* (ed. by T. Asada), University of Tokyo Press, Tokyo, pp. 175-216.

Ward W. J., III, Fleischer R. L. and Mogro-Campero A. (1977) Barrier technique for separate measurements of radon isotopes. *Rev. Sci. Instrum.* **48**, 1449-1441.

Zhang G. and Fu Z. (1981) Some features of medium- and short-term anamalies before great earthquakes. In *Earthquake Prediction* (eds. D. W. Simpson and P. R. Richards), *Amer. Geophy. Un.*, Washington D.C., pp. 497-509.

M. M. Monnin[1] and J. L. Seidel[1]

4.3 Radon and Volcanic Surveillance

4.3.1 Historical Introduction

Radon has been recognized for a long time as a detectable component of fluids associated with volcanoes (fumaroles, groundwaters, or soil gases). For instance, radon in hot springs has been mentioned as early as 1927 (Shiratoi, 1927), whereas its presence in fumaroles was made known more than half a century ago (Sicardi, 1940). Later, other studies (Cheminée, 1966; Iwasaki *et al.*, 1956, 1975; Leonard and Janzer, 1978; Wollenberg, 1975) have shown that thermal fluids typically have higher radon concentrations than cold groundwaters in the same geologic environment (Thomas, 1989) - particularly when associated with volcanic edifices. Accordingly, many investigations have been carried out in order to understand the structure and/or the behaviour of volcanoes based upon radon surveys whether in fumarole gases, in soil, or in groundwaters, as will be seen in greater detail below.

The time-dependence of radon concentration with volcanic activity in thermal springs was recognized for the first time by Chirkov (1975), who found that radon concentration would increase during seismic activity of the Karimsky, Russia, volcano leading to an eruption, as we will see later. Similarly, steep pre-eruption radon variations in ground air were observed. To our knowledge, the first of such events was measured on the Krafla volcano where up to a 20-fold increase in radon concentration occurred prior to the 1977 eruption. However, the first published result along these lines is due to Cox *et al.* (1980) and then Cox and Cuff (1981), who produced well-documented cases of radon variations correlated with volcanic activities of the Kilauea, Hawaii, volcano. Since then many time-dependent surveys have been carried out on some of the most studied volcanic edifices.

4.3.2 Technical Approach

Before going any further, one should pay some attention to the essential characteristics of radon and its consequent possibilites of movement, in order to understand the significance of radon measurements over volcanic structures.

Having a $6p^2$ $6p^6$ final P-electronic configuration, radon is a noble gas and thus does not undergo chemical reactions which could preclude its free movement within terrestrial materials. This statement is true, of course, for all isotopes of Rn. All Rn isotopes are daughters of radium. However, ^{226}Ra is the most abundant radium isotope (99% in the

[1]Université Montpellier II, Montpellier, France

natural abundance) and hence so is ^{222}Rn. In the words of Dongarra and Martinelli, 1993: "Although radium in rocks is generally related to the presence of uranium, secondary local accumulation may occur owing to its weak mobility in comparison with uranium. This reduced mobility is explained by the fact that radium forms insoluble phosphates and carbonates. In addition, radium is rather immobile in oxidizing environments and fairly mobile in reducing environments. Radium is also removed from the solutions by clays, organic matter, iron and manganese hydroxides. In some natural environments, chemical processes of dissolution-precipitation may concentrate radium so that much radon is produced in a small volume. Hot springs are an example". On the other hand, radon may be present even when very little radium can be found, because of the mobility of radon itself. Once radon is free to move, i.e. when it has left its original crystal lattice, it can migrate towards the earth's surface by different mechanisms. Diffusion is the simplest; but owing to its short half-life, radon cannot travel more than a few metres at most by mere diffusion. Convection is the second possibility. It comes into operation when a sufficient thermal gradient is available within the medium. It depends on many local parameters such as permeability, viscosity, ... Transport by a gas carrier is the third possibility for radon migration. Within a volcano, gases such as CO_2, H_2S, SO_2, H_2 are abundant candidates for being suitable carriers of radon. These gases may reach the surface at sometimes considerable rates. A strong temperature gradient and strong degassing are obviously common phenomena within a volcanic edifice. Therefore, both convection and transport migration lead to high radon readings on a volcano and must be taken into consideration when interpreting the gathered data.

When it comes to radon in groundwaters associated with volcanoes, one should remember that radon is moderately water soluble. Its solubility decreases with increasing temperature. Radon associates with water molecules through Van der Waals forces (clathrates). Low pH values destabilize clathrates and destroy them, thus releasing radon into the gaseous phase. Both heat and H^+ transfers are likely to occur in volcanic groundwaters and would contribute to radon release.

Performing radon measurements on a volcano is based upon the techniques described in sections 1.3, 2.2 and 2.3 of the present book. Differences may occur, depending on the type of the radon source at the different sampling sites.

4.3.2.1 Soil

The vast majority of in-soil radon measurements have been carried out using Solid State Nuclear Track Detectors (SSNTDs). The low-cost and ease of use favour this type of detector. Their rather long time of exposure (10 to 40 days on average) tends to smooth out the daily variations that could otherwise modulate the radon signal itself. They are placed in various "inverted cup" designs (see section 4.2). Conversely, the track detectors are unable to detect short intense burst-like radon signals, if any (or at best they see those in a very poor and weak manner). Thus real-time measurements are also performed using electronic data-storing devices. Finally, particularly for structure studies, sampling by pumping, storing and subsequent analysis of gas samples (either by ion chambers on the spot or by Lucas cells in the laboratory) have been commonly used.

4.3.2.2 Fumaroles

Because of the high temperature which is a common feature of fumaroles and of the often agressive nature of the fumarole gases, direct measurement of radon is ruled out. Sampling is achieved by different means such as vacuum pumping into glass vessels, or gas pumping into vessels filled with pure water or salt-saturated water. Samples are then brought to the laboratory for counting or gas separation and counting. In many cases, radon determination is based not on radon itself but on the growth rate of its daughters, ^{214}Bi and ^{214}Po.

4.3.2.3 Groundwaters

All the above possibilities can be carried out in groundwaters depending on accessibility. Techniques involving high length-to-diameter ratio inverted cups containing track detectors, as well as real-time electronic devices, water pumping and sampling, solvent (e.g. toluene) extraction, and so on, are to be found in the literature regarding radon measurements in groundwaters related to volcanoes.

4.3.3 Case Studies of Volcanoes Situated in Various Parts of the World

Before examining results obtained on some specific volcanoes, it is worth taking a look at studies in general, dealing with structural or behavioural investigations during quiescent phases.

In the first instance, investigations have been carried out in order to assess the intensity of geological radon sources. Such a process is exemplified by Wilkening's (1974) work over the Hawaii Island. The primary goal of the work was to assess the amount of radon released in the atmosphere by the volcanoes. It was found to be almost negligible, and most of the radon originated primarily from the deep volcanic soil rather than from the lava fields. This type of finding must be kept in mind when evaluating results of in-ground radon measurements.

Based upon the idea that fractured terrain or a fault should provide a preferential and easier path for transported radon, mapping of radon has been used to locate buried or hidden faults on volcanic sites (Aubert and Baubron, 1988; Cox, 1983; Cox and Cuff, 1981; Crenshaw *et al.*, 1982; Nielson, 1978; Stoiber, 1982; Williams, 1985; Wollenberg, 1975). In many cases radon measurements were coupled with other gas surveys. In the work of Crenshaw *et al.* (1982), for instance, mercury signatures were sought as well. Mercury is used because of its high volatility. They carried out transverse measurements accross areas of suspected faults. In some cases, radon and mercury peaks would nicely coincide with each other and with the fault trace (caldera fault scarp). In other instances, the radon peak location would differ by as much as 60 metres from the mercury's high-reading spot. Aubert and Baubron (1988) conducted a radon, carbon dioxide and helium study completed with SP (electrical self-potential) and temperature measurements. Several correlations were found, particularly between radon and CO_2 output. The authors offered a dynamical model in order to account for the various findings in which fluids including water could flow upwards, downwards and laterally, depending on their respective positions within the system.

Radon-based studies related to geothermal systems rely both on the existence of faults and a strong thermal gradient. Such studies have been carried out underwater (Koski *et al.*, 1985; Rosenberg *et al.*, 1988) and on land (Cox, 1980; Cox and Cuff, 1980; D'Amore and Sabroux, 1976; D'Amore *et al.*, 1978; Lombardi and Reimer, 1990; Lombardi *et al.*, 1984; Semprini and Kruger, 1984; Whitehead *et al.*, 1983; Whitehead, 1984; Williams, 1985). See section 4.5 of this book.

Let us now review some of the research work carried out by means of radon, often associated with the measurement of other parameters, over several volcanoes of the World.

4.3.3.1 Europe

One of the largest volcanoes of the world and the largest in Europe is Mount Etna, Italy. It is located in a populated area - the city of Catane lies at its foot. It often enters eruptive phases and is thus quite well studied.

Rn surveys have been carried out by several research groups. However, Etna is such a large volcano that some eruption may occur on some side of it without affecting in any way the Rn sensors placed far away. This is particularly true if the radon is transported and/or measured by means of an aquifer which may not be in any connection with the eruptive mouth. For instance, a northern eruption did not produce any effect at the Rn station located in the southern part of the volcano, but one where a peak outburst of radon was observed (Seidel and Monnin, 1982). On the other hand, performing measurements in two wells located at the foot of the Valle del Bove, Bottiglieri *et al.* (1987) observed a strong radon increase after the eruptive phase that lasted from october 1986 to february 1987 in this area. They concluded that the observed deep variation in radon could be ascribed to the vicinity of the eruptive vent localized on a fault. The fact that such an increase was observed after, and not before, the event is puzzling in the sense that one observes in general the opposite (see later, for instance the Rn behaviour after the El Chichon eruption).

During surveys carried out independently, similar data were obtained. Dall'Aglio *et al.* (1994) have analysed the chemical composition of groundwaters before and after the December 1991 eruption of Mt Etna. They were performing multi-element analysis. The physical parameters of the waters (temperature, pH, conductivity,...) were measured together with the compositon in the usual major elements. In addition, particular attention was paid to minor or trace elements, including radon. The overall chemistry of the waters showed an alkaline and alkaline-earth bicarbonate chemistry typical of waters that have been in contact with volcanic rocks. The plotting of Rn versus pCO_2 showed a definite, positive and significant correlation. The same is true also of relationships between temperature and pCO_2, and between conductivity and pCO_2. Accordingly, the basic geochemical process in water-rock interactions must imply a high input of carbon dioxide and heat into the groundwater, and to the subsequent leaching of cations from the volcanic minerals by carbonic acid. In this case, CO_2 flow seems to be responsible for the enrichment of Rn following the leaching of the radiogenic volcanites. Strong variations of radon (up to 70%) were observed. The authors concluded that the change in the concentration of He (He was also measured and found to decrease after the eruption) and Rn may be attributed to significant changes in fluidodynamic conditions governing the permeable conduits that allow them to ascend towards the surface and to spread into the water table. The work by D'Allessandro and Parello (1994) is interesting not only in itself but also in the sense that it supplements that of Dall'Aglio *et al.* (1994), which was a soil degassing study. The following gases were sought and their amounts determined: He, H_2, O_2+N_2, CO, CO_2 and Rn. Again, a very strong dependence of Rn on carbon dioxide was found. The role of carrier played by CO_2 in favour of the less mobile and stable Rn seems to be ascertained in this case. Before the December 1991 eruption, a clear increase was observed, followed by a sharp drop after the beginning of 1992. In a different place, where the dominant phase is methane, the authors again found an almost one-to-one correlation of the variations of concentration of Rn with those of CH_4. The methane was playing there the role of carrier for radon.

The Vulcano Island (which was supposed to be the home of Vulcan, the Graeco-Roman god of fire and from which derives the word "volcano") is part of the Aeolian archipelago which lies off the northern coast of Sicily, Italy. Del Pezzo *et al.* (1981) have carried out there a radon survey in a 10 m deep well plunging in a shallow aquifer. They observed a 10-fold increase in the radon readings a month before a seismic swarm started. The radon activity was rapidly decreasing when the seismic swarm reached its peak. However, the lag time between the radon peak and the time of increased seismic activity casts some doubt over the causal relationship between the two phenomena, particularly when taking into account the low magnitude of the seisms. The authors suggest that this radon anomaly - and possibly radon anomalies in general - over volcanoes, and the seismic activity could both be responses, with different delay times, to the same volcanic triggering effect, i.e. an increase of magmatic pressure at depth. This possibility was further substantiated on the Vulcano by an increase of the reducing power observed at the fumaroles simultaneously with the Rn increase.

Cioni (1990) and Cioni *et al.* (1984) have reported geochemistry surveys on the Vulcano, including some radon studies. The measurements were performed in fumaroles. In the beach fumaroles, the Rn content was found approximately constant with time (about 4 kBq l^{-1}). In the crater fumaroles, significant temporal variations were observed. However, a direct correlation beetween the radon content and the water fraction of the fumarole was found; thus indicating that radon is linked with the shallow fraction. The authors stress the fact that should a lateral boiling due to a pressure drop along fractures begin in an already Rn-rich aquifer, then a fluid even richer in radon would be produced; the steam/liquid water distribution coefficient being larger than one.

Another set of multi-element geochemistry survey was reported by Baubron *et al.* (1991) and by Toutain *et al.* (1992). The latter team has focused its attention on simultaneous measurements of CO_2, He, and Rn in the atmosphere of a water-well located at the base of the volcano. Over several days of recording, identical variations were found for the three gases. These periodic variations have a daily component (two daily maxima) and to a lesser extent they are modulated over a longer period of several days. When seeking a relationship between these gases, a direct correlation was found for both He and Rn with respect to carbon dioxide. The higher the CO_2 concentration of the atmospheric phase above the aquifer, the larger the concentration in He and Rn. Again, the carrier role played by CO_2 is pointed out. The finding suggests that radon is carried out of the water by bubbling CO_2. As a matter of fact, the role of micro-gas bubbles as a carrier was then experimentally and theoretically demonstrated by Varhegy *et al.* (1992). Following a radon extraction, by means of radon-free bubbles, (as demonstrated by Zagin and Saskina, 1966) this could play a determining role in radon transportation in aquifers situated below a volcanic structure. An important point stressed by Baubron and his co-workers (1991) is the fact that bulk concentration variations could probably not be usefully utilized as an assessment of the volcanic activity. On the contrary, these authors strongly advocate the use of correlated data such as concentration ratio between two species.

Baubron *et al.* (1991) have been sampling gas phase at a depth of 70 to 100 cm below ground level. Their overall conclusion is the same as above, partly because some of the reported results are identical. In addition, the authors set a sort of concomitant lower threshold for the three gases CO_2, He and Rn, above which such a gas anomaly could be considered as a signal of a possible resumption of volcanic activity. The set thresholds are 10% for the CO_2 concentration, 10 ppm for He, and 100 kBq m^{-3} for Rn.

The Krafla volcano is located in Iceland. It has also received some attention with the aid of Rn measurements. Hauksson (1981b) carried out a radon monitoring campaign on a fumarolic vent situated within the volcano caldera. Sampling was achieved at a frequency varying from 2 days during periods of expected tectonic activity to twice a week during the less uplift-active periods. In the mean-time, the water level of a nearby well was also monitored. The fissure extension and land elevation, as recorded by Bjornson *et al.* (1979) and continued afterwards, could be checked against Rn concentration variations. The most striking result of this work is the concomitant variations of radon emission (peak shaped variation), of water level in the well (peak shaped variation), of the land elevation (sharp drop followed by a regular but slower increase), and fissure extension variation (slow increase followed by a sharp drop at the time of common occurrence). The phenomenon repeated itself four times in a row. However, if they correlate well with the ongoing sequence of uplift and subsidence, the recorded radon data do not exhibit an obvious correlation with the seismicity. After a detailed analysis of their data in the light of existing models but primarily of that of D'Amore and Sabroux (1977), the authors calculated a radon velocity in the range 26 to 37 m s^{-1}. They pointed out that the simultaneous and correlated changes of radon and geothermal water-table level supported the conclusion that changes in the transport path between the water table and the surface fumaroles are the most likely mechanism for variation in radon emission at fumarole's level.

The 1975-1982 rifting activity in the Krafla volcanic centre was the first of its kind observed by scientists. Oskarson (1984) has reported fumarole discharge during this period of time. Among the monitored gases was Rn. The Rn activity was found to vary from 360 Bq m^{-3} up to 33 kBq m^{-3}, i.e. over a rather wide range. In order to check whether Rn came from water, from which it had been extracted by bubbles formed in the aquifer prior to and during actual boiling of the fluid, correlations were sought with the other inert gas flows and nitrogen. Such a trend was, indeed, found. It was also found that the petrochemical composition of the bedrock would control the radon abundance in the discharge. In a particular place, the radon data have confirmed that the studied fumarole belonged to the radial system and that its previous discharge was marked by the magmatic CO_2.

During the same rifting period, the Nordic Volcanological Institute laid a network of track detectors in the soil of the Krafla along a transverse that intersected the caldera rim. The surveys were carried out before and after the September 8th, 1977, eruption. A 5- to 20-fold increase in the Rn output was found before the eruption. In addition, the highest readings were found at the measuring sites located close to the trace of the caldera fault system, again underlining the major role played by fissures and faults in the transportation of Rn (Monnin and Seidel, 1991). The survey was again confronted with an eruption on November 18th, 1981. Before the event itself, a magmatic intrusion was well identified by severals means. It was preceded by a 3-fold increase, on average, in the measured Rn concentration. A subsidence phase followed, which was accompanied by a decrease in the radon concentration in the soil gas. Then another increase of the Rn concentration led ultimately to the eruptive phase. This set of events showed, for the first time, that radon measured in the ground could be taken as a precursory signal on some volcanic edifices, provided that the measuring sites were well chosen (Seidel, 1982).

4.3.3.2 North America and the Pacific

The so-called "fire belt" runs along the western coast of the American continent. A large number of volcanoes can be found there, and the largest set of them is located in Mexico. Several volcanoes have been monitored from the point of view of in-soil radon measurements.

The Paricutin is a young volcano born in 1943. On the hypothesis that this volcano emerged by the presence of a fissure, eight radon monitoring stations have been placed. Until now, the radon pattern has shown no particular feature - only a very good stability and reproducibility of data as a function of time.

The Colima volcano is the most active of the area with a predominant explosive activity. During the last 450 years, 29 explosive eruptions have occurred. The volcano is presently in a repose period and presents only an intense fumarolic activity. A radon network was placed inside the ancient crater. As for the Paricutin, the Rn counting pattern has been found to change very little during a two-year survey. It is worth mentioning that all changes at all the stations were correlated point-to-point. In addition, by mid-1982 a general rise of the Rn content was observed at all the stations. However, this general change could not be correlated either to fumarolic activity or to any other detectable change (Seidel and Monnin, 1984).

El Chichón volcano in the State of Chiapas in Mexico was known as an active volcano, but had been dormant for centuries. It erupted violently on March 28th, 1982. Two major eruptions followed on April 3rd and 4th, 1982. Five Rn stations were installed on the northern part of the volcano soon after the eruptive phase ended. The network was monitored for 11 months. Immediately after the eruptions, radon activity levels showed values one order of magnitude greater than those found several months later. It is known that atmospheric conditions can modify the Rn flow within the upper part of the soil. The data analysis with respect to climatic parameters suggests that the initial high values of Rn were unrelated to the weather. On the contrary, the smaller variations that followed were strongly correlated with rainfall. A Rn peak was observed in May, and it seems to be correlated with a flood that occurred at the end of the month when a pyroclastic deposit dam gave way. This produced a drastic change in the hydrological regime of the area, as was reflected by the Rn peaks. This result again supported the statement that Rn directly related to the eruption has been decreasing exponentially with time after the eruption, while the smaller variations observed later are related to climatic factors or to changes in the hydrological regime. Immediately (i.e. 33 hours) after the first eruption, samples of the distal ash deposit were taken to the laboratory. Contents of Rn daughters (^{214}Pb and ^{214}Bi) and thoron daughters (^{212}Pb and ^{212}Bi) were measured. They showed very high values. If Pb and Bi activities were in equilibrium with the reported values of U and Th content in juvenile ejecta of El Chichón, the expected activities should have been smaller by 3 orders of magnitude for the Rn daughters and by 6 orders for the thoron daughters. If one assumes that the radon released at the time of the eruption came from the magma, one cannot explain the high activities of its daughters. The authors (De la Cruz *et al.*, 1985) propose as an explanation that an important amount of the radon contained in the rocks surrounding the magma body was transported into it and dissolved in the melt. This argument can be supported by the fact that water above its critical point (374°C, 218 bar) and any noble gas would be perfectly miscible (Mitchell and Terrell, 1984). Such conditions are attained at depths below 800 m and the temperature of the very water-rich melt is reported to be around 800°C (Luhr *et al.*, 1984).

At the time of writing (winter 1995), the Popocatepetl, Mexico, volcano has been in an eruptive phase for more than a year. In-soil Rn measurements are carried out by means of electronic real-time detectors, but data are not yet available.

The Hawaii archipelago (in the Pacific ocean) results from a typical "hot spot". Kilauea is a young, basaltic shield volcano on the southeastern flank of the island of Hawaii. Its structure is dominated by a central summit caldera (3 km x 5 km) and major eastwards and southwestwards rift systems (Macdonald *et al.*, 1983).

A Rn survey was started in 1978 by Cox *et al.* (1980). Rn was monitored at 9 stations using LR 115 detector buried 30 cm below ground level. Most of them were located near faults. The survey started at a time of slow inflation of the summit. Some negative values of Rn degassing were found when subtracting the Rn activity background from actual counting. Such negative readings were attributed to downward gas flows. A minor eruption occurred on November 16th-17th, 1979. Before and during the eruption, the soil radon concentration exceded by 0.5 to 20.5 times the background as measured during the quiescent phases. The highest concentrations were found near an eruptive fissure formerly active six years earlier. Another strong Rn outgassing was found close to an area of steaming grounds related to an intrusive event that occurred in 1938. These findings again support the idea that thermal release and easy pathways lead usually to high rates of Rn transportation towards the surface. Three of the Rn stations were kept in operation. In February 1980 they exhibited a smaller increase in Rn; which was followed by two seismic swarms (March 2nd and 9th). The authors attributed the increase in outgassing either for the eruptive or the seismic events to movement of the magma towards the surface and associated deformation of the country rock. They further thought that the rise of magma induces temperature increases that would facilitate the volatilization of Rn from the magma, and fracture permeability that in turn would facilitate the transport of radon toward the surface. The decrease in Rn after the event is then attributed to the exhaustion of the available Rn by flushing in the strong outgassing.

However, further monitoring of the Kilauea summit and flanks (Cox, 1983; Thomas and Konyanagi, 1986; Thomas *et al.*, 1986) demonstrated that intrusions and eruptions could produce substantial radon anomalies at stations so far away from the volcanic activity that a different mechanism needed to be proposed. The studies first confirmed that meteorological conditons did not produce important variations in the soil Rn content of the Kilauea owing to the mild seasonal temperature variations and to the high surface permeability of the topsoil, which usually prevents saturation. Several high radon-activity phases were observed. The first one was correlated to seven intrusions, one eruption and 13 earthquakes. Another period of higher activity was linked to a renewal of the summit activity with caldera eruptions. A further set of high Rn activity occurred corresponding to the onset of a sequence of eruptions. To account for the findings the authors proposed a mechanism which would be that the "deformation of the volcanic edifice associated with inflation and deflation of the summit magma chamber, and intrusion of magma into the flanks of the volcano was sufficient to generate microfracturing in near-surface rocks. The resultant increase in the release rate of radon was believed to be sufficient to account for the radon anomalies observed".

Another piece of information regarding the Rn release mechanism is due to Gill *et al.* (1985). Activities of Rn and ^{210}Po were measured in a basic andesite lava sawed off the lava bed one day after the eruption. In-growth pattern showed that all Rn was volatilized during the eruption and that Po degassing was also complete. If this could be generalized to all volcanic eruptions, it would again support De la Cruz *et al.*'s hypothesis (1985).

4.3.3.3 Central America

Rn measurements were also performed on basic andesite lava from the Arenal volcano in Costa Rica. This volcano has been erupting such lava flows almost continuously since 1968. The sample was collected 2 or 3 days after it was discharged. As found on the Kilauea, Rn was completely degassed during exit from the vent. However, ^{210}Po had been removed only up to 84%. Five in-soil Rn stations have been set on this volcano. Although from 1981 to 1984 it has emitted 6 block lava flows, radon monitoring has shown very few appreciable changes. The result is attributed to poorly selected sites.

The Poas volcano is also located in Costa Rica. When the first Rn monitoring started, it was in a repose period with fumarolic activity from the principal crater, with temperatures up to 970°C. The Rn stations were placed along an approximately straight line that started from one rim of the crater, crossed it, and then went down the flank. The Rn concentration pattern in the soil shows a very strong Rn degassing inside the crater with respect to measurements achieved outside it. In the mean-time, measurements of Cl^-, SO_4^{2-} and F^- anions were performed on the water of a summit lake. During quiet phases, no correlation could be found between these parameters and Rn. However, the volcano has now entered a more active phase. Comparison of data shows a direct correlation again between Rn and the concentration of the above-mentioned ions in the water.

The Irazu volcano is located near San Jose, the capital city of Costa Rica. The radon survey started there during a phase of quiescence. For as long as it lasted, little variation in Rn in the soil was observed. The Rn track detectors were then changed for three real-time Rn probes. The probes were placed before the December 8th, 1994, eruption. In tropical latitudes, much care should be taken to account for precipitations when assessing Rn data. For the full duration of the months of October, November and December the rainfall was 390 mm, 270 mm and 28 mm, respectively. Heavy rainfalls are known to induce a sharp decrease of Rn in the topsoil, presumably by closing up the available pores. During October, the three stations exhibited slight deviations from the average activity. At the station that produced the most striking Rn anomaly, the average concentration was about 300 kBq m^{-3}. During November the Rn content gradually increased, reaching its peaks (i.e. 2.5 times the average background) around November 15th. Yet the rainfall had not been less severe than during October. The Rn concentration then decreased and kept decreasing during the month of December for the three stations. It then started again to increase slightly for one of them. The eruption that occurred on December 8th was of a phreatic nature. It is then likely that during its build-up, large amounts of dry vapour had been emitted, which transported Rn toward the surface. The relatively weak blocking due to heavy rainfall could not prevent its motion. Afterwards, continuous flushing right before and during the eruption accounts for the rapid decrease in Rn concentration. The slight increase observed at one station would then be attributed to the renewed build up of radon in the upper layers of the bedrock.

4.3.3.4 Indian Ocean

Two islands in the Indian ocean are of volcanic origin. Mauritius is one of them, but no volcano is any longer active there. The island of La Réunion, France, which is believed by some to be the second daughter of the same hot spot that produced Mauritius, houses the Piton de la Fournaise, a fairly active volcano.

The Rn monitoring of the Piton was started, using SSNTDs. Seven measuring sites were selected after a seismic and topographic survey, and the detectors were placed in the most active zones based on eruptive and volcano-tectonic criteria. Data recorded from 1983 to 1986 were reported by Seidel *et al.* (1988a, b). They showed large Rn peaks. Some of them occurred before eruptive phases, others not. The Fourier transform analysis of the rain and radon data showed a strong correlation corresponding to a twice-yearly wave-length: which links Rn data to tropical heavy-rains effects. However, for higher frequencies the correlation with rainfall is not clear. Some of correlations are clearly related to heavy rain, but on the contrary the two highest anomalies could not find their origin in any meteorological factor. They are definitely correlated with the volcanic activity which was revealed by the December 3rd, 1983, and June 14th, 1985, eruptions. The remaining radon peaks offer the double characteristics to be possibly attributed to rainfall as well as to volcanic activity. As with other volcanoes, some consideration may be given to the relationship between Rn response and the location of the measuring sites. Three stations are located close to the main fracture zone of the June 1985 eruption, and the data they produced reflect the proximity of the intrusive phenomenon.

The track detector network was later complemented with automatic probes. A week before an intrusive phase, very high Rn peaks were observed in 1991 at several stations. However, whether such anomalies were related to the volcano's activity could be questioned, because the signal-to-noise ratio was as high as 10^4: much higher than any that had ever been recorded. A new set of 20 automatic probes, protected from moisture effects, were then set in place. Unfortunately (for the researchers!), the volcano has entered a phase of quiescence since the beginning of 1992. The data recorded since then exhibit temporal variations that are connected with meteorological conditions, mostly rainfall.

4.3.3.5 Far Eastern Asia

In 1991, Notsu *et al.* reported the hydrological and geochemical changes related to the 1989 seismic and volcanic activities of the Izu peninsula, Japan. They have concentrated on a hot spring in Naka-Izu, where the water is self-spouting from a 700 m deep well. Radon fluctuations were found to be small in summer and large in winter, but oscillating only from a rather constant baseline. Starting at the beginning of July 1989, the radon concentration increased steadily. Practically at the same moment the seismic swarm activity became intense. From then on, Rn decreased and increased almost in phase with seismic activity. Eventually the number of earthquakes decreased. At this time volcanic tremors began to occur when the Rn concentration again increased, until it dropped just before a submarine eruption took place about 3 km off the eastern coast of the Izu peninsula. This combination of events is of considerable interest, particularly when no appreciable Rn variation was observed at a groundwater station located only 1 km away from the previous one. The importance of a judicious choice of measuring sites could not be better emphasized.

To pay tribute to the pioneering work of Chirkov, we have kept to the end the very first report on Rn concentration variations related to volcanic activity. Chirkov (1975) reported having regularly measured Rn concentrations in thermal springs on the Karimsky volcano, Kamchatka peninsula, Russia. The survey started in 1966 and was continued until 1972. After an eruption that occurred in 1966, the Rn concentration decreased until mid-1968. It then increased slightly for approximately a year. In October 1969 the Rn concentration increased dramatically, and at its peak reached a signal-to-background ratio of up to about 7. In the mean-time, the local seismic activity also increased, until another eruption occurred. When the lava flow ceased, the Rn concentration in the thermal spring also decreased. It further decreased for several months, and ultimately reached a practically constant activity level. The study was continued through a second eruptive cycle (Flerov *et al.*, 1986). Radon concentrations were again found to be well above background during intense seismic and eruptive activity, and then decreased as the eruption rate fell. These initial findings of Chirkov and co-workers were the starting point of a new series of international investigations into precursory signals of volcanic eruptions.

4.3.4 Mechanisms

We would now like to briefly review the mechanisms that can be invoked in order to explain the field data, provided that these do reflect some internal behaviour of geophysical significance. In other words, data processing must be carried out in order to extract genuine "signals" from variations due to local conditions. An interesting paper has been recently published along these lines (Pinault and Baubron, 1996) for soil Rn concentration.

4.3.4.1 Fumaroles

Hauksson (1981b) attributed the Rn activity variations to changes in the travel time of the steam carrier from the water table to the surface discharge point. The calculated Rn concentration increases were consistent with the field data. However, according to this mechanism a delay in the Rn activity should be present; but this has not been found in the field data. Cioni *et al.* (1984) have proposed a mechanism in which two different sources provide the fumarole fluids: a shallow high-Rn-bearing vapour, and a deep magmatic gas with a less important Rn charge. Changes in the mixing ratio of these fluids seem to be able to account for the observed behaviours.

4.3.4.2 Soil

The first type of mechanism invoked for Rn soil observation is that an increased heat flow or steam discharge could increase top soil Rn concentrations (Cox, 1980; Cox and Cuff, 1981; De la Cruz *et al.*, 1985). This model is reasonably conceived in the light of the processes that occur during an intrusion or an eruption. However, the reported data from Kilauea show that some anomalies might be recorded at a relatively large distance from intrusive or eruptive activities.

The second type of mechanism one can think of is that increased Rn activities in the soil are due to the collapse of pore volume and upflow of deeper radon-rich ground gas. This was first proposed by King (1979, 1986). However, Rn anomalies substantially larger than those that the model allows for have been observed. A transient model proposed by Monnin and Seidel (1988) suggests that higher Rn anomalies could be generated by an increase of the velocity of the upward gas column that would sweep the pre-existing radon in front of it.

The final model proposed is that stress-induced microfracturing increases the rate of radon release from the rock matrix, and this seems to be consistent with laboratory and field data (Thomas *et al.*, 1986). However, the levels of stress that are required are higher than those known to occur at some distance from an eruptive vent or an intrusion. On the other hand, the stress distribution on a volcano is far from uniform and local hyper-stresses could occur.

4.3.4.3 Groundwater

The first possible mechanism, as proposed by Chirkov (1975), was that stresses induced by the intrusion of magma into the shallow magma chamber cause micro-fracturing, allowing radon to be released from rocks. Laboratory experiments support this mechanism (Holub and Brady, 1981; Honda *et al.*, 1982); but it is not clear whether the degree of fracturing one could expect in nature would explain the magnitude of changes which are observed in the field.

Another attempt to explain Rn anomalies in groundwater was to consider that micro-fracturing could generate ultrasonic vibration, which in turn would liberate Rn from its parent rocks (Flerov *et al.*, 1986; Thomas, 1988). However, the amount of vibrational energy in laboratory experiments is several orders of magnitude higher than what micro-fracturing could produce.

Gasparini and Mantovani (1978) and Del Pezzo *et al.* (1981) have suggested that the influx of magmatic gas into the shallow groundwater could explain increases in the Rn concentrations. However, in order to produce anomalies, Rn would have to be rapidly transported throughout the water system before decaying away. The micro-bubble transportation mechanism suggested by Varhegyi *et al.* (1992) could offer the additional means of transportation.

This section owes much to Thomas (1990), who also proposed the aquifer breaching/fluid mixing model (Thomas, 1988) which, though proposed for earthquake-induced geochemical anomalies, could possibly apply to volcanoes.

4.3.5 Concluding Remarks

The examination of the available data strongly suggests that volcanic activity does, in many different ways, influence the Rn values that can be detected in fumaroles, groundwaters and soils over or near a volcanic edifice. However, because of the very complexity of such systems, the correlations sought are not yet clearly and firmly established. Increasing the number of case studies is certainly the key to an improved understanding and a more efficient approach to the problem. The need for the highest possible temporal resolution is a clear conclusion that one could draw from the existing data. (A good cost-to-efficiency compromise could be automatic revolving SSNT detectors.) However, both real-time measurement instruments and extensions of monitoring networks are costly undertakings. A good compromise could be the selection of a small number of representative and active volcanoes that could be intensively monitored. Finally, Rn measurements should be definitely supplemented by measurements of other physical or chemical parameters.

Under such circumstances, knowledge of the geochemistry of volcanoes could rapidly increase in the immediate future.

References

Aubert M. and Baubron J. C. (1988) Identification of a hidden thermal fissure in a volcanic terrain using a combination of hydrothermal convection indicators and soil-atmosphere analysis. *J. Volcanol. Geotherm. Res.* **35**, 217-225.

Baubron J. C., Allard P., Sabroux J. C., Tedesco D. and Toutain J. P. (1991) Soil gas emanations as precursory indicators of volcanic eruptions. *J. Geol. Soc. London* **148**, 571-576.

Bjornsson A., Johnsen G., Sigursdsson S. and Thorbergsson G. (1979) Rifting of the plate boundary in North Iceland 1975-78. *J. Geophys. Res.* **84**, 3029-38.

Bottiglieri L., Condorelli D., Pece R. and Tedesco D. (1987) Misure di radon in aree vulcaniche italiane (Campi Flegrei, Vesuvio, Etna, Vulcano). *GNV Bull.*, 159-174.

Carbonnelle J., Dajlevic D., Zettwoog P. and Sabroux J. C. (1982) Gas output measurements from an active volcano. *Bull. Volcanol.* **45**, 3, 267-268.

Cheminée J. L. (1966) Nouveaux résultats sur la teneur en radon des fumerolles des volcans italiens. C. R. Somm. *Soc. Geol. Fr.* **10**, 399-401.

Chirkov A. M. (1975) Radon as a possible criterion for predicting eruptions as observed at Karimsky volcano. *Bull. Volcanol.* **37**, 126-131.

Cioni R. (1990) Geochemistry applied to volcanic surveillance: the case of Vulcano island. In *Proc. Int. Workshop on Radon Monitoring in Radioprotection, Environmental Radioactivity and Earth Sciences* (eds. L. Tommasino *et al.*), Trieste, 1989, World Scientific, Singapore, pp. 315-339.

Cioni R., Corazza E. and D'Amore F. (1984) Radon in fumarolic gases from Vulcano island (Sicily, Italy). *Geothermics* **13**, 4, 385-388.

Cox M. E. (1980) Ground radon survey of a geothermal area in Hawaii. *Geophys. Res. Lett.* **7**, 4, 283-286.

Cox M. E. (1983) Summit outgassing as indicated by radon, mercury and pH mapping, Kilauea volcano, Hawaii. *J. Volcanol. Geotherm. Res.* **16**, 131-151.

Cox M. E. and Cuff K. E. (1980) Rn and Hg surveys: Geothermal exploration in N.E. Maui, Hawaii. *Trans. Geotherm. Resour. Counc.* **4**, 451-454.

Cox M. E., Cuff K. E. and Thomas D. M. (1980) Variations of ground radon concentrations with activity of Kilauea volcano, Hawaii. *Nature* **288**, 74-76.

Crenshaw W. B., Williams S. N. and Stoiber R. E. (1982) Fault location by radon and mercury at an active volcano in Nicaragua. *Nature* **300**, 345-346.

Cuff K.E. and Thomas D.M. (1985) Soil gas radon changes at Kilauea volcano, Hawaii. *Eos. Trans. Am. Geophys. Un.* **18**.

Dall'Aglio M. (1993) Geochemistry of natural fluids: The link between natural waters management and natural disasters surveillance. In *Sci. Meeting on the Seismic Protection*, Venice, pp. 133-136.

Dall'Aglio M., Pagani F. and Quattrocchi F. (1994) Geochemistry of the groundwaters in the Etna region before and after the paroxysmal phase of the eruption of December 1991: Implications for the geochemical surveillance of Mt Etna. *Acta Vulcanol.*

D'Alessandro W. and Parello F. (1994) Soil degassing from the lower flanks of Mt Etna. *Environ. Geochem. Health* **16** (suppl.), 189-201.

D'Amore F. and Sabroux J.C. (1976/77) Signification de la présence de radon-222 dans les fluides géothermiques. *Bull. Volcanol.* **40**, 2, 106-115.

D'Amore F., Sabroux J. C. and Zettwoog P. (1978/79). Determination of characteristics of steam reservoirs by radon-222 measurements in geothermal fluids. *Pure Appl. Geophys.* **117**, 253-261.

De la Cruz S., Mena M., Segovia N., Chalot J. F., Seidel J. L. and Monnin M. (1985) Radon emanometry in soil gases and activity in ashes from El Chichón volcano (Mexico). *Pure Appl. Geophys.* **123**, 410-421.

Del Pezzo E., Gasparini P., Mantovani M. S. M., Martini M., Capaldi G., Gomes Y. T. and Pece R. (1981) A case of correlation between Rn anomalies and seismic activity on a volcano (Vulcano Island, Southern Tyrrhenian Sea). *Geophys. Res. Lett.* **8**, 9, 962-965.

Dobrovolsky I.P., Zubkov S.I. and Miachkin V.I. (1979) Estimation of the size of earthquake preparation zone. *Pure Appl. Geophys.* **117**, 1025-1044.

Dongarra G. and Martinelli G. (1993) Migration processes of radon towards the earth surface: implications for the prediction of seismic and volcanic events. In *Sci. Meeting on the Seismic Protection*, Venice, pp. 141-147.

Flerov G. N., Chirkov A. M., Tretyakova S. P., Dzolos L. V. and Merkina K. I. (1986) The use of radon as an indicator of volcanic processes. *Izvestya Earth Phys.* **22**, 213-216.

Gasparini P. and Mantovani M. S. M. (1978) Radon anomalies and volcanic eruptions. *J. Volcanol. Geotherm. Res.* **3**, 325-341.

Gill J., Williams R. and Bruland K. (1985) Eruption of basalt and andesite lava degasses ^{222}Rn and ^{210}Po. *Geophys. Res. Lett.* **12**, 17-20.

Hauksson E. (1981a) Radon content of groundwater as earthquake precursor: evaluation of worldwide data and physical basis. *J. Geophys. Res.* **86**, 9397-9410.

Hauksson E. (1981b) Episodic rifting and volcanism at Krafla in North Iceland: radon emission from fumaroles near Leirhnjukur. *J. Geophys. Res.* **86**, 11806-11814.

Holub R.F. and Brady B.T. (1981) The effect of stress on radon emanation from rock. *J. Geophys. Res.* **86**, 1776-1784.

Honda M., Kurita K., Hamano Y. and Ozima M. (1982) Experimental studies of He and Ar degassing during rock fracturing. *Earth Planet. Sci. Lett.* **59**, 429-436.

Iwasaki I., Katsura T., Shimojima H. and Kamada M. (1956) Radioactivity of volcanic gases in Japan. *Bull. Volcanol.* **18**, 2, 103-123.

Iwasaki I., Ozawa T., Yoshida M. and Kamada M. (1975) Forecast of volcanic eruptions by chemical methods. *Bull. Volcanol.* **39**, 1, 91-103.

King C.Y. (1979) On radon emanation as a possible indicator of crustal deformation. *Tectonophys.* **52**, 120.

King C.Y. (1986) Gas geochemistry applied to earthquake prediction: An overview. *J.*

Geophys. Res. **9**, 12269-12281.

Koski R.A., Normark W.R. and Morton J.L. (1985) Massive sulfide deposits on the southern Juan de Fuca ridge: Results of investigations in the USGS study area, 1980-83. *Mar. Min.* **5**, 147-164.

Leonard R.B. and Janzer V. J. (1978) Natural radioactivity in geothermal waters, Alhambra hot springs and nearby areas, Jefferson county, Montana. *J. Res. U.S.G.S.* **6**, 4, 529-540.

Lombardi S., DiFilippo M. and Zantedeschi L. (1984). Helium in Phlegrean Fields soil gases: July 20th-26th - September 19th-25th, 1983. *Bull. Volcanol.* **47**, 259-265.

Lombardi S. and Reimer G. M. (1990) Radon and helium gases in the Phlegraean fields, Central Italy. *Geophys. Res. Lett.* **17**, 6, 849-852.

Luhr J.F., Carmichael I.S.E. and Varekamp J.C. (1984) The 1982 eruption of El Chichon volcano, Chiapas, Mexico: mineralogy and petrology of the anhydrite-bearing pumices. *J. Volcanol. Geotherm. Res.* **23**, 69-108.

Macdonald G.A., Abbott A.T. and Peterson F.W. (1983) *Volcanoes in the Sea: the Geology of Hawaii.* University of Hawaii Press, Honolulu.

Martinelli G., Albarello D. and Mucciarelli M. (1995) Radon emissions from mud volcanoes in Northern Italy: possible connection with local seismicity. *Geophys. Res. Lett.* **22**, 15, 1989-1992.

Matsuo S., Ossaka J., Hirabayashi J., Ozawa T. and Kimishima K. (1982) Chemical nature of volcanic gases of the Usu Volcano in Japan. *Bull. Volcanol.* **45**, 3, 261-264.

McCabe W.J., Barry B.J. and Manning M.R. (1981) Radioactive tracers in geothermal underground water flow studies. *N.Z. Inst. Nucl. Sci. Rep.*, INS-R-298.

Mitchell J.G. and Terrell D.J. (1984) Noble gas solubility in supercritical water: implication for inert gas studies and geochronology. *Geof. Int.* **23**, 483-490.

Monnin M. and Seidel J.L. (1988) Sur une hypothétique émission intense de radon avant un événement géophysique majeur: une analyse théorique. *C. R. Acad. Sci. Paris*, **307**, ser. II, 1363-68.

Monnin M. and Seidel J. L. (1991) Radon in soil, air and groundwaters related to major geophysical events: recent advances. *Nucl. Instr. Meth.* **A314**, 316-330.

Navarro M., Canales O., Strauch W., Conway F. M., Lafemina P. and Hill B. E. (1995) Cerro Negro: Small lava flows in main crater; ash eruptions end in mid August. *Global Volc. Net. Bull.* **20**, 9, 9-10.

Nielson D.L. (1978) Radon emanometry as a geothermal exploration technique; theory and an example from Roosevelt hot spring. KGRA, Utah, *Rept. Earth Sci. Lab., Univ. Utah Res. Inst.*, ESL-14.

Notsu K., Wakita H., Igarashi G. and Sato T. (1991) Hydrological and geochemical changes related to the 1989 seismic and volcanic activities of the Izu Peninsula. *J. Phys. Earth* **39**, 245-254.

Oskarson N. (1984) Monitoring of fumarole discharge during the 1975-1982 rifting in Krafla volcanic center, North Iceland. *J. Volcanol. Geotherm. Res.* **22**, 97-121.

Pinault J.L. and Baubron J.C. (1996) Signal processing of soil gas radon, atmospheric pressure, moisture, and soil temperature data: A new approach for radon concentration modeling. *J. Geophys. Res. - Solid Earth* **101**, B2, 3157-171.

Rosenberg N.D., Lupton J.E., Kadko D., Collier R., Liley M.D. and Pak H. (1988) Estimation of heat and chemical fluxes fropm a seafloor hydrothermal vent field using radon measurements. *Nature* **334**, 604-609.

Seidel J. L. (1982) Radon emanometrie appliquée á la geophysique interne. Ph.D., Université Clermont II, pp. 1-81.

Seidel J. L. and Monnin M. (1982) Some radon activity measurements of geophysical significance. In *Proc. 11th Int. Conf. on Solid State Nuclear Track Detectors* (eds. P. H. Fowler and V. M. Clapham), Bristol, 1981, Pergamon Press, Oxford, pp. 517-523.

Seidel J. L. and Monnin M. (1984) Mesures de radon-222 dans le sol de l'Etna: 1980-1983. *Bull. Volcanol.* **47**, 4, 1071-1077.

Seidel J. L., Monnin M., Cejudo J., Chalot J. F., Segovia N., De la Cruz S., Mena M., Malavassi E. and Fernandez E. (1984) Radon emanometry in active volcanoes. *Nucl. Tracks Radiat. Meas.* **8**, 1-4, 411-414.

Seidel J.L., Bonneville A. and Lenat J.F. (1988a) Mesures de radon en relation avec l'activité du Piton de la Fournaise, Réunion, de 83 à 87. *C.R. Acad. Sci. Paris*, **306** (ser.2), 89-92.

Seidel J.L., Monnin M., Boneville A., Delorme H., Lenat J.F. and Segovia N. (1988b) Radon in soil measurements related to the activity of the Piton de la Fournaise volcano, La Réunion Island, France. *Nucl. Tracks Radiat. Meas.* **15**, 1-4, 567-570.

Semprini L. and Kruger P. (1984) Relationship of radon concentration to spatial and temporal variations of reservoir thermodynamic conditions in the Cerro Prieto geothermal field. *Geothermics* **13**, 1-2, 103-115.

Shiratoi K. (1927) The variation of radon activity of hot springs. *Sci. Rep. Tohoku Imperial Univ.* **16** (Ser.3), 614- 621.

Sicardi L. (1940) Il recente ciclo dell'attivita fumarolica dell'isola di Vulcano. *Bull. Volcanol.* **7**, 85-149.

Stoiber R. E. (1982) Emission of Hg and Rn from Nicaraguan and Costa rican volcanoes. *Eos Trans. Am. Geophys. Un.* **63**, 45.

Thomas D.M. (1988) Geochemical precursors to seismic activity. *Pure Appl. Geophys.* **126**, 241-266.

Thomas D. M. (1990) Radon gas as a tracer for volcanic processes. In *Proc. Int. Workshop on Radon Monitoring in Radioprotection, Environmental Radioactivity and Earth Sciences* (eds. L. Tommasino *et al.*), Trieste, 1989, World Scientific, Singapore, pp. 295-314.

Thomas D. M., Cuff K. E. and Cox M. E. (1986) The association between ground gas radon variations and geologic activity in Hawaii. *J. Geophys. Res.* **91**, B12, 12186-12198.

Thomas D. M. and Koyanagi R. Y. (1986) The association between ground gas radon concentrations and seismic and volcanic activity at Kilauea volcano. *Eos Trans. Am. Geophys. Un.* **67**, 905.

Toutain J. P., Baubron J. C., Le Bronec J., Allard P., Briole P., Marty B., Miele G., Tedesco D. and Luongo G. (1992) Continuous monitoring of distal gas emanations at Vulcano, Southern Italy. *Bull. Volcanol.* **54**, 147-155.

Varhegyi A., Hakl J., Monnin M., Morin J.P. and Seidel J.L. (1992) Experimental study of radon transport in water as a test for a transportation microbubble model. *J. Appl. Geophys.* **29**, 37-46.

Whitehead N.E. (1984) Geothermal prospecting by ground radon measurements. *J. Volcanol. Geotherm. Res.* **20**, 213-229.

Whitehead N.E., Gingrich J.E. and Fisher J.C. (1983) A New Zealand test of the track-etch method of prospecting for geothermal steam. *J. Volcanol. Geotherm. Res.* **15**, 339-354.

Wilkenning M. H. (1974) Radon-222 from the Island of Hawai: Deep soils are more important than lava fields or volcanoes. *Science* **183**, 413-415.

Williams S. N. (1985) Soil radon and elemental mercury distribution and relation to magmatic resurgence at Long Valley Caldera. *Science* **229**, 551-553.

Wollenberg H.A. (1975) Radioactivity of geothermal systems. In *Proc. Snd. U.N. Symp. on the Development and Use of Geothermal Resources,* U.S. Gov. Printing Of., Washington D.C., pp. 1283-92.

Zagin B.P. and Saskina N.N. (1966) Transfert du radon dans la chambre d'émanantion à l'aide de gaz carbonique. *Radiochimie* **8**, 125-126.

H. A. Khan[1], A. A. Qureshi[1] and I. E. Qureshi[1]

4.4 Radon and Mineral Exploration

4.4.1 Introductory Remarks

Uranium (^{238}U and ^{235}U) and thorium (^{232}Th) decay, over the geological time periods, through their respective well known decay series, generating a wide spectrum of daughter products. Uranium and thorium, as well as all their daughter products, except ^{222}Rn, ^{219}Rn and ^{220}Rn, are solids, and therefore, do not easily migrate from one place to another under normally prevailing conditions. The radioactive daughters ^{222}Rn, ^{219}Rn and ^{220}Rn, being gaseous, are mobile and can thus move away from their respective source materials (^{238}U, ^{235}U and ^{232}Th), through the process of forced and/or natural migration. They may also be carried away owing to the convectional flow of fluids in the earth's crust.

The radon isotope ^{219}Rn is insignificant because of the low abundance of ^{235}U, while ^{220}Rn is of less importance due to its very short half-life (55 s). So the radon isotope of our concern in this section will be ^{222}Rn. In the following subsections any mention of radon means ^{222}Rn, unless otherwise specified.

The isotopes ^{238}U and ^{232}Th are present everywhere in the earth, though in low concentrations in most cases. Average values of ^{238}U concentrations in common igneous rocks and earth's crust as a whole are given in Table 4.4.1 (Davis, 1947; Evan and Goodman, 1941; Koevil, 1938; Senftle and Koevil, 1947).

Table 4.4.1: Average concentration of ^{238}U in common rocks and earth as a whole.

Rock Type	^{238}U (ppm)
Ultramafics	0.03
Basic Rocks	2.00
Gabbros	0.94 - 0.46
Intermediate Rocks	1.4 - 3.0
Granites	2.8 - 4.0
Alkaline Granites	10.0 - 100
Earth's Crust	0.2

Owing to the ubiquitous presence of trace amounts of uranium, its daughter radon is widely distributed in rocks and soil gas. This chemically noble gas is radioactive and decays by alpha activity. It can diffuse through the soil and rocks and can be detected

[1]Pakistan Institute of Nuclear Science and Technology, Islamabad, Pakistan

down to a level of 10-20 atoms/million atoms. The importance of radon for mineral exploration rests on three features: (i) its radioactivity, (ii) its relation to uranium, and (iii) its inertness. Because of the radioactivity, radon can be measured in minute concentrations. The connection with uranium makes it useful in the search for uranium ores. Owing to its inertness it can travel appreciably long distances within the earth's crust, which makes it an effective tracer for understanding geophysical processes. The 'mobility' of radon can be helpful in providing useful information concerning the possible existence of sub-surface uranium mineralization. Therefore, if there exists some uranium mineralization at a certain depth, a fair amount of radon will be produced in time, which will have the tendency to migrate away from the source. A fraction of radon so produced succeeds in reaching the earth's surface before being lost in the process of natural decay. If some convenient and accurate method is employed for the measurement of radon gas, an area showing anomalously high radon signals may indicate the presence of uranium mineralization. The earliest measurements of radon were made by Behounek (1927) as a possible means of uranium prospecting on the basis of field experience around uranium deposits in Czechoslovakia. Further developments of field methods were made in the USSR during the early thirties, based on the determination of radon in soil gas (Kirikov *et al.*, 1932). Whereas the high radon levels are indicative of uranium deposits, the anomalously low radon values in some areas also have a certain significance. The oil/gas reserves are usually capped by impervious rocks, which hamper the normal flow of radon gas, and an area of low radon level is developed over oil/gas deposits. So, by the use of the method of radon measurement under favourable conditions, existence of petroleum and/or natural gas can also be predicted. The measurement of radon activities in soil gas and ground-water has also been made to map faults and to characterize geothermal sources. Experience shows that the high temperature present underneath the radon source causes the movement of the radon through cracks and pores in the earth's crust. Therefore, the observation of exceptionally high radon levels may also indicate the possible existence of a geothermal energy source lying deep underneath the earth's surface. The presence of a geological fault also facilitates the movement of radon. Thus, higher radon emission along a linear structure in earth crust indicates the existence of a fault. It shows that the technique of radon measurement in soil gas can yield a lot of information regarding the subsurface geological features and the presence of mineral, oil and gas reserves.

Over the years, a large number of techniques and methods have been developed to measure radon concentration in the 'soil gas' in selected areas of interest. These techniques can be classified into two categories namely, (a) Active or Instantaneous techniques, and (b) Passive or Time-Integrated techniques. The 'Instantaneous' radon measurement techniques are based upon methods in which grab samples of the soil-gas are collected at an instant of time, followed by measurement of radon concentration through its alpha radioactivity. On the other hand, the time-integrated radon measurement techniques are based upon radon measurements in the soil-gas over a cumulative period, varying from 4 to 6 weeks. In this time-integrated method, detection of alpha particles resulting from the decay of radon isotopes as well as of their daughter products is made. A number of time-integrated radon measurment techniques have been established by now. It has been experienced that all the known radiometric techniques based on the measurement of γ-activity have serious limitations when used for locating deeply buried uranium ore bodies. Under these conditions, extensive drilling and gamma logging methods are employed, which are expensive and require complicated instrumentation. The technique of radon measurement for exploration/prospecting of uranium mineralization using the Etched Track Detector (ETD) method has been quite successful. The required methodology for the successful application in mineral exploration has been developed during the last two decades or so.

In the following sections, the passive methods of radon detection using etched track detectors is described in detail. Our discussion on etched track detectors will include the principle of the technique, effects of tube size, position of the detector in the tube, exposure time and the environmental conditions. Finally, some successful applications of the technique, merits and demerits of the etched track detectors, will be discussed. In the end, application of this technique to locate oil and gas reserves will be discussed.

4.4.2 Migration of Radon

The distances to which radon can manage to migrate are of great interest when using the technique of radon measurement for uranium exploration. In this regard, it is important to study the effect of radon flow on its concentration in the ground. Fig. 4.4.1 summarizes the pattern of radon concentration variation as a function of depth in the ground. The dashed-line curve represents the conditions when the flow results only from diffusion and the vertical flow is assumed to be zero. The solid curves on its right and left have been calculated on the assumption that the net upward and downward flows, respectively, have been taking part in the movement of radon. Drastic shifts in the radon concentration profiles take place owing to the effect of some net upward and downward flows (in this figure the flow velocity of $\pm 7\sqrt{D/\tau}$ per unit porosity has been used, where D is the 'diffusion coefficient' and τ is the mean life of radon). If the flow was only due to diffusion, the typical value of $\sqrt{D\tau}$ in the soil would be only 1.2 m. If it is assumed that the ratio of the radon emanating power of the ore to that of the surrounding matrix is 10^4, the excess signal will exceed the background radon concentration only up to a distance of eleven metres. However, uranium has been detected at depths of more than 200 metres using radon-based measurements. This clearly indicates that diffusion is not the only cause of radon flow. It is thus quite obvious that a vertical flow of gases in the upper part of the earth is greatly responsible for transporting radon to the surface region where the radon concentration measurements are made.

4.4.3 Methods of Radon Measurement

A large number of techniques and methods are available to measure radon concentration in soil gas (Andrew and Wood, 1972; Chirkov, 1976; Fleischer *et al.*, 1975; Iwasaki *et al.*, 1956; King 1979; Shiratoi, 1972; Sempirini and Kruger, 1984; Soonawala and Teleford, 1975; Tanner, 1964, 1980; Tidjani, 1984). All techniques for the measurement of radon and its daughters are based on the detection of particles or radiations emitted as a result of their natural radioactive decay. Most of these methods rely on the detection of alpha particles, some on detection of gamma emissions, and only a few on the measurement of beta decays. In addition to the type of radiation that is detected, there are other characteristics which dictate the choice of the measurement technique. The objective of the investigation, i.e. whether only radon concentration is to be monitored or its decay products are also to be recorded, is one such characteristic. Another major feature, of which there are three broad classes, is the time duration. In grab-sample techniques, the radioactive content of discrete samples of air collected over a short time from a specific place is analysed. Continuous techniques are designed to provide information on

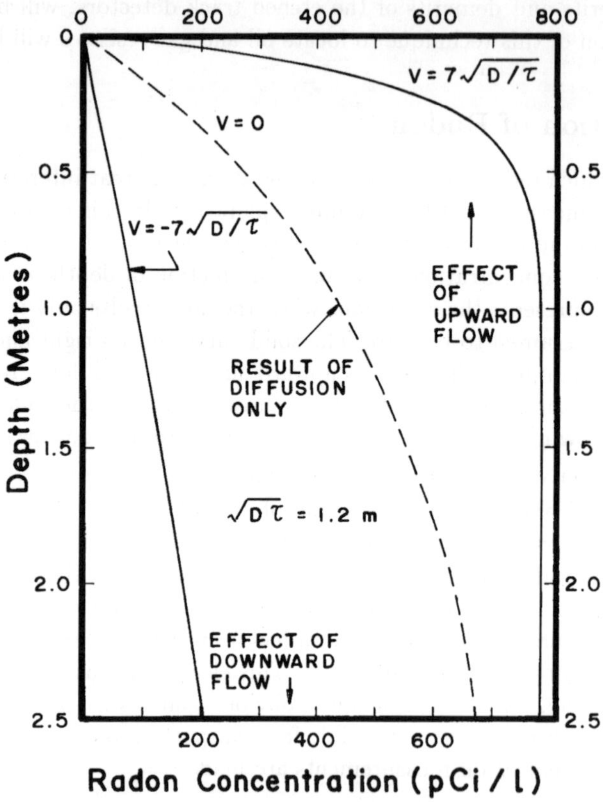

Figure 4.4.1: Value of radon concentration versus depth calculated for zero vertical flow and flow per unit porosity of $\pm$ 7 $\sqrt{D/\tau}$. The mean diffusion distance in the ground is 1.2 m, and the mean diffusion time of radon is 5.5 days (mean life of radon). The horizontal scale is calculated to correspond to typical field values at a site in Thoreau, New Mexico, USA (Fleischer, Hart and Mogro-Campero, 1983).

the time-dependent concentrations. Time-integrated techniques provide a single readout concentration averaged over a period of a few days to a year or so. The integrating techniques are most commonly used for survey work or to determine the long-term radon and/or its decay product concentation at a specific place.

4.4.4 Etched Track Detector Technique Used in Mineral Exploration

A considerably successful method of radon detection for uranium exploration is based on the use of etched track detectors. This system of charged particle detection has been described in detail in Chapter 2 of the present volume. The etched track detector technique for radon measurements now offers the opportunity of inexpensively detecting uranium mineralization buried several hundred metres deep. Such ore bodies are usually not traceable by conventional exploration techniques. The details of this technique are presented in the following subsections.

4.4.4.1 Principles and methods

In common with other methods, the principle of this technique is that a detectable amount of radon, above the background level, reaches the earth's surface as a result of the natural decay of a subsurface uranium ore body. The α-particles produced by radon decay are registered in the detector in the form of latent damage trails and can be counted in an etched detector with the help of an optical microscope. The number density of α-tracks is directly related to radon concentration. In practice, many measurements are made in a grid pattern over a suspected ore body, and radon density contour maps are prepared. The ore body is expected to be present under the area of relatively higher radon concentration values.

In the field applications of this method, the etched track detectors are used in containers of convenient geometries such as tubes or cups (Fleischer *et al.*, 1975; Gingrich and Fisher, 1976; Khan, 1980, 1992; Khan and Akber, 1976; Khan *et al.*, 1991). These containers have dimensions which are suitable for installation in holes drilled in the field. Previously, cellulose nitrate films were mostly used for the detection of radon and its daughters. However, during the last decade or so a polyallyl carbonate detector (CR-39) has become much more common for such applications. This new film is not only relatively easier to instal and handle in the rough field conditions, but it also offers certain unique advantages during the final processing and evaluation stages.

The tubes containing any of the above-mentioned track detectors (cellulose nitrate or CR-39) are used with or without 'membrane filters' which are fixed on their open ends. These filters not only help in eliminating the interference of water vapours and soil, but also prove to be fairly effective in reducing the contribution of thoron (^{220}Rn): Presence of a membrane at the open end of the tube causes a delay in the entry of thoron into the tube. During this delay, the short-lived thoron decays into its solid daughters which are not capable of entering into the tube to emit any α-particles to create tracks. The filters are essential if the area under study is known to have some thorium deposits and if thoron contribution is desired to be eliminated. This method has been very successfully applied in the field during the last couple of decades (Gingrich and Fisher, 1976; Khan and Akber, 1976).

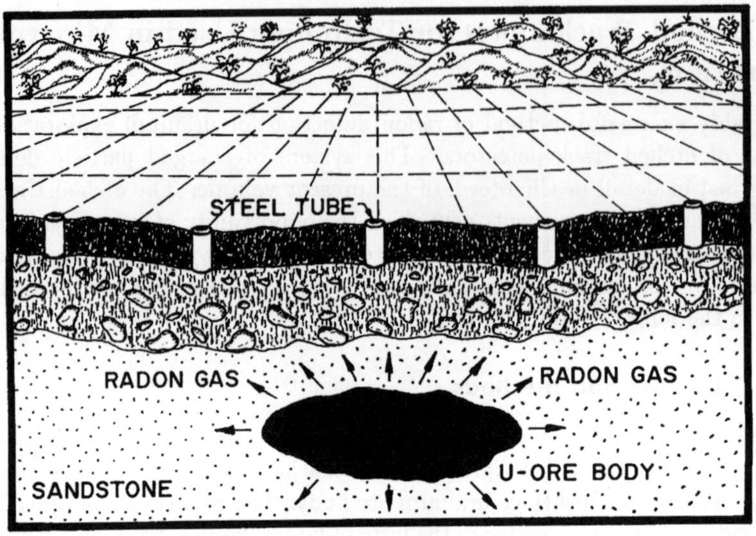

Figure 4.4.2: A geographical survey grid laid on the ground for the burial of steel tubes containing track detectors for the study of uranium mineralization. Similar grid pattern is followed in laying the track detectors containing tubes to locate oil/gas reserves and other geological studies based on radon measurements.

In a typical uranium exploration programme, steel tubes containing detectors are placed in holes about 60 cm deep, drilled in a grid-like pattern, over the area being explored, as shown in Fig. 4.4.2. The grid lines are between 25 and 1000 m apart depending upon the stage of exploration, size of the area being explored, the dimensions of the expected ore body, and the local geology. At the prospecting stage the tubes are installed at distances as wide as one km apart. The grid is narrowed to 25 m or so when the exploration is in the final stages and the three-dimensional shape of the ore body is to be evolved for final grade/tonnage calculation. After the tubes are placed, the holes are covered with soil and left undisturbed for 4-6 weeks. By leaving the tubes in the ground, the detectors in the tubes continuously accumulate the tracks produced by the α-decay of the radon gas coming from the subsurface uranium ore. The α-count is indicative of the long-term average value of radon concentration. Since the short-term measurements (1-2 weeks) have a significant variability, therefore, a long-term exposure is preferable. After retrieval from the field, the detectors are chemically etched in an alkaline solution for a specific time at a specific temperature depending upon the nature of the detector that is being used. After etching, the α-tracks are counted by using an ordinary microscope to determine the number of decays of radon accumulated in the tubes. To obtain the maximum amount of information, the track data is usually presented in the form of radon track-density contour maps.

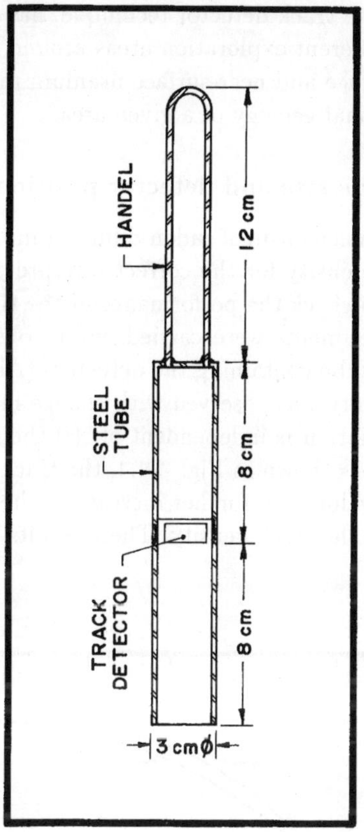

Figure 4.4.3: A sketch of the steel tube used for radon measurement near ground surfaces to study uranium mineralization, oil/gas reserves and other geological studies based on radon measurements.

The tubes, whether made of PVC or of stainless steel (Fig. 4.4.3) have no mechanical or electrical parts, so that there are no components which may cause problems in severe environmental conditions. This factor is very important in surveying rugged terrain and in less developed areas where transport and logistical facilities are limited. Moreover, even semi-skilled manpower can be trained to apply this technique. The system is usable for measuring radon levels in very wet surface conditions and also under the snow, since radon can penetrate many media. The technique can be applied even under water by placing the tubes in inverted position. A special application of the system has also been developed for measuring radon in deep drill-holes by lowering these tubes into the holes.

The etched track detector technique, like other techniques, requires an accurate determination of the background level of radon for the area being surveyed. Thus a minimum of one hundred sampling stations should normally be set up to statistically determine the general background level of radon for the area. Experience with a large number of exploration programmes, using the track detector technique, has shown that the background levels can greatly vary in different exploration areas around the world. These variations are due to differences in surface and near-surface uranium mineralization and due to the differences in local and regional geology of a given area.

4.4.4.2 Effect of the tube size and detector position in the tube

In order to understand the mechanism of radon diffusion into the tubes and the development of a reasonable track density for the correct interpretation of subsurface uranium ore bodies, it is necessary to check the performance of the tube-detector system. In this connection, a series of experiments were carried out to observe the variation in track density with the size of the tube containing the detectors (Akbar *et.al.*, 1980). A specific relative variation in the density was observed with change in the tube radius. It was observed that this relative variation is independent of: (i) the radon source concentration, and (ii) the exposure time. As shown in Fig. 4.4.4, the track density increases with tube size; but, beyond a certain value, any further increase in the tube radius does not result in an appreciable increase in the track density. These results were found to be helpful in

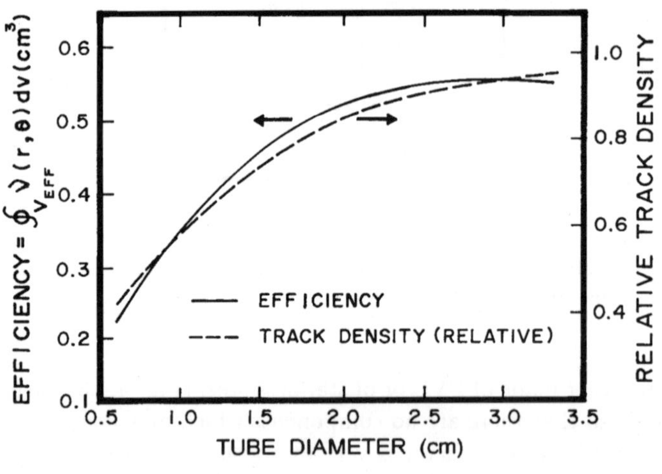

Figure 4.4.4: The variation in the efficiency of the tube-detector systems with the tube radius. The dashed line is the curve of relative track density, which is in excellent agreement with the theoretically expected results (Akber *et al.*, 1980).

designing a 'tube detector' system with optimum performance for uranium exploration. The variation of track density with tube dimension in the effective volume V_{eff} refers to the region about a certain point in the tube such that the α-particles produced as a result of any disintegration within that volume possess a non-zero probability of making etchable damage at that point. This volume is the region bounded by the intersection between a hemisphere centred at a point P with radius R (i.e. the α-range in air) and the cylindrical walls of the tube (Fig. 4.4.5a).

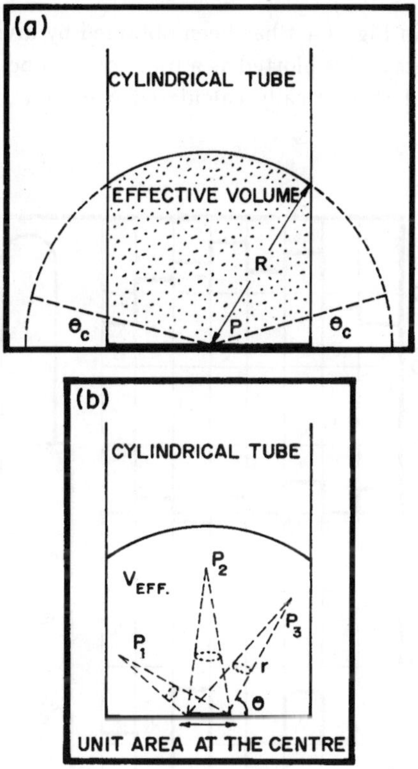

Figure 4.4.5: (a) The effective volume of a gas (radon) about a certain point P in the cylindrical tube designed for Rn measurement. The effective volume is the intersection of a sphere of radius R with the inner boundaries of the tube. R is the etchable range of the particles emitted, in this case through the decay of radon. (b) A sketch to illustrate the variation in the relative probability of registration of tracks within the effective volume. The relative probability of track registration depends upon the solid angle subtended at the area of detection The efficiency of the system is an integral of the relative probability over the entire effective volume (Akber *et al.*, 1980).

This region is further reduced due to the effect of the critical angle of etching θ_c. So that the net effective volume is the dotted region in Fig. 4.4.5a. Even within the effective volume, the probability of α-particle registration varies as a function of distance r (from the point of α-production to the point of incidence on the target) and angle θ (between the vector r and the detector surface) (Fig. 4.4.5b).

The efficiency of the system can be obtained by integrating the relative registration probability ν (r, θ), over the entire effective volume V_{eff} i.e.

$$Efficiency = \int_{V_{eff}} \nu(r,\theta)dv \qquad (4.4.1)$$

The efficiency shown in Fig. 4.4.4 has been obtained by dividing the effective volume into small volume segments and is plotted as a function of tube radius. The experimental relative track density and theoretically calculated efficiency show a strong correlation (Fig. 4.4.4).

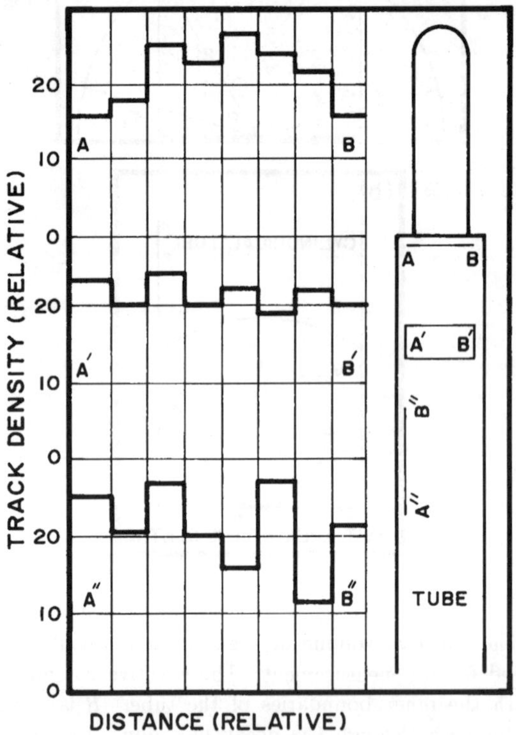

Figure 4.4.6: The figure shows the variation in the track registration efficiency due to changes in the position of the track detector placed at various positions in the tube. The detector marked A′-B′, placed as a loop in the tube, shows a uniform track distribution, whereas the other detectors, A-B and A″-B″, show large variations (Akber *et al.*, 1980).

The contribution of the α-emitting radon daughters may be important in the above-mentioned experiment. The radon daughters, being charged and chemically reactive, have a tendency to stick to the walls of the tube and to the dust particles, and consequently their distributions is not the same as that of the radon gas itself. It has been observed that within the limits of the tube sizes (0.5 cm diameter upwards) this factor has not introduced much error in the readings. This may be due to one of the two reasons: (i) the relative probability for registration from the boundaries of the tube is much less than that from the rest of the effective volume, and (ii) their distribution is not much perturbed in a generally dust-free atmosphere.

The track density recorded by a track detector, placed in a tube, is a function of the solid angle subtended by the effective gas volume on the detector. Therefore, the position of the detector in the tube plays an important role in the α-track accumulation. In this connection some experiments were carried out to determine the position of the detector in the tube that would obtain optimum track density. For this purpose the detectors were positioned in the tube at different locations, e.g. (i) along the length of the wall of the tube, (ii) on the inner top flat face of the tube, and, (iii) in the form of a ring in the centre of the tube as shown in Fig. 4.4.6. The tube was kept in radon atmosphere for a few weeks and then the detectors were analysed. It was observed that:

(a) In the case of the detector placed along the length of the tube, the track density was found to be maximum somewhat in the middle of the detector; it decreased towards the ends of the detector, as shown by the detector position A″-B″ in Fig. 4.4.6.

(b) The track density was again maximum in the central part of the detector and lowest on the ends of the detector placed on the inner top surface of the tube, as shown by detector position A-B in Fig. 4.4.6.

(c) Maximum and uniform track density was noted on the detector positioned in the form of a ring or a loop in the central part of the tube, as shown by the detector position A′-B′ in Fig. 4.4.6.

The maximum values so observed were found to increase with the tube size up to the tube diameter of about 3.5 cm. So a tube size, of inner diameter 3.5 cm and length 16 cm, with the detector fixed as a cylindrical loop along the inner walls of the tube, was found to have an ideal efficiency for radon measurement using the etched track detector technique.

4.4.4.3 Effects of the exposure time, environmental conditions and etching conditions

It is important to measure the track accumulation rate as a function of the exposure time in order to find out the time interval required for producing a steady-state or saturation rate during the exposure. Both, plastic track detectors and silicon surface barrier detectors were employed, and the following conclusions were drawn from the results obtained (Akber *et al.*, 1980; Khan and Akber, 1976).

(i) An exponential type of pattern is observed in the accumulation rate vs. exposure time curve during the initial stage;

(ii) The saturation or steady-state value is obtained in about 4 weeks.

Usually, the 'tube-detector' system is left in the field for 4 weeks. The experiments conducted to study the effect of temperature indicate that the cellulose nitrate detectors can retain α-tracks even if they are heated at 40-50°C for a few weeks. It has been estimated that the temperature at the depths of 60 cm at which the detectors are buried does not exceed 45°C. Therefore, they are fairly safe as regards the danger of annealing of tracks.

Eperiments conducted to study the effects of increased humidity on the track registration properties of track detectors show a slight enhancement in the efficiency of the detector with humidity. The exact contribution of humidity on detectors lying in the field should be evaluated for the particular type (and the batch) of the detector to be employed for the exploration work.

During the detection of α-particles from radon gas, α-particles of a continuous energy spectrum are recorded. If etched track detectors are employed in these measurements, the tracks due to α-particles of various energies will show different behaviour towards etching. Therefore, the role of etching conditions becomes important for such a detection system. For example, we need to employ etching conditions so that the latent trails due to α-particles incident even at the highest energies are revealed. The optimum etching conditions may therefore depend upon the tube size and detector position as well. Taking this observation into account it was found that for CN-85 cellulose nitrate detectors, etching carried out in 6M NaOH at 50°C for 70 minutes represents the most suitable etching conditions. In CR-39 the α-tracks are best revealed by etching in 6M NaOH at 70°C for 180 min.

4.4.4.4 Geology of the area and radon movement

Geology of the area under investigation plays an important role in understanding radon measurement results. It has been observed that loosely cemented coarse grained and fractured sandstones are likely to give higher radon yields, since the structural characters of such rocks provide an easy path for radon movement. Radon studies have, however, been found to be equally valid for other types of terrains such as the sedimentary basins of Western U.S.A., vein-type deposits of Front Range Colorado and metasedimentary deposits of Northern Canada.

Ground-water movements may also have some effect in producing halos around an ore body. Periodic rises in the general level of water table may displace ores of uranium to a higher level than the original place of the deposit. The new mineralized zone can act as another radon source and may produce two zones of radon concentration over the same ore body. In this case a very careful interpretation of subsurface orc body is required. In very alkaline waters, such as in active karst areas, ^{226}Ra can also be easily transported. It is sometimes moved under the same conditions as those that cause calcium, barium and strontium to be transported. Under such circumstances, the radon measurement technique may give erroneous results.

In all scenarios of soil-gas transport mechanisms, it may be remembered that radon is an extremely small component of soil gas (37.4 Bq m^{-3} = $0.7 \cdot 10^{-8}$ litres Rn/litre). Therefore, radon must travel as a very minor constituent of soil gas as compared to oxygen, nitrogen, carbon dioxide, etc. It has been estimated that the radon transport velocity may not be more than $(4\text{-}6) \cdot 10^{-3}$ cm s^{-1} or approximately 3 - 5 m per day (Andrew and Wood, 1972; Soonawala and Teleford, 1975; Tanner, 1964). Using the lower value of 3 m per day and 1 ppm of uranium in soil overlying an ore body having a uranium concentration of 2000 ppm and buried as deep as 120 m, an anomaly 3 times the background should be produced over the area of the ore-body. Using a transport velocity of 5 m per day under the same conditions, the anomaly could be as high as 5 times the background.

4.4.4.5 Corrections to be applied

The detectors received from the field are etched in the laboratory and the α-tracks counted visually under an optical microscope. The density of α-tracks obtained in this way gives only rough information about the radon concentration in soil gas over the ore body. It is better if they are analysed in the light of different factors, such as approximate depth of the ore body, geology of the area, structural features, climatic variations, and the surface radiometric pattern. A brief account of these factors is given below:

The movement of radon through the rocks depends largely upon the geology of the transmitting medium. Factors which affect the passage of radon include lithology, compaction, porosity and cementation (Borreto *et al.*, 1972). A very compact crystalline rock would allow less penetration of radon gas than a porous sandstone or a fractured rock. A thin bed of shale in the path of radon may slow down the gas transmission or change its route along passages of easier access. As a precaution, due consideration should be given to these factors while interpreting the radon data.

Some local structural features such as faults, joints or fractures tend to create errors in radon measurements. This was evident in an experiment in which some tubes were planted (a) on a solid rock containing high uranium, (b) over a fault zone, and (c) on a drill hole. The results indicated track densities of the order of 10^2 cm^{-2} over the solid rock, 10^3 cm^{-2} over the fault zone and 10^5 cm^{-2} over the drill hole. Therefore, as a precaution, the tubes should not be planted in close proximity to any of the structures mentioned above.

A thorough checking of the local sources of radioactivity in the proximity of the hole where a detector tube is to be planted is also necessary. If some alpha-emitting source (uranium or thorium) is present near the hole, it would tend to supply abnormal amounts of radon or thoron to it, which will, in turn, yield erroneous results on the general radon levels. So before planting the tubes, field counters may be used to register γ-activity, which would indicate the local presence of uranium.

The radon gas produced in the uranium mineralization travels in all directions. However, its movement is very slow and it takes a long time to cover small distances. When it has to travel long distances, most of it decays and a very small fraction of it reaches the top surface. Therefore, the depth is a very vital factor while interpreting the radon measurement data.

Climatic changes such as rainfall or pressure variations affect the radon escape tremendously. The rain water, while percolating through the pore spaces also pushes the escaping radon back into the earth. Similarly, pressure changes also retard or escalate the radon escape rate. To avoid these errors, the tubes may be planted in a dry season. The pressure effect is averaged out by exposing the detectors for long emplacement times of about 4 weeks.

4.4.4.6 Merits and demerits of the ETD-technique

The following merits of the Etched Track Detector (ETD) technique have made it more useful as compared to other conventional techniques available for uranium exploration (Gingrich and Fisher, 1976; Khan, 1992; Khan *et al.*, 1991).

1. It is a sensitive technique for uranium reconnaissance /prospection and has a fair potential for a follow-up exploration programme, especially for deeply buried deposits undetectable by the conventional methods.

2. The detectors can be installed by one to two individuals, in 60 cm deep holes.

3. Exposure duration of 4 to 6 weeks would ensure an integrated long-term average radon concentration read-out, which eliminates variations experienced in some of the other techniques.

4. The method is most effective in areas having permeable overburden or fractures in rocks, which would act as conduits to the uranium plumbing system and hence facilitate the migration of radium, and consequently of radon, to the near-surface zones.

5. The system is portable, inexpensive and is extremely convenient in rugged and remote areas even without vehicular access.

Some of the demerits of the technique (Gingrich and Fisher, 1976; Khan, 1992; Khan *et al.*, 1991) are listed below:

1. Mobility of parent nuclides (^{226}Ra and/or ^{238}U), and leaching of uranium along the dip, usually gives more than one radon anomaly at the surface, which may lead to erroneous results.

2. Excessive porosity and permeability of the lithology results in additional release of radon, and its retention in the denser, less permeable units. Therefore, a good geochemical/geological model is necessary to evaluate most of these anomalies, and supportive sampling may be necessary for more accurate interpretation of the results.

3. The method requires determination of the radon background level for a specific area, which necessitates a number of measurements to be made in order to establish the mean background level.

4. The results are not immediately available for on-the-spot guidance during the pattern drilling.

5. Radon emanations and exhalations from a local source may sometimes cause apparent radon anomalies.

6. The dispersion of radon in soils, water and rocks is related to, and reflects, the concentration of radium. In certain cases, the radium is mobilized and deposited away from the original source of ^{238}U. In such cases it is difficult to locate the ^{238}U deposit.

4.4.4.7 Some successful applications of the ETD-technique

The etched track detector technique has been employed in over 300 uranium exploration programmes in a wide variety of geological environments. Early surveys were carried out in the sedimentary areas of Western USA and in the vein-type deposits of Australia. Recently, several successful programmes have been carried out in Canada and Africa. The following results are typical examples from some of these programmes (Gingrich and Fisher, 1976).

(i) Western United States Sedimentary Basins: The pattern drilling techniques have been used for uranium exploration in many sedimentary basin areas of the Western United States, since there are no surface or air radioactivity anomalies found in this area. One of these areas in the Grants Mineral Belt of New Mexico, located in the West Water Canyon and consisting of sandstone member of the Morrison Formation, was explored with the etched track detector technique. The detector tubes were planted in a 50 m wide grid pattern covering an area of 1.6 km x 0.5 km. As a result of this study three regions of anomalous radon levels were detected, where the track counts were more than three times the background. During exploratory drilling, based on these findings, the third drill hole showed the first signs of mineralization. The fourth drill hole intersected the mineralized zone at a depth of 100 m, which had a thickness of 3 m and a grade of 0.34% U_3O_8. Subsequently, the complete ore body was mapped and found to contain several hundred tonnes of uranium ore.

(ii) Front Range Colorado - Vein-type deposits: The Front Range of the Rocky Mountains in Colorado contains high-grade uraninite veins in breccia fillings. The Schwartzwalder Mine in this region has produced more than 5000 tons of U_3O_8 since 1956, with an average grade of 0.7%. Conventional techniques are generally ineffective in locating uranium deposits in this area because of the intense leaching of rocks and the presence of colluvial cover. An extensive comparative study was made in this area, encompassing airborne and ground radioactivity measurements, soil geochemistry, geobotany, biogeochemistry, SO_2 gas geochemistry, radon emanometry and track detector techniques. Most of these techniques were found to be largely unproductive for several reasons. The radon emanometry

technique showed extreme variations (10-15 times the background) in the radon in soil gas due to the effects of barometric pressure changes, soil temperature variations, and other factors.

A comparison of the various methods showed that the etched track detector technique was the most effective in this area. All the major uranium deposits were successfully mapped with this technique on the basis of an observed radon anomaly of 4 to 10 times the background value. In one of the studies, a very pronounced anomaly of 88 times the background was observed in an area where conventional techniques had shown a minor potential for uranium mineralization.

(iii) Metasedimentary deposits of Northern Canada: In order to determine the applicability of the etched track detector technique in a glaciated terrain, a survey was conducted in the Cluff Lake "N" deposits located in Northern Canada. In this region, the uranium deposit consisted of pitchblende-bearing lenses associated with fractured zones ranging in depth from 10 to 120 m in paragneisses and granites of Archaean or Aphebian age. The detector tubes were placed in a 70 m wide grid pattern extending well beyond the boundaries of the known ore body. A number of exceptionally high readings, more than 50 times the background, were delineated on the basis of track etch studies over the main ore body. The highest readings were obtained from the area where the high ore-grade mineralized lenses are found at subsurface level.

This area was resurveyed using the track detector technique during the winter by placing the tubes under the snow cover. The winter survey also showed a similar pattern of anomalies, although the readings were considerably lower - perhaps owing to a combination of lower radon migration rates and the fact that the detectors were covered with a layer of ice crystals.

(iv) The sedimentary and paleochannel deposits of Australia: In Australia a number of exploration programmes based on etched track detector were conducted in the Alligator Rivers area of the Northern Territory, the Fromme Lake Embayment, and several other sedimentary areas. In the early phase, surveys were conducted over a known ore body in the East Alligator River uranium district. Track detectors placed on a 30 m x 150 m grid clearly located an anomaly over the ore body, with the highest reading more than 40 times the background. The ore body was at a depth of 75 m and the water table at 3 to 10 m. The signal was not substantially attenuated by the thick water cover. Results indicate that an enhancement of the signal through a thick water table may take place.

In another Australian test programme, in the Fromme Lake Area of South Australia, a survey was made over an ore body about 4 m thick with 0.25% grade uranium in Lower Tertiary paleochannel sand which was at a depths of 110 m. The whole area is overlain by impermeable Upper Tertiary lacustrine clays. The etched track detector method identified several low-order (2 to 3 times the background) anomalies over a known ore body. The anomalies were displaced about 100 m from the ore body, for which the reasons were not fully understood. The low order of the anomalies probably reflects the large thickness of the overburden and the low permeability nature of the clay cover.

(v) Remobilized deposit of Isakhel, Pakistan: In one of the areas in Pakistan (Isakhel), uranium mineralization is associated with the water table at a depth of 104-110 m. The

shale underlying the ore body and the water table is holding the ore body in its present position. The mineralization level in Isakhel has been changing in the past owing to water-table fluctuations (Qureshi *et al.*, 1988, 1993). The latest mobilization is very recent, since the deposit has not yet attained equilibrium, so that the gamma counters and logging units indicate a false uranium anomaly due to the gamma emitting daughters now present there.

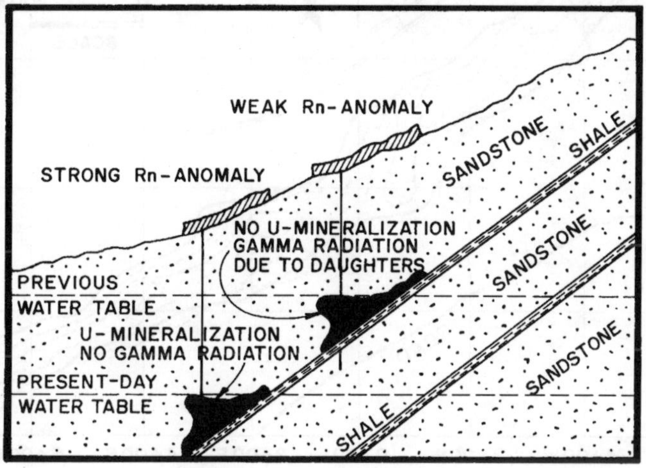

Figure 4.4.7: A hypothetical model developed on the basis of surface and sub-surface studies for uranium mobilization in Isakhel area (Pakistan) due to water-table fluctuation. The previous site of mineralization contains only gamma-emitting uranium daughters, which produced a weak radon anomaly at the earth's surface horizontally close to the shale exposure. The present site of mineralization is associated with the present-day water table at a site deeper than the previous site of mineralization. This site contains radon-emitting uranium, which has produced a strong anomaly on the earth's surface horizontally away from the shale exposure (Qureshi *et al.*, 1993).

Radon measurements using the etched track detector technique, applied in this area, indicated two impressions of the ore body: a weaker impression horizontally, close to the shale exposure, and a stronger impression, away from the shale, as shown in Fig. 4.4.7. Subsequent drilling indicated the downward mobilization of the uranium ore body due to water-table fluctuations. The impression close to the shale exposure indicates the previous subsurface site of mineralization, and the second impression away from shale exposure is the site of the present-day mineralization. The ore body has now been established away from shale on the basis of an analysis of drill cores. To test the reliability of the radon measurement technique, tubes containing etched track detectors were installed using a closer grid over an ore body proven by drilling. An astonishing match of the subsurface ore body and the surface track density contour map was found:

ninety percent of the productive drill holes were within the high track-density contours, as shown in Fig. 4.4.8.

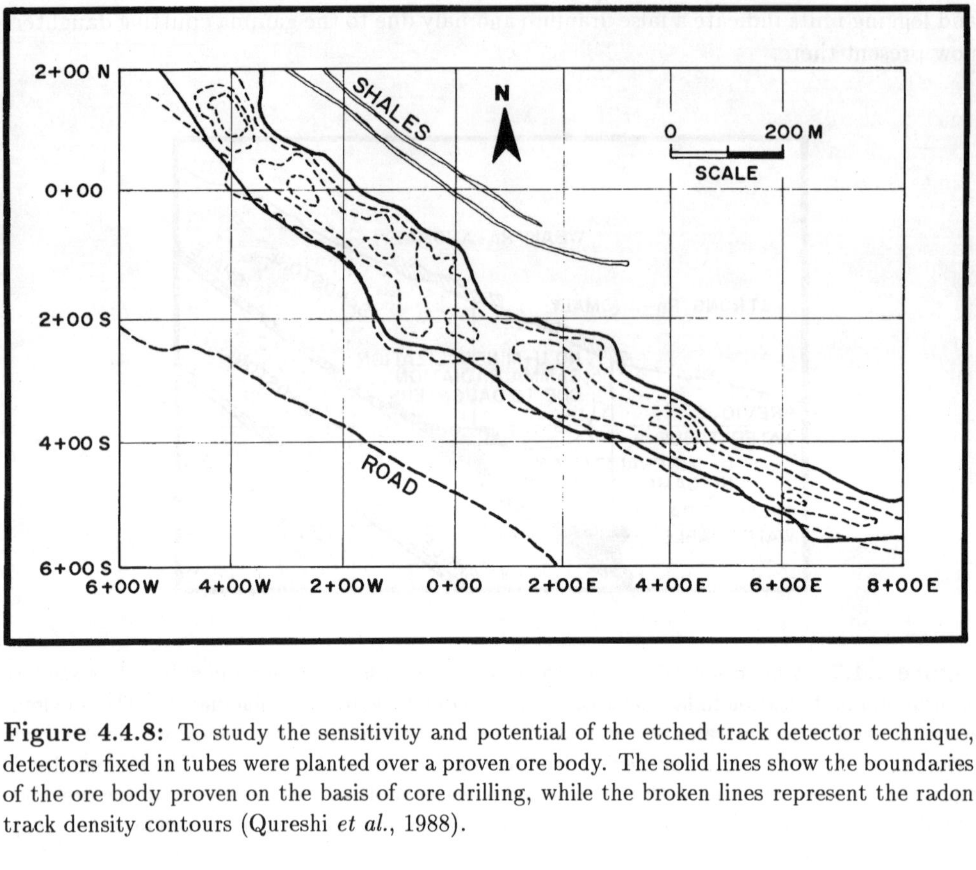

Figure 4.4.8: To study the sensitivity and potential of the etched track detector technique, detectors fixed in tubes were planted over a proven ore body. The solid lines show the boundaries of the ore body proven on the basis of core drilling, while the broken lines represent the radon track density contours (Qureshi *et al.*, 1988).

4.4.4.8 Location of oil and gas deposits

The utilization of radon signals for the exploration of oil and gas is in its initial stages. However, the limited work carried out to-date has provided very useful information in this field. In this section a brief account is given of the scanty results reported on this topic.

Recent experiments have shown that in certain favourable cases radon anomaly can provide useful information for locating the deeply buried oil and gas reserves. Radon is constantly emitted from the interior of the earth into the atmosphere at a fairly constant rate, usually in small amounts. Anomalously large radon emissions from uranium deposits, geothermal sources, geological faults, volcanoes, and during the period of an earthquake formation process have been known for a long time. More recent work has shown a negative anomaly of radon emanation (i.e. radon level deficiency) from a region of proven oil and gas reserves (Fleischer and Turner, 1983, 1984; Ishankuliev and Tretyakova, 1991; Makarov *et al.*, 1977).

Oil and gas deposits are geochemical anomalies, and structures containing hydrocarbons and barriers to stop oil migration within the ground are geophysical anomalies (Fleischer and Turner, 1983). So, in the search for hydrocarbons, both geophysical and geochemical methods of mapping may be employed. Fleischer and Turner (1983) carried out three types of surveys in the Pine Bush Area in Albany and Schenectady, New York state. Firstly, a set of $\delta^{13}C$ measurements were made on the soil samples ($\delta^{13}C=1000[(^{13}C/^{12}C) - (^{13}C/^{12}C)s] / (^{13}C/^{12}C)s$), where $(^{13}C/^{12}C)s$ is the carbon-13 to carbon-12 ratio in standard material). The low values of $\delta^{13}C$ ($<$-20) are characteristic of hydrocarbon deposits. Deposits of marine origin lie largely in the range $>$-24 and $<$-20, and terrestrial values typically are below -24; but significant exceptions do exist (Galimov and Kodina, 1982). Although the quantitative nature of anomalies expected near hydrocarbon deposits is subject to dispute, the relative values could provide maps that are useful in identifying anomalies.

Secondly, a set of measurements of non-carbonate total carbon in soil samples (referred to as "ΔC") was made using the carbon outgassed in the temperature range 500°-600°C during continuous heating. These are the 'avoiding temperature ranges', where the carbonate minerals are dissociated and hydrocarbons are released that may have migrated from a reservoir such as the one sketched in Fig. 4.4.9 (Duchscherer, 1991).

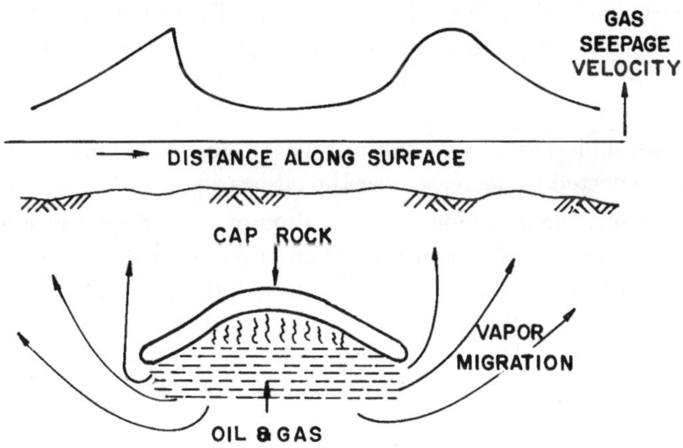

Figure 4.4.9: Sketch of upward flow of ground gases (containing radon) from an oil/gas reserve. The cocentration of radon at the earth's surface is shown in the graph (Fleischer and Turner, 1983).

The third method is based on the measurement of radon emanation from the ground.

In this case, the expectation is that the flow of gases will alter the near-surface gradients in radon concentration, as indicated in Fig. 4.4.9. Thus, a possible locally decreased emanation can be identified over oil/gas reserves as compared to the surroundings.

For the purpose of measurements, soil samples for $\delta^{13}C$ and ΔC determination were taken at depths of 1.2 m. Using a separate hole, 1-1.5 m away from the sampling site for $\delta^{13}C$ and ΔC, radon measurements were made at a depth of 0.6 m using the etched track method, for an exposure period of 64 days. Samples from these holes were also taken for the study of radon emanation properties of the soil.

The results of the measurements for $-\delta^{13}C$ are shown in Fig. 4.4.10a. The obvious characteristic is the single high value of 48 for $-\delta^{13}C$ at one location, and a monotonic decrease around that location. Measurements of ΔC given in Fig. 4.4.10b show the extreme value of 81 at the same position as that where $-\delta^{13}C$ has its maximum value.

The radon data have their lowest value (0.6), at the same position where $-\delta^{13}C$ and ΔC have maximum values. The area having the lowest radon value is surrounded by an annular region of high radon values, as shown in Fig. 4.4.10c. This is the configuration expected for the simple model sketched in Fig. 4.4.9. In short, all three data sets place an anomaly at the same site.

The above data for carbon isotopic composition $\delta^{13}C$ in soil, for carbon released ΔC at 500°-$600^o C$ from soil, and a radon-based indicator of upward gas transport, all display an anomaly at the same location in a portion of the Eastern Overthrust Belt, Albany and Schenectady, New York. The cause of the anomalies is unidentified (Fleischer and Turner, 1983), but they may be related to subsurface oil/gas-bearing deposits.

Latest tectonic movements and their correlation with the older geological structure are of great importance in oil geology (Ishankuliev and Tretyakova, 1991). In West Turkemanistan, the oil and gas deposits are associated with the dome structures of Pliocene - Quaternary time, which represent the latest tectonic events in the region. Studies carried out in this area show a decrease in γ-activity over the geological structures which contain oil and gas reserves underneath. The upward flow of deeper ground gases such as CH_4, CO_2, etc., is also expected to decrease over the oil-bearing structures.

In order to find out the behaviour of emanation of these deeper ground gases (CH_4, CO_2, etc.) studies were carried out in the region of oil and gas reserves of West Turkemanistan over Kum-Dag, Kuydguk, Kyzil-Kum and the Kobak domed structures by Ishankuliev and Tretyakova (1991). The upward flow of these gases was studied by monitoring the radon concentrations in 2m deep holes using etched track detectors. The deeper ground gases have concentrations of the order of 15-16 magnitudes higher than those of radon.

A few results of radon profiles in terms of α-tracks cm^{-2}, normalized as per day exposure, are shown in Fig. 4.4.11. It can be seen that the curve of sub-soil radon concentration has two pronounced maxima at the edges of the dome, and a depression in the central part. The difference in the radon concentration across the dome agrees with the model presented by Fleischer and Turner (1983) in which radon gas concentration increases at the edge of the dome structure. This has been accounted for as being due to the absence of cap rock and/or an increase in the number of cracks on the edge of the

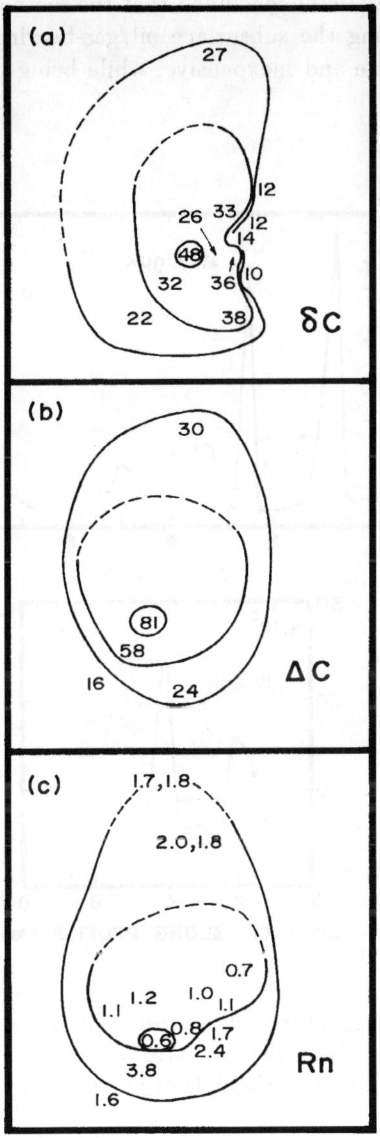

Figure 4.4.10: Values of -δ^{13}C, ΔC and Rn concentrations according to the survey carried out in the Pine Bush Area, New York, are shown in figures (a), (b) and (c). The highest values of -δ^{13}C (48), ΔC (81), and the lowest value of radon emanation Rn (0.6), all superimpose each other, which is indicative of some anomaly of unknown significance (Fleischer and Turner, 1983). This could possibly be a subsurface structure containing oil and gas reserve.

dome due to tectonic upheaval and increased oil/gas pressure below the dome structure. Ishankuliev and Tretyakova (1991) concluded that the method of etched track detection can be helpful in recognizing the subsurface oil/gas-bearing structures, and that the technique is relatively simple and inexpensive, while being more effective compared to γ-spectrometry.

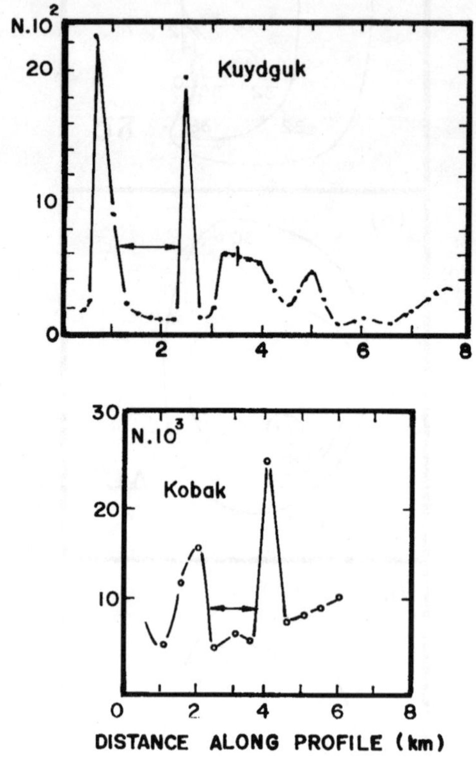

Figure 4.4.11: Radon measurement studies carried out in Kuydguk and Kobak area in the former Soviet Union. The area demarcated by arrows consists of dome structures which may contain oil and gas deposits (Ishankuliev and Tretyakova, 1991).

Makarov *et al.* (1977) carried out experiments to identify oil-saturated beds using the radon measurement technique. They estimated the relative amplitudes of anomalies by employing laboratory-based and bore-hole investigations. The radon anomalies recorded over the oil-bearing strata were opposite to those observed over the water-bearing strata.

4.4.5 Conclusions

One may draw the following conclusions from the results and discussions noted above:

1. The etched track detector technique has been found to be fairly successful for mineral exploration under varied conditions. It can be fruitfully employed at the reconnaissance level, though it is frequently applied to areas where other methods and favourable geological criteria have identified them as possessing potential for sub-surface ore. Therefore, it is appropriate for the follow-up and detailed level of exploration for reserve estimations.

2. In view of the problems that may crop up owing to disequilibrium conditions and dispersion of ^{226}Ra, it is advisable, in general, to use radon measurement methods in conjunction with other radiological methods and geological surveys.

3. Owing to the migration of radium and re-channeling of radon itself, the uranium deposit may not lie vertically below the location of a radon anomaly detected with the help of the etched track detector technique.

4. Encouraging results are obtained in situations where the overburden comprises a few tens of metres of soil or permeable rocks overlying the subsurface ore.

5. For detailed interpretation of radon anomalies, a working knowledge of the over-all geological, structural and stratigraphic set-up is necessary. In case the anticipated ore body is below the water table, acquaintance with the geohydrology and ground water regime is also required.

6. By using different radon monitoring systems, the information obtained about the subsurface geology can be used not only for uranium exploration but also for the exploration of other minerals and oil/gas reserves.

References

Akber R.A., Khan H.A., Ahmad I. and Jamil K. (1980) Some important considerations in the use of solid state nuclear track detectors for radon gas concentration measurements. *Nucl. Instr. Meth.* **173**, 183-189.

Andrew J.N. and Wood D.F. (1972) Mechanism of radon release in rock matrices and entry into ground water. *Trans. Min. Metall. Sect. B* **81**, 198-209.

Behounek F. (1927) On the proportion of radioactivity in the uranium mining district of St. Joachimsthal in Bohemia, Radioactivity of spring, ground and mine air, and the atmosphere. *Z. Phys.* **28**, 333-342.

Borreto P.M.C., Clark R.B. and Adams J.A.S. (1972) Physical characteristics of radon-222 emanation from rock, soil and minerals: Its relation to temperature and alpha dose. *Natural Radiation Environment* **11**, 731-740.

Chirkov A.M. (1976) Radon as a possible criterion for predicting eruptions as observed at Karimsky Volcano. *Bull. Volcanol.* **39**, 126-131.

Davis G.L. (1947) Radium contents of ultramafic igneous rocks: Laboratory investigations. *Am. J. Sci.* **245**, 677-693.

Duchscherer W., Jr. (1991) Non gasometric geochemical prospecting for hydrocarbons with case histories. *Oil and Gas J.*, 312-327.

Evan R.D. and Goodman C. (1941) Radioactivity of Rocks. *Geol. Soc. Am. Bull.* **52**, 459-490.

Fleischer R.L. and Turner L.G. (1984) Correlations of radon and carbon isotopic measurements with petroleum and natural gas at Cement. *Oklahoma Geophysics* **49**, 810-817.

Fleischer R.L. and Turner L.G (1983) Geophysical and Geochemical Anomaly in NE New York. *Technical Information Series of General Electric Company, New York 12345, No. 83 CRD 248.*

Fleischer R.L., Price P.B. and Walker R.M. (1975) *Nuclear Tracks in Solids: Principles and Applications.* University of California Press, Berkeley.

Fleischer R.L., Hart H. R. Jr. and Mogro Campero A. (1980) Radon emanation over an ore body: Search for long distance transport of radon. *Nucl. Instr. Meth.* **173**, 169-181.

Galimov E.M. and Kodina L.A. (1982) On the factors controlling carbon isotope composition of organic matter of sedimentory rocks. *5th Int. Conf. on Geochronology, Cosmochronology and Isotope Geology*, Nikko, Abs., pp. 109-110.

Gingrich J.E. and Fisher J.C. (1976) Uranium exploration using the track etch method. In *IAEA Proc. of a Symposium on Exploration for Uranium Ore Deposits*, IAEA-SM-208/19, pp. 213-227.

Ishankuliev J. and Tretyakova S.P. (1991) Radon measurements using SSNTD in the region of oil and gas deposits of West Turkemanistan. *Nucl. Tracks Radiat. Meas.* **19**, 1-4, 329-331.

Iwasaki I., Katsura T., Shimojima H. and Kamada M. (1956) Radioacivity of volcanic gases in Japan. *Bull. Volcanol.* **18**, Series 2, 103-123.

Khan H.A. (1980) Track registration and development efficiencies of solid state nuclear track detection. *Nucl. Inst. Meth.* **173**, 43-54.

Khan H.A. and Akber R.A. (1976) The measurement of radon by alpha sensitive plastic track detectors for use in uranium exploration. In *Proc. 9th Int. Conf. on Solid State Nuclear Track Detectors* (eds. F. Granzer *et al.*), Neuherberg/Muenchen, Pergamon Press, Oxford, pp. 803-814.

Khan H.A. (1992) Usefulness of radon measurements in earth sciences. *Nucl. Tracks Radiat. Meas.* **22**, 1-4, 355-364.

Khan H.A., Qureshi A.A. and Brandt R. (1991) Mineral exploration based on radon measurements. In *Proc. 2nd Workshop on Radon Monitoring in Radioprotection, Environmental and/or Earth Sciences* (eds. G. Furlan and L. Tommasino), Trieste, 1991, World Scientific, Singapore, pp. 472-485.

King C.Y. (1979) On radon emanation as a possible indicator of crustal deformations. *Tectonophysics* **52**, 120-126.

Kirikov A.P., Tverskoy P.N. and Grammakov A.G. (1932) On the question of the emanation method of searching for radioactive deposits. *Vses. Geol. Razved. Ob"edineniye Izv.* **51**, 1269-1274.

Koevil N.B. (1947) Thorium-uranium ratios of rocks and their relation to lead ore veins. *Econ. Geol.* **33**, 685-696.

Makarov M.S., Pinkenzon D.B. and Kalinin V.G. (1977) Identification of oil-saturated beds using radon tracer technique. *Vypusk* **59**, 165-169.

Qureshi A.A., Beg M.A.S., Ahmed F. and Khan H.A. (1988) Uranium exploration in Pakistan using alpha sensitive plastic films (ASPF). *Nucl. Tracks Radiat. Meas.* **15**, 1-4, 735-739.

Qureshi A.A., Hussain G., Mahmood K., Beg M.A.S. and Khan H.A. (1993) Use of alpha sensitive plastic film (ASPF) technique to locate re-mobilized uranium ore bodies in sandstones. *Nucl. Tracks Radiat. Meas.* **22**, 1-4, 431-433.

Sempirini L. and Kruger P. (1984) Relationship of radon concentration to spatial and temporal variations of resevoir thermodynamic conditions in Cerro Prieto geothermal field. *Geothermics* **13**, 12, 103-115.

Senftle F.E. and Koevil N.B. (1947) Thorium-uranium ratios in the theory of genesis of lead ores. *Trans. Am. Geophys. Union* **28**, 732-738.

Shiratoi K. (1972) The variation of radon activity of hot springs. *Sci. Res. Tohoku Imperial Univ.* **16**, 614-621.

Soonawala N.M. and Teleford W.M. (1975) Diffusion of ^{222}Rn and interpretation of techniques. *75th Meeting Soc. Exploration Geophysicists*, Denver, Colorado.

Tanner A.B. (1964) Radon migration in the ground: A review. In *Proc. Symp. Natural Radiation Environment* (ed. J.A.S. Adams), Houston, Texas, Univ. of Chicago Press, Chicago, pp. 161-190.

Tanner A.B. (1980) Radon migration in the ground. A supplementary review. In *Proc. Symp. Natural Radiation Environment III.* (eds. F.F. Gesell and W.M. Lowder), Houston, 1978, Rep. CONF-780422, U.S. Dept. of Energy, Washington, D.C. 1, pp. 5-56.

Tidyani A. (1984) Etudes de sondes pour l'analyse de processes geophysiques-radon-emanometrie. Thése No. 779, Université de Clermont-Ferrand 11, France.

M. Balcázar[1]

4.5 Radon and Geothermal Energy Production

4.5.1 Geothermal Energy: A Contribution to the World Energy Needs

The world geothermal electrical power at the end of 1994 was 6,272 MW (Hiriart *et al.*, 1995), which represents 0.2% of the total electrical power produced by all other means. Although it is generally recognized that geothermal energy is not a viable alternative source for satisfying the world's energy demands, it is nevertheless true that for some developing countries like Philippines the geothermal energy production reaches 17%; it reaches 19% in El Salvador, 9% in Nicaragua and 8% in Kenya. In the USA the geothermal electrical power represents only 0.3% of the total installed capacity; however, this 0.3% production equals twice the total geothermal power generated by the four countries mentioned above. Table 4.5.1 shows how the geothermal installed power is distributed amongst the twenty-one producing countries. Important economic factors arise in analysing specific points, for example, the three geothermoelectric sites in Mexico comprise 2.4% of the total installed power (Hiriart *et al.*, 1995); but those had an average availability of 85% in 1994, producing 4.2% of the total power. This means a saving of more than 10 millon barrels of oil for Mexico.

The future contribution of geothermal energy to the generation of electricity in the world is estimated (Barbier, 1991) to be 12,000 MW in the year 2000; 17,000 MW in 2010 and 25,000 MW in 2020. Two major developments concerning the use of geothermal energy are still to come in the near future: an increased efficiency of energy production and the use of low enthalpy fields for applications other than energy generation.

In the first case, the efficiency of energy production by geothermoelectric power is about three times lower than the efficiency of nuclear and fossil-fulled plants. This is due to the low temperature of the avalaible steam, typically of the order of 150°C. There are some proposed combined cycle plants (Hiriart *et al.*, 1995; Holland *et al.*, 1992; Mulas *et al.*, 1985), to either increase the efficiency production or even to make use of hot water at temperatures above 85°C.

In the second case, geothermal water at temperatures of up to 90°C has been successfully used (Facchini *et al.*, 1993) for heating greenhouses, drying forage and maize and for aquaculture tanks, amongst others uses. Because the temperature of the available heat is determined by the geothermal reservoir and not by the use to which it is put, careful integration into the industrial processes is required.

---

[1]Instituto Nacional de Investigaciones Nucleares, Mexico, D.F., Mexico

Table 4.5.1: Geothermal Installed Power.

Country	Capacity (MW)[a]	% of total power installed[b]
USA	2,594	0.3
Phillippinas	1,053	14
Mexico	753	2.4
Italy	637	1
Japan	298	0.1
New Zealand	285	2.2
Indonesia	256	1.4
El Salvador	105	19
Nicaragua	70	8.9
Costa Rica	55	
Iceland	50	
Kenya	45	7.8
China	31	
Turkey	20	
Rusia	11	
Other six countries	9	
Total	6297	

[a] Hiriart y Gutiérrez (1995)
[b] Barbier (1991)

The underground thermodynamic activity in a geothermal field brings to the earth's surface gases, whose analysis provides: geological structure information, fluid temperature in the reservoir, fluid-rock chemical interaction and evolution of the hydrothermal system. The radioactive nature of the natural by occurring radon and radium, make them suitable tracers for geothermal properties analysis of the field.

4.5.2 Geothermal Energy Sites

The inner core of the earth reaches maximum temperatures of about 4,000^{o}C at a depth of 6,000 km. This temperature is gradually reduced to 1,000^{o}C in the semifluid mantle located at a depth of between 35 and 300 km; the crust, thus, has an average temperature gradient of 30°C km^{-1}.

There are some selected areas on the earth where temperature gradients higher than the average are present. Those areas are usually found in young vulcanism sites (Barbier, 1991; Gutiérrez *et al.*, 1984; Hiriart *et al.*, 1995), close to the margins of major crustal plates where molten rocks from the earth's mantle come nearer the surface: there, the heat is transfered upwards by conduction until it reaches the permeable rocks containing aquifer reservoirs. Then heat transfer takes place by convection in the reservoir, producing super-heated steam in the upper part with temperatures greater than even 300°C. Impermeable rocks at the top of the reservoir prevent the trapped heat from escaping.

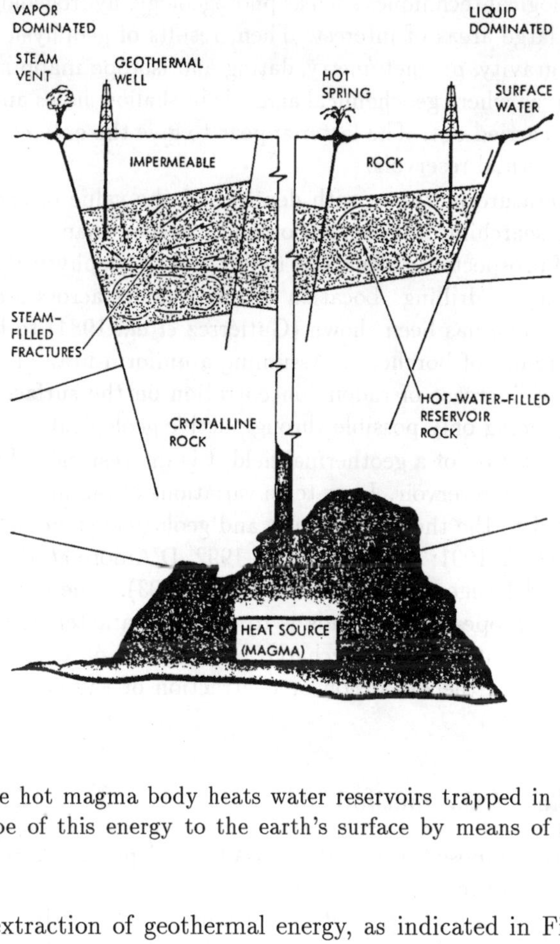

Figure 4.5.1: The hot magma body heats water reservoirs trapped in impermeable rocks, leading to the escape of this energy to the earth's surface by means of steam vents or hot springs.

The practical extraction of geothermal energy, as indicated in Fig. 4.5.1, requires four conditions: a heat source, either the magma body or hot rocks; a heat-carrier fluid to the surface; a reservoir formed by fractured or permeable rocks; and, lastly cap rocks providing an impermeable and insulating cover. A geothermal field is classified according to the specific enthalpy of the fluid realised or extracted at the earth's surface. Low-temperature fluids (at 30°C to 100°C) are low-enthalpy fields of 0.03 to 0.4 MJ kg^{-1}; medium enthalpy are fields producing fluids between 100 and 140°C; and high-enthalpy fluids yield 0.5 to 3 MJ kg^{-1}. The heat trapped in high and low enthalpy fields can be used for electricity production if a hole is drilled through the insulating material, deep enough to release the heat-carrier fluid from the reservoir. Low-enthalpy fields are good energy suppliers for plant cultivation, animal husbandry and house heating.

4.5.3 The Use of Radon in Geothermal Surveys

The exploration of geothermal energy fields involves several steps: firstly, rapid low-cost reconnaissance geological techniques such as photogeology, hydrogeology and vulcanology are used to define large areas of interest. Then, results of geophysical methods such as electric resistivity, gravity, magnetometry, dating and isotope mapping are cross-checked to define smaller areas where geochemical analysis in shallow holes and deepwell drillings for flow testing are carried out. The latter appear to give the only conclusive evidence of an exploitable geothermal reservoir.

Since the temperature increases with depth, and the value of energy increases with temperature, most searching is done at boreholes deeper than 1 km, consequently increasing the cost of prospection. Thus, one major goal of geophysical measurements is to improve the efficiency of drilling. Location of active faults across a geothermal field by means of radon mapping has been shown (Gutiérrez *et al.*, 1984) to be a good indicator for reducing the number of boreholes. Assuming a uniform uranium content in a given geothermal field, any increase of radon concentration on the surface is due to a higher transport rate, this being only possible through active geological faults.

During the exploitation of a geothermal field it is interesting to know the thermodynamical evolution of the reservoir. Long-term variations of radon and radium isotopes in wellheads are related to the thermodynamics and geological structure of the geothermal resevoir (Balcázar *et al.*, 1991; Battaglia *et al.*, 1992; D'Amore *et al.*, 1978-79; Segovia *et al.*, 1991; Seprini and Kruger, 1984; Stewart *et al.*, 1993). The objective of monitoring natural radioactive isotopes in wellheads is to obtain characteristics of the geothermal reservoir, such as the enthalpy, the recharge process, the potential of the field for increasing the production, the expansion or contraction of the two-phase boiling zones, etc.

4.5.3.1 Principles of the method

The crust is mainly composed of granitic, basaltic and perioditic rocks, which contain small quantities of radioactive nuclei, particularly uranium, thorium and potassium. However, granite, the main component of the crust, is relatively rich in uranium and thorium, which gives rise to three natural radioactive decay chains of particular importance for geothermal studies: ^{238}U, ^{235}U and ^{232}Th. Amongst the radioactive nuclei present in those decay series are radium isotopes (^{226}Ra, ^{228}Ra, ^{224}Ra, ^{223}Ra) and radon isotopes (^{222}Rn, ^{220}Rn, ^{219}Rn), both subject to being transported by geothermal fluids. The very short half-lives of ^{220}Rn (55.6 s) and ^{219}Rn (3.96 s) exclude them from geothermal applications.

As indicated in Table 4.5.2, radon and radium isotopes in geothermal fields have been used during two stages: exploration and exploitation.

Applications during the exploration stage are described below first.

4.5.3.1.1 Areas of thermal significance associated with rift structure. These areas are defined by radon mapping in geothermal fields under exploration. Assuming the absence of high uranium-content minerals, and as a result of the high convection temperature processes underground in geothermal fields, the high radon activity is associated with fractures and faults through which radon is transported by geothermal fluids (Balcázar *et al.*, 1991, 1993a,b; Cox, 1980; De la Cruz *et al.*, 1986; Lombardi *et al.*, 1992; López *et al.*, 1987). In practice, this relationship is not straightforward, because the amount of ^{222}Rn detected from the radon-bearing rock depends on an important internal property: direct ejection of the radon atom by recoil in the alpha emission, which is a function of the total quantity of radium in the rock, its distribution in it and the grain size; as well as on external properties such as: the pressure gradient, the physicochemical characteristics of the fluids, and the stress field in the parent rock.

4.5.3.1.2 Fluid-rock chemical interaction. Measurements of ^{222}Rn and radium isotopes (^{223}Ra, ^{224}Ra, ^{226}Ra and ^{228}Ra) in springs and shallow wells are useful to identify reservoirs and to asses the dynamics of such geothermal systems (Battaglia *et al.*, 1992; Roldán-Manzo, 1992); the technique is based on measuring the ratios ^{228}Ra/^{226}Ra and ^{224}Ra/^{223}Ra, which are expected to represent the Th/U ratio in the host rock and thus to provide information on the origin of the geothermal system. Radon and radium content in geothermal fluid samples are measured using scintillation counters and γ-spectrometers, after chemical treatment of the sample (Santoyo *et al.*, 1991).

Applications during the exploitation stage require continuous measurements of radon and radium isotopes in wellheads; Table 4.5.2, summarizes some of the applications briefly described below.

4.5.3.1.3 Evolution of hydrothermal system. The goal is to establish a correlation between radon concentration and the thermodynamics and geological structure of the geothermal reservoir (D'Amore *et al.*, 1979). Since radon is a short-lived isotope, its concentration at the wellheads indicates reservoir properties and thermodynamic conditions within transport times of less than 30 days prior to fluid production (Semprini *et al.*, 1984). Two processes govern the wellhead radon concentration in geofluids: (a) radon emanation into the geofluid from the reservoir rock; (b) convective transport of the geofluid through the reservoir. Then, changes in radon concentration during periods of field exploitation are related with the thermodynamic behaviour of the reservoir, such as expansion or contration of the two-phase boiling zones.

4.5.3.1.4 Mixing between fresh water and old geothermal brine. The activity ratios ^{224}Ra/^{223}Ra and ^{228}Ra/^{226}Ra appear to be suitable indicators of host rock composition; Battaglia *et al.* (1992) have proposed a transport model for describing the physicochemical interactions between the radionuclei in the fluid and in the rock phase. From their work the following conclusions are obtained: (a) concentrations of ^{224}Ra and ^{228}Ra in a production test well give information on mixing between fresh water and old geothermal brine; (b) ^{223}Ra and ^{226}Ra in in conjunction with ^{222}Rn in the fluid phase provide information on the time-scale of water-rock interactions in the thermal reservoir prior to expansion.

Table 4.5.2: Uses of radon and precursors in geothermics.

Exploration Stage		
Application	Radioisotopes measured	References
Areas of thermal significance associated with rift structure	^{222}Rn (Radon mapping in the ground)	Balcázar *et al.* 1990, 1993a,b,c; Cox, 1980; De la Cruz *et al.*, 1986; Lombardi and Pinti, 1992; López *et al.*, 1987
Fluid-rock chemical interaction	$^{224}Ra/^{223}Ra$; $^{224}Ra/^{228}Ra$; $^{228}Ra/^{226}Ra$; ^{222}Rn	Battaglia *et al.* 1992; Roldán-Manzo, 1992

Exploitation Stage		
Application	Radioisotopes measured	References
Evolution of hydrothermal system	^{222}Rn (long term variation in a wellhead)	Balcázar *et al.*, 1991; D'Amore *et al.*, 1978-79; Segovia *et al.*, 1991; Semprini and Kruger, 1984; Stewart *et al.*, 1993
Mixing between fresh water and old hydrothermal brine	^{224}Ra and ^{228}Ra (measured at producing-test wellhead)	Battaglia *et al.*, 1992

4.5.3.2 Geostatistics for optimizing the number of detectors in the field

Radon concentration distribution in a given geothermal field is subject to variation in space due to the differences in soil and geology; variations in time occur according to processes of recharge and explortation. The time-dependence in a geothermal field under prospection is less affected than in a field under exploitation where the reservoir is subject to a greater thermodynamical activity. The small time-dependence in a geothermal field under prospection due to rain-fall and atmospheric pressure changes is eliminated by placing all detectors at the same time in the area of study.

In the case of México, radon mapping has been performed in four geothermal fields (Balcázar *et al.*, 1990) under contract with the Comisión Federal de Electricidad (CFE, national agency for electric power supply). Fig. 4.5.2 shows the location of the fields: Los Azufres, Tres Virgenes, Derrumbadas and El Ceboruco; in the states of Michoacán, Baja

California, Puebla and Nayarit respectively. The areas of interest for radon mapping in each field were chosen by the geothermal prospecting department of the CFE, taking into account geological and preliminary geochemical considerations.

Figure 4.5.2: Location of main geothermal fields under study and in production in Mexico. Radon mapping is used as a tool for detailed prospection.

The non-regular shape of the regions studied had an average area of 50 km^2; square graticules were marked in each geothermal region with the help of a magnetic compass. The sizes of the squeres were between 500 x 500 m^2 and 400 x 400 m^2 for preliminary studies, and down to 150 x 150 m^2 when choosing detailed areas of geothermal influence. Detectors placed at the corners of the squares varied in number from one field to another but a minimum of 160 detectors were used in each case.

Radon measurements are carried out by means of samplers, each composed of a thin LR 115 plastic detector placed at the upper end of a 30 cm long PVC tube, inside a second 1 m long PVC tube of a bigger diameter. The length of the inner tube was enough to allow the undesirable ^{220}Rn from thorium to decay before reaching the detector. The samplers were buried in such a way that the detectors were at 70 cm below the earth's surface. Tracks of alpha-particles from radon decay, registered as latent trails in the detectors during 20 days of exposure time, were chemically developed as holes through the thin plastic detectors. The number of holes, evaluated by means of a spark counting device (Balcázar *et al.*, 1984; Cross and Tommasino, 1970), are proportional to radon concentration.

A typical spatial histogram is shown in Fig. 4.5.3, corresponding to an area of 4.4 x 6 km^2 in the Los Azufres geothermal field; the heights of the bars are proportional to radon concentration, after correcting the detector readings for residual thickness and after applying the calibration factor to convert the hole density into Bq m^{-3}. The frequency distribution of radon concentration in the four geothermal fields displayed in Fig. 4.5.4 follows a log-normal distribution.

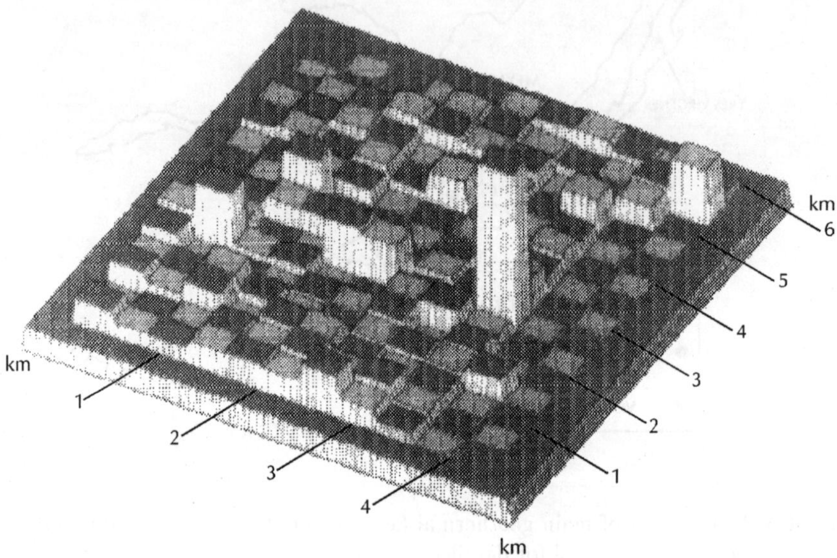

Figure 4.5.3: Radon concentration values obtained from plastic detectors, placed in the field under study, are fed into a computer program to draw a map of the area. Software permits one to rotate the maps for easy analysis.

Measurements of the variable radon concentration are often spatially correlated; this means that measurements close to each other are more likely to be similar than measurements at large distances from each other. A spatial correlation of the variable is described by the semivariogram, which is defined as a function of the distance between measurement locations (Badr *et al.*, 1993). The maximum distance between two observation points in a geothermal field is such that the corresponding measurements begin not to be correlated; this maximum distance is defined by analysing a semivariogram function $\Gamma(h)$ (Balcázar and Flores, 1992), which is obtained by plotting $\Gamma(h)$ as a function of the distance h in a given direction. Analysis of semivariograms in different directions provides information on the possible anisotropic characteristics of the field.

The average radon concentration of the geothermal fields in Fig. 4.5.4 for Las Derrumbadas is similar to the one for El Ceboruco; however, it is bigger for Tres Virgenes and Los Azufres. It is believed that these differences in the mean values are mainly related to differences in soil characteristics, and, to a minor extent, to geothermal properties. Although the average radon concentration varies from one field to another, it has been possible to define zones of interest for detailed geothermal prospection.

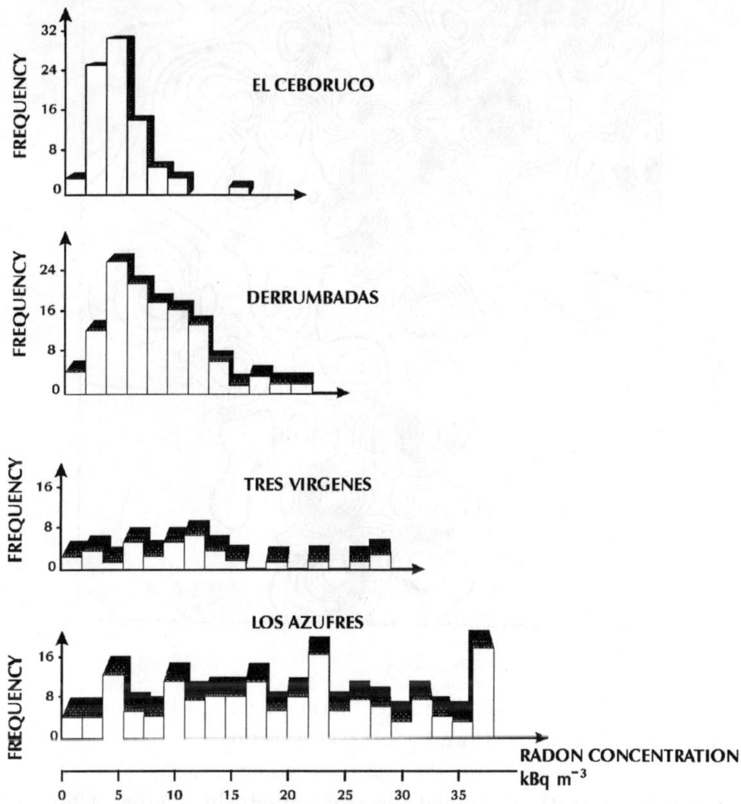

Figure 4.5.4: The radon concentration distribution in geothermal fields follow a log-normal distribution. Differences in mean values are mainly due to differences in soil characteristics.

4.5.3.3 Radon anomalies: geophysical-geochemical relationship studies

The Instituto Nacional de Investigaciones Nucleares (ININ), México, was asked to participate in a geothermal energy prospection project in El Salvador, Central America. GEOCEL, the national company in El Salvador, dealing with geothermal energy production, and the Instituto de Insvestigaciones Eléctricas (IIE), a Mexican electric research institute, conducted geochemical and geophysical investigations to select potential energy sites in El Salvador (Balcázar *et al.*, 1993a).

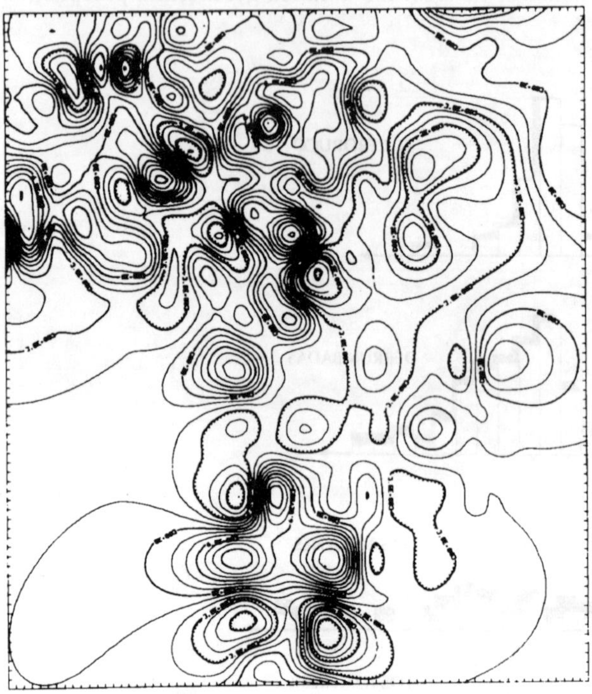

Figure 4.5.5: The Ahuachapan-Chipilapa geothermal field in El Salvador, Central America. Several faults are located in the northwest-southeast direction.

Results of such investigations defined the area called Ahuachapan-Chipilapa as potentially promising for detailed studies. This area is located towards the west of El Salvador, near the Guatemala border, 90 km to the notheast of the capital of the country: El Salvador City. Ahuachapan-Chipilapa is an area of 200 km^2, geographically located between 13° 57' 24" and 13° 50' 00" north latitude and 89° 49' 56" to 89° 42' 11" west longitude. The work is part of the project named "Geoscientific and engineering reservoir studies for the feasibility project of the Chipilapa geothermal field".

An area of 8.74 km^2 in the Chipilapa region was chosen by geologists from GEOCEL and IIE for radon mapping. The area and its location, shown in Fig. 4.5.5, is to the south of San Lázaro hill and to the north of Laguna Verde (Green Lake). There are a number of faults in the northwest-southeast direction; among them are, from east to west, Paraíso (Par), Río Nuevo (RNUE), N1, N2, Cuyanausul (Coy), Agua Shuca (Shu), N4 and N3. There is one fault in the southwest-northeast Escalante (ESC) which intersects N4.

One hundred and sixty LR 115 plastic detectors were put in the area, in an array-like grid, and approximately 200 m apart from each other. The shape of the radon concentration is displayed in Fig. 4.5.6 as isodose curves in the field, and as frequency distribution in Fig. 4.5.7. A few detectors had a mean deviation greater than 10%, but in all cases it was smaller than 20%. The mean radon value obtained in the field was 4.6 $\pm$ 1.8 kBq m^{-3}; detectors that were either damaged or wet were not taken into account for calculating this mean value. A three-dimensional radon concentration distribution, similar to that shown in Fig. 4.5.8, for the Ahuachapan geothermal energy field, provides an easy interpretation of radon anomalies by rotating them in all directions.

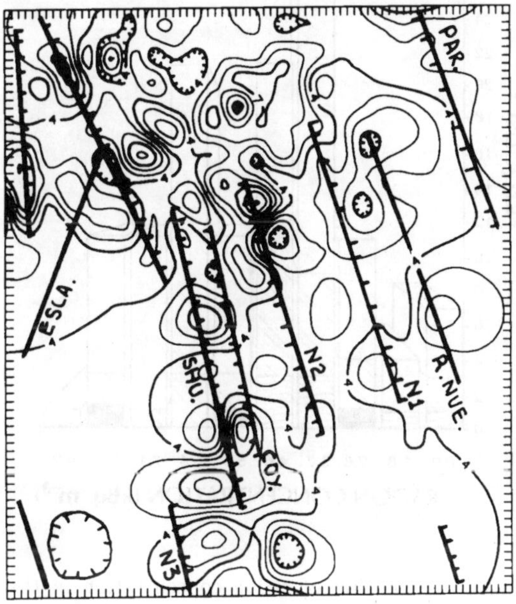

Figure 4.5.6: Radon isodose values clearly show anomalies in the southern part of the field and the northwest of the region. Two producing wells were successfully located in the south of the area.

There are two main areas of high radon activity shown in Fig. 4.5.5 and Fig. 4.5.6. The one in the south corresponds to faults N3, SHU, COY and N2, defining a region of structural anomaly; the area in the northwest was identified as a hydrothermal discharge of the geothermal field. In both areas, there is high thermodynamic activity for which radon acts as a tracer, and is brought to the surface by other gases moving through faults in the first case, and by the discharge fluids in the second case. Analysis of geochemical, geophysical and radon results suggested a potential source for electricity production in the area of study. As a consequence of the analysis, two wells 2,100 m deep, named CHA and CHB, were drilled. CHA is located between faults SHU and COY; and CHB is to the north of N3 and to the west of SHU. Temperature of geothermal fluids in both wells is above 240°C, making them suitable for energy production (Balcázar *et al.*, 1993b).

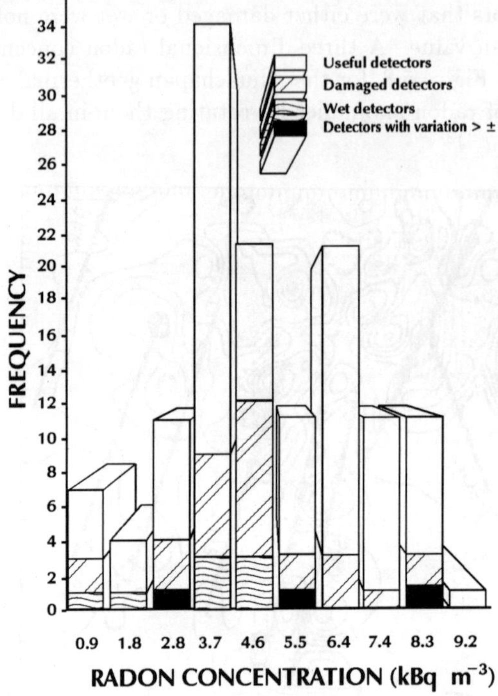

Figure 4.5.7: Frequency distribution of radon concentration in the Chipilapa geothermal field, Republic of El Salvador. The average concentration was 4.6 ± 1.8 kBq m^{-3}, and values higher than 6.4 kBq m^{-3} were considered as anomalies.

4.5.3.4 Radon-enthalpy relationship in geothermal energy production

Among the thermodynamic properties needed for geothermal production and reservoir engineering calculations is the enthalpy (i.e. the energy per unit mass from the well), for which it is necessary to measure the specific volume i.e. the vapour fraction, a parameter closely related with the enthalpy. Then, the purpose of the work is to correlate radon content in geothermal fluids from heat producing geothermal wells with their enthalpy, a key parameter for energy production.

A radon-enthalpy study in the already working geothermal field of Los Azufres, México, was performed during five years by Santoyo *et al.* (1991). Los Azufres is located in the so-called Mexican volcanic belt. It is an area where there are springs, fumaroles and several active faults showing geothermal activity. The wellheads are situated between 2,700 m and 3,200 m above the sea level and from 1,750 m to 3,000 m deep below the earth's surface. Radon studies in soil and in geothermal fluids were carried out in two areas: Maritaro in the northern part of the field where geothermal fluids from wells show less vapour-fraction content, in contrast to the Tejamaniles area in the southern part where high vapour-fraction content in geothermal fluids is present (Nieva *et al.*, 1987).

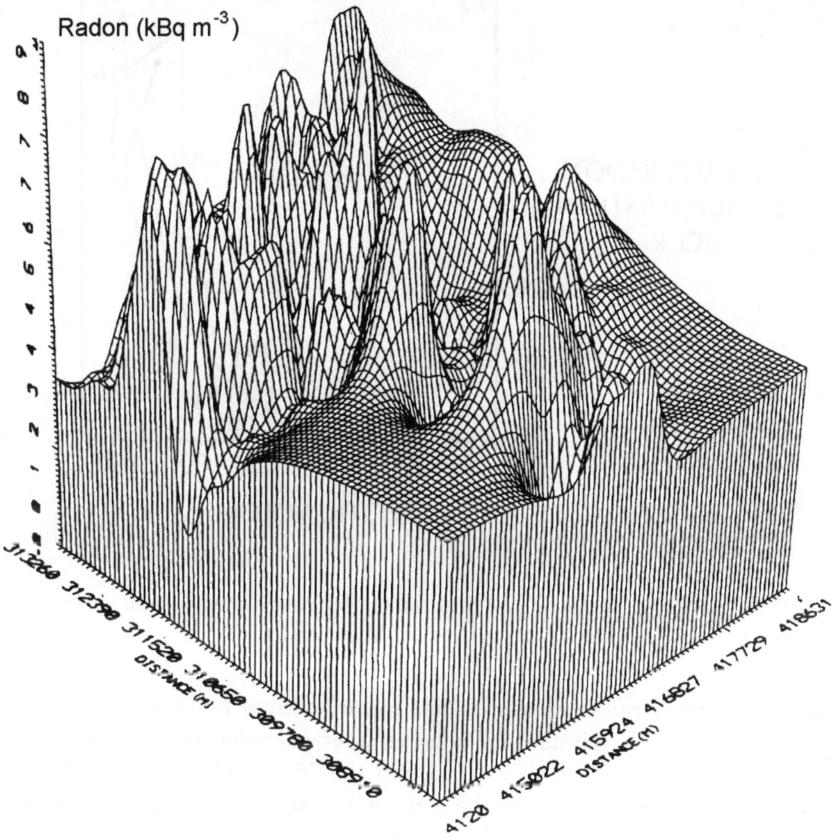

Figure 4.5.8: A three-dimensional radon map for the same values as displayed in Fig. 4.5.6. Rotations allow one to observe anomalies of interest.

Radon samplings in geothermal wells are made in two steps. Firstly, a "T" is connected to the main vapour output pipe, to drain geothermal fluids, which are then passed through a condenser. Samples are obtained in an evacuated stainless steel cylinder from a second "T" fitting, which purges the line prior to sampling; this second "T" is placed between the cylinder and the condenser. Secondly, in the laboratory, radon separation is made by sweeping the radon from the cylinder with helium into an extraction system.

The radon evaluation is made by a scintillation counter; a correction for the elapsed time between the sampling and counting is made.

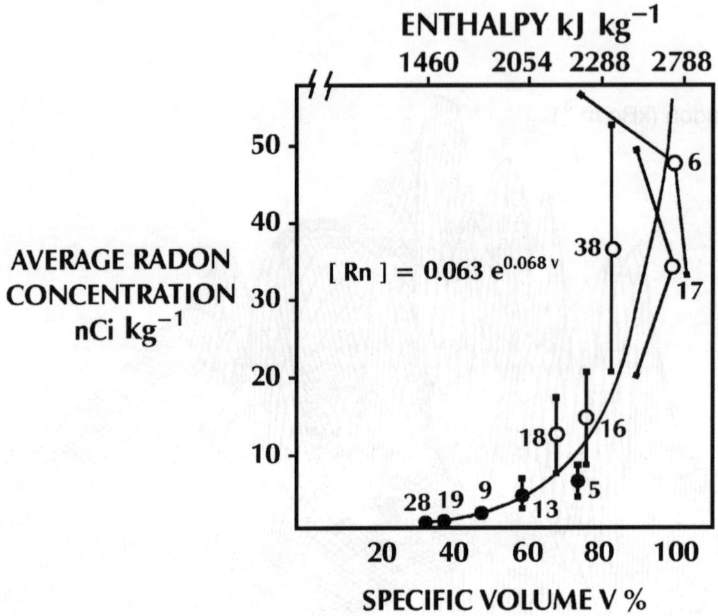

Figure 4.5.9: The average specific activity of radon observed during several years in geothermal fluids from several wells shows a correlation with the enthalpy.

Results of long-term radon measurements in geothermal fluids from nine wells are displayed in Fig. 4.5.9. A relationship between average radon concentration |Rn| and specific volume v of the type |Rn| = 0.063 exp (0.068v) is observed (Balcázar *et al.*, 1991). Numbers on the curve refer to the well-code-number, black dots represent those wells situated in the northern part of the field having low specific volume, whereas open circles correspond to wells in the south, with higher specific volume. A difference in Rn concentration amounting to as much as ten times can be observed for the two sets of wells. Radon sampling was carried out from 1983 through 1988, with no systematic periodicity. However, it is observed that the higher the specific volume the higher the deviation in average radon concentration. We shall further consider this point later. The specific volume is closely related to the enthalpy i.e. the amount of energy released per unit mass by a geothermal well. Thus, it seems that radon concentration is a good indicator of the well potentiality.

The following question now arises: Is radon emanation from soil near geothermal wells related to the specific volume? Radon exhalation from the soil near to a low enthalpy well A9 and near to a medium enthalpy one A16 (see Fig. 4.5.9) were observed during one year at ten points on a line joining the two wells at distances of 50 m apart along the line (Balcázar *et al.*, 1991). Radon emanation was consistently two to ten times higher in the medium enthalpy area. Rain-fall is considered to be the same in the areas of both wells; however, a reduction in radon emanation, with an increase of rain fall, in the area of low vapour fraction is not observed in the area of high rainfall.

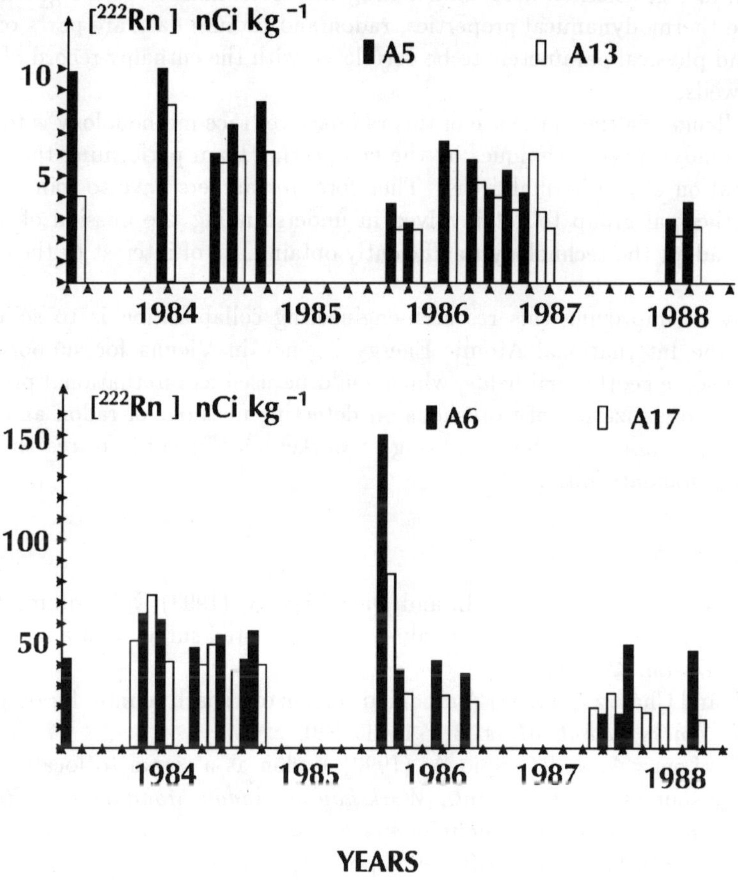

Figure 4.5.10: Geothermal fluids in wells A6 and A17 have a high specific volume, and therefore higher enthalpy values, than well A5 and A13. Minimum values of |Rn| in wells A5 and A13 and maximum values of |Rn| in well A6 and A17 correspond to the 1985 earthquake in Mexico.

Finally, Fig. 4.5.10 shows the long-term radon-concentration measurements from wells A5 and A13 located in the low enthalpy area, and from wells A6 and A17 in the high enthalpy area. During the earthquake that occurred in México on 19 September 1985, an increase in radon concentration was observed for the high-enthalpy wells, in contrast to a decrease in radon concentration for the low-enthalpy wells. This point requires further study (Balcázar *et al.*, 1993).

4.5.4 Concluding Remarks

At the present stage the high radon transport-rate through active faults permits us to locate them in a geothermal field. Concerning the determination of the geothermal field through the thermodynamical properties, radon and radium data are parts of the set of chemical and physical parameters to be correlated with the enthalpy record of individual producing-wells.

The challenge that faces the use of this radioactive trace methodology is to match the demand of ready-to-use techniques by the engineering team performing the exploration and exploitation of geothermal areas. Therefore, researchers have to join a multidisciplinary geothermal group to get involved in understanding the mission of the project and quickly adapt the technology to efficiently obtain data of interest to the engineering team.

One way of improving this research-engineering collaboration is to solicit the assistance of the International Atomic Energy Agency in Vienna for supporting model programs in some geothermal fields, which could be used as international projects. Additionally, the organization of workshops on detecting methods of radon and radium in geothermal fields would, we believe, bring to workers in this subject-area the benefit of the necessary standardization.

References

Badr I., Oliver M. A., Hendry G. L. and Durrani S. A. (1993) Determining the spatial scale of variation in soil radon values using a nested survey and analysis. *Radiat. Prot. Dosim.* **49**, 433-442.

Balcázar M. and Chávez A. (1984) A modified version of a spark counter for α-spectroscopy. *Nucl. Tracks Radiat. Meas.* **8**, 1-4, 617-620.

Balcázar M., López A. and Cuapio A. (1990) Radon as a signal to locate geothermal energy sources. In *Proc. Int. Workshop on Radon Monitoring in Radioprotection, Environmental Radioactivity and Earth Sciences* (eds. L. Tommasino *et al.*), Trieste, 1989, World Scientific, Singapore, pp. 463-468.

Balcázar M., González D., Santoyo E. and González E. (1991) Radon measurements in heat-producing geothermal wells. *Nucl. Tracks Radiat. Meas.* **19**, 1-4, 283-287.

Balcázar M. and Flores J. H. (1992) Statistics applied to radon mapping in geothermal fields. Unpublished results.

Balcázar M., González E. and Arellano V. M. (1993a) Anomalias de radón en la zona geotérmica de Chipilapa, El Salvador C. A. *Revista Mexicana de Geotérmia* **9**, 231-241.

Balcázar M., González E., Ortega M., Flores J. H., Faz P., Renteria D. and Torres V. (1993b) Geothermal energy prospecting in El Salvador. *Nucl. Tracks Radiat. Meas.* **22**, 1-4, 273-276.

Balcázar M., Segovia N., Monnin M., Chávez A. and Flores J. H. (1993c) The need of real-time devices for measuring volcanic and seismic signals. International Atomic Energy Agency, Vienna TECDOC-726, pp. 130-134.

Barbier E. (1991) Geothermal energy: its role in the generation of electricity and its environmental impact. International Atomic Energy Agency, Vienna, 600, pp. 163-176.

Battaglia A., Ceccarelli A., Ridolfi A., Fröhlich K. and Panichi C. (1992) Radium isotopes in geothermal fluids in central Italy. *International Atomic Energy Agency*, Vienna, 789, pp. 363-383.

Cox M. E. (1980) Ground radon survey of geothermal area in Hawaii. *Geophy. Res. Lett.* **7**, 283-286.

Cross W.G. and Tommasino L. (1970) Rapid reading technique for nuclear particle dammage tracks in thin foils. *Radiat. Eff.* **5**, 85-89.

D'Amore F., Sabroux J. C. and Zettwoc P. (1979) Determination of characteristics of steam reservois by ^{222}Rn measurements in geothermal. *Pageoph.* **117**, 253-261.

De la Cruz S., Isabelle D. B., Mena M., Armienta M. A., Monnin M., Seidel J. L., Romero M., Segovia N. and Pialoux P. (1986) Radiation emanation related to geothermal Fauls. *Nucl. Tracks Radiat. Meas.* **12**, 1-6, 875-878.

Faccini U., Magnoni S. and Sordelli C. (1993) Rodigo 1, northern Italy a geothermal complex for agriculture. *Geothermics* **22**, 2, 135-147.

Fleischer R. L. (1988) Radon in the environment - oportunities and hazards. *Nucl. Tracks Radiat. Meas.* **14**, 4, 421-435.

Gutiérrez L., López A. and Balcázar M. (1984) Application of dating for searching geothermic sources. *Nucl. Tracks Radiat. Meas.* **8**, 385-389.

Hiriart G. and Gutiérrez L. C. A. (1995) Calor del subsuelo para generar electricidad. *Ingenieria Civil* **313**, 13-22.

Holland F. A. Heard C. L., Nieva D. and Best R. (1992) Development in geothermal energy in Mexico - part forty: The future for geothermal energy in Mexico. *J. Heat Recovery Systems* **12**, 6, 451-456.

Lombardi S. and Pinti D. L. (1992) ^{222}Rn behavior at the latera geothermal field (Northern Latium Italy). *J. Radioanal. Nucl. Chem.* **161**, 2, 365-375.

López A., Gutiérrez R., Razo A. and Balcázar M. (1987) Radon mapping for locating geothermal energy sources. *Nucl. Instr. Meth.* **A255**, 1-2, 426-429.

Mulas P., Nieva D. and Holland F. A. (1985) Development in geothermal energy in Mexico-part one: general considerations. *J. Heat Recovery Systems* **5**, 4, 277-283.

Nieva D., Verma M., Santoyo E., Barragán R. M., Portugal E., Ortiz J. and Quijano L. (1987) Chemical and isotopic evidence of steam up-flow and partial condensation in Los Azufres reservoir. *Proc. XII Workshop on Geothermal Resources Engineering,* pp. 253-257.

Roldán-Manzo A. R. (1992) Investigaciones geoquímicas realizadas en los campos geotérmicos de Zunil y Amatitlán, Guatemala. International Atomic Energy Agency, Vienna, 457, pp. 279-305.

Santoyo E., Verma S. P., Nieva D. and Portugal E. (1991) Variability in the gas phase composition of fluid discharged from Los Azufres geothermal field (México). *J. Vulcanology and Geotherm. Resources*, **47**, 161-181.

Segovia N., De la Cruz-Reyna S., Mena M., Seidel J. L. and Monnin M. (1991) Radon-222 as an indicator of geothermal reservoirs behaviour. *Nucl. Tracks Radiat. Meas.* **19**, 1-4, 401-404.

Semprini L. and Kruger P. (1984) Relationship of radon concentration to spatial and temporal variations of reservoir thermodynamic conditions in the Cerro Prieto geothermal field. *Geothermics* **13**, 1-2, 103-115.

Stewart M. K., Burnett W. C. and Whitehead N. E. (1993) Survey of some natural decay-series isotopes in the wairakei geothermal area and possible residence-time applications. *Institute of Geological and Nuclear Science Ltd.* **93/38**, pp. 1-16.

5.

CONCLUSION

S. A. Durrani[1] and R. Ilić[2]

5.1 Concluding Remarks and Future Outlook

Any reader who has perused the whole of this book with some care will, we believe, be struck by the great breadth and depth of the studies that have been, and are being, carried out on the whole gamut of radon research. These range from geological considerations and investigations that form the essential background to the causes and rates of emission of radon, through economically beneficial applications of the gas such as mineral exploration and search for geothermal energy sources, all the way to indoor and outdoor surveys of radon concentration levels, and critical appraisals of the health hazards of excessive exposure to radon and the measures that can be taken to minimize any deleterious effects - whether to the general public or to those working in mines and caves. And all through the book runs the unifying thread, linking its various parts: the theme-line of Solid State Nuclear Track Detection (SSNTD). This, as we stated in the Preface, is the technique that is most commonly used today for studying the many and various aspects of radon measurements and applications. Several sections of the book have, therefore, been devoted to giving the reader a broad background to the SSNTD discipline and also providing him or her with detailed descriptions of all the various types of instrumentation and devices (both "passive" and "active") used in applications ranging from simple determination of radon concentration levels in homes to earthquake prediction and volcanic surveillance - not to forget the important subject of calibration and standardization of the detectors.

It would be futile and unnecessary - and indeed an unavailing exercise - for us to attempt to summarize here all the topics covered in this monograph. So we shall desist from making any such attempt. Instead, a few general remarks are all that we shall allow ourselves the luxury of making by way of an overview - and perhaps venture to put forward some comments about the future of radon studies: as we perceive them.

The first general comment is that much still remains to be done in several fields. Take one example: effects of radon on human health. There exists today a whole spectrum of opinion on this subject. Views range from the assertion that "a little bit of radiation (and hence of radon) is good for you *(hormesis)* " to the other extreme that maintains that "*all* radiation is bad for you - however small in quantity and of whatever type or form". Without wanting to arbitrate on these beliefs, let us merely remark that further research is needed before the truth can be established (and it is likely that there will always be grey areas where the positive and the negative will indistinguishably merge).

Another example that comes forcefully to mind is the whole question of using radon as a signal of an impending earthquake. Here, it is still early days in this field of in-

[1]University of Birmingham, Birmingham, UK

[2]J. Stefan Institute, Ljubljana, and University of Maribor, Maribor, Slovenia

vestigation. As R. L. Fleischer puts it in his review (section 4.2): "Presently, routes to earthquake prediction are at most in their infancy; the use of radon shares in being at an early stage of development". Man has endeavoured from time immemorial to understand the causes of earthquakes (whether believing, as in ancient India, that they are caused by a subterranean bull shifting the mass of the earth from one horn to the other; or the Greek god Atlas carrying out a similar manoeuvre from under the high mountains). But although we may pride ourselves in having gained some deeper understanding of the phenomenon in terms of colliding tectonic plates and the molten nickel-iron core of the earth, we are no closer (we regret to have to admit) to predicting the time, the place and the intensity of the next earthquake than the ancient Chinese sage observing the behaviour of cats and dogs and birds, and the water-level in his wells. (This, we hasten to add, is our personal non- specialist's belief - not necessarily shared by the specialist or the active researcher in the field.) Much further work in this subject area, it is clear, still needs to be done - and we feel sure that, like the multiparameter problem of weather-forecasting, which has made "quantum leaps" in accuracy over the last half-century or so, the efficacy of interpreting the radon signals, in conjunction with other pointers, for earthquake prediction will undergo a sea change in the years to come.

A further change that is bound to gain momentum in the coming years is the automation of measurements and instruments in radon work and related studies. More real-time and electronic-cum-track detector devices and computer-controlled data- gathering and -reduction systems are bound to come on the market.

All in all, we feel that radon studies are in their "adolescent" stage in most areas - i.e. they have passed their infancy but not as yet come into full-bodied and vigorous adulthood. This is a good, and indeed an exciting, phase in the development of any scientific discipline. Feelings of lethargy, resignation, and even ennui have not yet set in - even though the compensating richness of wisdom and experience of old age is also at this stage missing. It may, thus, not be a bad idea for us to leave the reader poised in animated suspense at this historical juncture. He, or she, may engage themselves in speculating, with eager anticipation, about the veils of mystery yet to be lifted by those nearest to the action in this eternal drama involving Man and Nature as the protagonists. And may we hope that the reader himself/herself will, by having read the preceding pages of this volume, feel impelled to join this battle which, we believe, is of some "cosmic" significance (if that does not signify excessive delusions of glory)? Good hunting, in any case, dear reader! We wish you all success.

6.

APPENDIX

R. Ilić[1], **S. A. Durrani**[2], **M. Sohrabi**[3], **G. Åkerblom**[4], **T. K. Ball**[5] **and R. L. Fleischer** [6]

6.1 Glossary of Terms in the Context of Radon Measurements by Etched Track Detectors and their Applications

The following glossary is intended as a conveniently accessible source of definitions and explanations of technical and specialist terms used in the text of the present book, with special reference to radon in the context of radiation protection, earth sciences and the environment. The aim is to provide concise definitions, and no attempt has been made to be exhaustive: for greater insight the reader is referred to standard texts in the relevant subject areas. Only terms actually used in the book have been included. Cross references in the definitions are indicated by **bold type**.

absorbed dose, the fundamental dosimetric quantity D, energy absorbed per unit mass. The SI unit is **joules** per kilogram (J kg^{-1}), termed **gray** (Gy). 1 Gy = 1 J kg^{-1}.

action level, the level of **dose** rate or **activity** concentration above which **remedial action** or **protective action** should be carried out in chronic **exposure** or emergency exposure situations. Also colled reference level in some countries.

activated charcoal, a term used to describe charcoal as a material that exhibits a high degree of porosity and extremely large suface area, so as to adsorb large quantities of gases such as **radon**. During activation (heating) in the absence of oxygen, pores are created in the solid matrix of starting material.

active dosimeter, an instrument for measuring the **exposure** or **activity** concentration in real time, in which air is usually pumped through a filter.

activity, the quantity A expressing the decay rate of a radionuclide, defined as $A = dN/dt = \lambda N$, where λ is the **decay constant** (s^{-1}) and N is the number of nuclei undergoing spontaneous disintegration. The SI unit of activity is **becquerel** (Bq). 1 Bq

[1]J. Stefan Institute, Ljubljana, and University of Maribor, Maribor, Slovenia
[2]University of Birmingham, Birmingham, UK
[3]Atomic Energy Organization of Iran, Tehran, Islamic Republic of Iran
[4]Swedish Radiation Protection Institute, Stockholm, Sweden
[5]British Geological Survey, Nottingham, UK
[6]Rensselaer Polytechnic Institute, Troy, New York, USA

= 1 disintegration s^{-1}. The activity concentration is usually expressed in Bq m^{-3}.

advection, transfer of heat by horizontal flow of air or gas.

aerosol, colloidal particles dispersed in the air, that can serve as nuclei to which the solid radon **daughter products** can attach themselves.

alluvial deposits, deposits of material of any grade which has been carried in suspension by water.

alpha particle, the nucleus of a ^{4}He atom, with an atomic mass of 4. It is emitted in one of the principal modes of radioactive decay of unstable **isotopes** of high-Z elements.

alum shale, a Scandinavian uranium-rich black **shale** of Upper Cambrian age (550-500 Ma). The alum shale unit is usually about twenty metres thick and covers large areas in Sweden and Norway. The uranium content of the alum shale is 50 to 400 ppm U (**activity**, 600 to 5,000 Bq kg^{-1}).

annual limit on intake (ALI), the intake by inhalation, ingestion or through the skin, of a given radionuclide in a year by the reference man, which would result in a **committed dose** equal to the relevant **dose limit**. The ALI is expressed in units of **activity**.

atomic mass number (A), mass of a **nuclide**, i.e. total number of protons and neutrons in it. A is the nearest integer to its precise atomic mass.

atomic mass unit (amu), one-twelfth of the mass of ^{12}C, based on a scale where the latter has a mass of 12. This makes 1 amu = 1.6606 x 10^{-27} kg.

atomic number (Z), nuclear charge number, i.e. number of protons in the nucleus of atoms of an element. The atomic number defines the chemical properties of the element.

attached fraction, that part of the solid radon **daughters** that gets attached to **aerosol** particles.

barn (b), a unit of **cross section** for nuclear reactions. 1 b = 10^{-28} m^2.

basalt, a fine-grained **basic** igneous **rock**.

basic rocks, igneous rocks in which the silica content is between 45 and 55%.

becquerel (Bq), the SI unit of **radioactivity**. 1 Bq = 1 disintegration per second.

black shale, a **shale** that contains organic material decomposed to particles of coal or **kerogen**, and often rich in sulphides.

boulder clay, a glacial deposit consisting of a stiff tenacious clay containing sub-angular stones of various sizes.

Bragg curve, a plot of linear rate of energy loss (dE/dx) against residual range or energy, as a charged particle slows down in a medium. The 'Bragg peak' is the region near the end of the particle's range where the loss rate is the maximum (beyond which the effective charge of the heavy **ion** declines as it picks up electrons from the medium, and hence dE/dx falls).

breccia, a cemented fragmental rock consisting of angular fragments of other rocks.

caldera, a volcanic crater of great size caused by expulsion of a large volume of **magma** and subsequent collapse of the **magma chamber**.

cancer, condition in which there is malignant tumour from uncontrolled growth of tissue, tending to spread and recur when removed.

cap-rock, a rock that resists the flow of fluid. It may, e.g., form a cap over an anticline, the core of which is permeable.

chemical etching (CE) of tracks, developing a **latent track** by an etching solution that selectively removes the material of the **latent track**, thus enlarging it to a point usually observable under an **optical microscope**.

clay gouge, a thin seam of clay separating ores or ore and rock.

collective dose, an expression for the total radiation **dose** incurred by a population, defined as the product of the number of individuals exposed to a source and their average radiation dose. The collective dose is expressed in man-**sieverts** (man-Sv).

collector of radon products, a thin metal foil which collects radon **daughter products** by an electrostatic field. See also **plate-out**.

committed dose, the time-integrated **equivalent** or **effective dose** following the intake of radioactive material by the body. The time is normally expressed in years (e.g. 50 for adults and 70 for children).

concentration of radon, i.e. its **activity** concentration; often referred to as radon level; its SI unit is Bq m^{-3} (the volume refers to that of the surrounding air). Also expressed in Bq (or Ci) per litre of a liquid. 1 pCi l^{-1} = 37 Bq m^{-3}.

Coulomb explosion, origin of the atomic collision cascade in solids, caused by the passage of a rapidly moving **ion** in a **dielectric medium**, and leading to displaced atoms and ultimately to a **latent track** if certain conditions are satisfied.

counting system, semi- or fully-automatic system for the measurement of **track density** and related parameters.

CR-39, trade name of a **track detector** made of polyallyl diglycol carbonate.

critical angle of etching, the half-cone angle of etch pits, defined as $\theta_c = \sin^{-1} V_B/V_T$ for the given detector. V_T stands for the velocity of etching along the **latent track** and V_B for the velocity of etching of bulk material. If a **latent track** makes a (dip) angle with the detector srface which is less than θ_c, it cannot be revealed by etching.

cross section of a nucleus, the measure of the probability of a particular process or reaction involving the nucleus, usually expressed as a geometric area σ (cm^2), termed 'microscopic cross section' of the nucleus. If N is the number of target nuclei per cm^3, then the 'macroscopic cross section' is $\Sigma = N\sigma$, (cm^{-1}).

curie (Ci), old unit of radioactivity. One curie corresponds to 3.7 x 10^{10} nuclear transformations or desintegrations per second. 1 Ci = 3.7 x 10^{10} Bq.

damage trail, see **track, latent**.

daughter product(s), substance(s) formed by the radioactive decay of another (the 'parent' or progenitor); also called the **progeny**.

decay constant, the probability of radioactive decay per unit time. Thus, if λ is the decay constant, then out of N unstable nuclei λN will decay per unit time. It is also the reciprocal of the **mean life** τ.

deposition, see **plate-out**.

diatomaceous earth, a friable earthy deposit composed of nearly pure silica and consisting essentially of the frustules of microscopic plants called diatoms.

dielectric medium, an insulating material, i.e. one that can sustain high electric fields.

diffusion coefficient/constant, D, characterizes the transport of (radon) **ions** through normal thermal agitation. Its dimensions are m^2 s^{-1}.

diffusion distance, mean, distance traversed by a radon **isotope** over its **mean life**. It is defined as $\sqrt{D\tau}$, where D stands for **diffusion coeficient** and τ for **mean life**.

dike/dyke, a tabular body of igneous rock relatively thin compared with its lateral extent, that cuts across the structure of adjacent rocks or cuts massive rocks. See also **sill**.

distal ash, fine-grained **pyroclastic** rocks deposited from a volcano.

dome, a structural body characterized by local uplift approximately circular in outline.

dose, a measure of the radiation received or absorbed by a target.

dose equivalent/equivalent dose, has the SI unit **sievert** (Sv). 1 Sv = 1 J kg^{-1}. The concept includes a weighting (or quality) factor of the radiation type.

dose limit, the value of the **effective dose** or the **equivalent dose** to individuals from controlled practices that shall not be exceeded, as laid down by some regulatory authority.

dose response, quantity representing the effect in a medium (e.g. tracks cm^{-2}), or in a device (e.g. total counts), of a given **dose** of radiation.

dosimeter/dosemeter, a device, instrument or system which can be used to measure or evaluate any quantity related to either **absorbed dose** or **equivalent dose**.

earth: **core, mantle** and **crust**: the crust extends from the surface to a variable depth but usually about 35 km; the mantle extends from about 35 km to 2,900 km; the core extends from 2,900 km to the centre of the earth.

effective dose, the quantity E, defined as a summation of the tissue-equivalent **doses**, each multiplied by the appropriate tissue weighting factor. The SI unit of effective dose is J kg^{-1}, termed the **sievert** (Sv).

electret, an electrical analogue of a permanent magnet, and carrying a permanent electric charge.

electrochemical etching (ECE) of tracks, developing **latent tracks** by using an ECE chamber, usually consisting of two compartments filled with a suitable etching solution of various molarities, and insulated from each other by a **polymer track detector**, across which a high alternating electric field is applied. This results in electrical breakdown of the **polymer** at the pointed tip of the etched track. The resulting **etch spots** are often visible to the naked eye.

emanation, the movement of a radon atom from a mineral grain to the pore void, usually as a **recoil** in the **alpha**-decay of the parent radium atom.

enthalpy, total heat content of a system.

equilibrium equivalent concentration of radon (EEC$_{Rn}$), for a non-equilibrium (i.e. $F \neq 1$) mixture of short-lived Rn **daughters** it is that (smaller) **activity** concentration of Rn in secular radioactive equilibrium with its **daughter products** that has the same **potential alpha energy (PAE)** concentration.

equilibrium factor (F), the ratio of the activity of a given **radon daughter** to that of the parent (radon) itself in a given volume; e.g. for a short-lived daughter i, $F_i = A_i/A_{Rn}$, where A is the **activity**. When the daughters are in secular equilibrium with radon, $F = 1$.

equivalent dose, the quantity H, defined as $H = D_T W_R$, where D_T is the absorbed **dose** of radiation type R in organ T, and W_R is the weighting factor (formerly called the quality factor Q) for the radiation type R. The SI unit is **sievert** (Sv). 1 Sv = 1 J kg^{-1}.

eskers, long serpentine ridges, sometimes of considerable height, more than 50 metres above the surrounding terrain, and consisting of gravel sand and waterborne rounded stones. Eskers are formed at the edge of the continental ice sheet or glaciers by material rapidly transported by water.

etch pit, the pit, or mouth-opening, of a charged-particle **track** as a result of etching the detector with an appropriate chemical reagent.

etch spot, the two-dimensional section of a spark (resembling a Lichtenberg figure with craggy edges), resulting from electrical breakdown of the **polymer** during the **electrochemical etching** of a charged particle **track** in a plastic detector.

etched track detectors, see **solid state nuclear track detectors**.

exhalation, emision of radon from a surface, for example the ground surface or a wall surface, by diffusive or convective movement through pore spaces.

exposure, the act or condition of being subject to irradiation. It can be either external (irradiation by sources outside the body) or internal (sources inside the body). It is often expressed as Bq m^{-3} h, where h is the number of hours of exposure to a given **activity**

concentration.

fluence, total number of particles incident per unit surface area. It is the time-integrated flux (particles crossing a unit area at right angles to them per unit time).

fumaroles, small vents on the flanks of a volcanic cone, or in the crater itself, from which gaseous products emerge.

gabbro, a **basic** plutonic medium-to-coarse grained **rock** usually black or greenish dark grey. Main minerals are calcic plagioclase, pyroxene and olivine.

geometric mean (GM), when the observed values of a quantity differ widely, it is often more meaningful to represent them by a geometric, rather than an arithmetic, mean. The GM of n quantities x_1, x_2,.....x_n is given by: GM $= (x_1 \cdot x_2....x_n)^{1/n}$.

geometric standard deviation (GSD), is a dimensionless quantity which indicates the range over which $\cong$ 68.3% of all readings, distributed **lognormally,** fall. If, say, GSD is 2.0 and GM is 22 Bq m^{-3}, then the range of the quantity (**activity** concentration) within 1 GSD on either side of GM (i.e. GM $\times/\div$ 1 GSD) is 44 - 11 Bq m^{-3}. The GSD is based on fractional errors $(\delta x/x)$ in the readings. Thus ln GSD $= (1/n)[\delta x_1/x_1 + \delta x_2/x_2 + ... + \delta x_n/x_n]$.

geostatistics, a statistical formalism that allows one to estimate or interpolate values of a geological property at unsampled locations from known values at sampling points.

geothermal, pertaining to the earth's internal heat.

gneiss, a coarse-grained rock formed by **metamorphism** in which bands rich in granular minerals alternate with bands in which schistose minerals predominate.

grab sample, sample, e.g. that of air containing radon, collected over a short time-interval from a given location.

granite, a coarse-grained igneous rock, normally with greater than 65% silica. It is usually relatively uranium-rich, and hence a strong source of radon.

gray (Gy), SI unit of **absorbed dose**. 1 Gy = 1 J kg^{-1}.

half-life ($\tau_{\frac{1}{2}}$), time in which the atoms of a radionuclide decay to half their initial number (and **activity**). It is related to **mean life** (τ) as: $\tau_{\frac{1}{2}} = (ln2)$ x τ.

head, an unstratifield or poorly stratified accumulation of rock fragments of local derivation. Formed under cold conditions by **solifluxion.**

Henry line, a straight line of **lognormal distribution**, defined by mean value and its standard deviation.

hypabyssal rocks, igneous rocks that have risen form the depths as **magma** but solidified mainly as minor intrusions such as **dikes** and **sills** before reaching the surface.

intermediate rocks, igneous rocks in which the silica content is between 55 and 65%.

intervention, any action intended to reduce or avert **exposure** or the likelihood of exposure to sources which are not part of a controlled practice or which are out of control as a consequence of an accident. See also **protective action, action level**.

intrusive rock, a rock that consolidated from **magma** beneath the surface of the earth.

ion, an atom which has lost its orbital electron(s) and has thus become positively charged. Most fast-moving particles traversing a medium, or those involved in radioactive decay, become ionized.

ionizing radiation, any radiation (including charged particles, γ-rays, etc.) that forms ions through Coulomb or photoelectric, etc., interactions with the atoms of a medium.

isotopes, nuclides of an element having the same **atomic number** Z, but different **atomic mass numbers** A.

isotopic abundance, the number of atoms of a particular isotope, expressed as a fraction of all the atoms of the element (e.g. those in nature).

joule (J), SI unit of energy. 1 J = 1 kg m^2 s^{-2}. 1 J = 10^7 erg = 6.2418×10^{18} eV. Also 1 J = 1 Ws.

karst, areas in which the rock has suffered solutional activity with the development of caves and enlarged fractures.

kerogen, a solid, bituminous mineraloid substance in oil **shales**.

lacustrine, accumulated in lakes.

latent track, see **track, latent**.

laterite, red residual soil developed in humid tropical and subtropical regions of good drainage.

leaching, the process by which mineral salts are washed out of rocks by percolating water.

LET, linear energy transfer. Energy transferred per unit path length by an **ion** or other radiation moving in a medium.

leukaemia/leukemia, disease in which there is an excess of white corpuscles in tissue and usually in blood.

Lexan, trade name of a **track detector** made of a bisphenol-A polycarbonate.

life-time risk, risk (i.e. probability of a defined deleterious outcome), integrated over a whole life-time (e.g. 70 years for a child).

limestone, a sedimentary rock consisting mainly of calcium carbonate with or without magnesium carbonate.

limit, the value of a quantity used in certain specified activities or circumstances that must not be exceeded, as laid down by a regulatory authority.

lithology, the character of a rock expressed in terms of visible attributes that, in aggregate, impart individuality to the rock.

lithophile elements, elements enriched in the silicate **crust**. Elements with a greater free energy of oxidation, per gram atom of oxygen, than iron.

litosphere, the solid **crust** that envelops the central **core** of the **earth**.

loess, a sediment, commonly nonstratified and unconsolidated, composed predominantly of silt-size particles, deposited initially by the wind.

lognormal distribution, a distribution in which the logs are distributed 'normally', (i.e according to the Gaussian distribution).

LR 115, trade name of a **track detector** made of cellulose nitrate.

lung cancer, see **cancer**.

magma, molten fluid within the earth's interior from which igneous rocks are formed by solidification.

magma chamber, a void filled with **magma**, usually a term given to a sub-volcanic feature.

Makrofol, trade name of a **track detector** made of a bisphenol-A polycarbonate.

mean life (τ), time interval over which the number of radionuclei falls to 1/2 of its initial value through radioactive decay. It is related to the **decay constant** (λ) and the **half-life** ($\tau_{1/2}$) as follows: $\tau = \lambda^{-1}$; and $\tau = \tau_{1/2} / \ln 2$.

median value, that value in a series of observed values which has exactly as many readings above it as below.

metamorphism, process by which consolidated rocks are altered in composition, texture, etc. Pressure, heat and the introduction of new chemical substances are the principal causes, often yielding new minerals.

mole (M), molecular weight of a compound in grams, based on the **atomic mass numbers** A of its constituents. A mole contains the Avogadro's number N_A of molecules. $N_A = 6.0220 \times 10^{23}$ molecules.

monitoring, the measurement of **dose** or contamination for reasons related to the assessment or control of **exposure** to radiation or radioactive substances, and the interpretation of the results.

moraine, a trail of rock-waste left either at the margins or at the base of a mass of ice.

natural exposure, an **exposure** delivered by natural sources.

Northampton Sand Formation, sandstone containing beds of calcareous iron stone.

nuclide, a species of atoms characterized by their **atomic mass number** A, and **atomic number** Z. Unstable nuclides that decay by the emission of α's, β's or γ's are termed radionuclides or radioactive nuclides.

overthrust belt, a mass of rock which has been thrust over another along a low-angled reverse fault.

paleochannel deposit, a sedimentary in-fill of a fossil stream channel.

paragneiss, a high-grade metamorphic rock derived from a sedimentary rock.

passive dosimeter, an instrument for measuring the **exposure** or **activity** concentration, where air diffuses to the detector or radiation falls naturally on it, e.g. a **solid state nuclear track detector**. Such detectors usually integrate the response over a period of time.

periglacial, the effect of cold dry conditions, characterized by **solifluxion** deposits.

permeability, the capacity of a medium to transmit gas or fluid. Fracture-permeable rocks are those in which the fluid is transmitted along joints and fissures, etc.

placer deposit, an alluvial or glacial deposit composed of sand or gravel and containing particles of heavy valuable minerals.

plate-out, deposition of solid decay products of **radon** onto solid surfaces.

Pliocene, a period during the Tertiary era extending from 7 to 2 million years ago.

podsol/podzol, a highly bleached ashlike soil that is low in iron and lime. It consist of a **leached** upper layer followed by largely nonweathered and nonleached layers.

polycarbonate(s), (range of) thermoplastics based on the action of phosgene (carbonyl chloride) on dihydroxy organic compounds such as diphenylol propane.

porphyry, an acid extrusive or **hypabyssal rock** with phenocrysts of alkali feldspar or quartz.

potential alpha energy release (PAER), the total α-energy ultimately released by the decay of the short-lived **daughters** of ^{222}Rn down to but not including ^{210}Pb, when in secular equilibrium with the progenitor ^{222}Rn.

precursor, in **radioactivity**, a substance from which another is formed by decay.

primary ionization, the ionization produced by the primary particle traversing a medium and removing orbital electrons from its atoms by the Coulomb force, in contrast to total ionization, which includes the secondary ionization produced by the delta rays emitted in the process.

progeny, in **radioactivity**, substance(s) formed by the decay of another (the 'progenitor'); also called **daughter product(s)**.

protective action, an **intervention** intended to avoid or reduce **doses** to members of the public or personnel of nuclear facility, etc., in chronic or emergency **exposure** situations.

pyroclastic deposit, the consolidated accumulation of solid fragments ejected from **volcanoes**.

Quaternary, the most recent era of the geological succession extending from 2 million years ago to the present.

radiation biology, the study of radiation effects in living matter.

radiation, high-LET, radiation that loses energy per unit length at rates much higher than do electrons or X-rays. Examples are heavy charged particles (including **α-particles**), and neutrons.

radiation, low-LET, examples are fast electrons, X-rays and gamma-rays.

radioactivity, the property of certain unstable **nuclides** (termed radionuclides) of emitting radiation or particles(chiefly α, β, or γ's) during their spontaneous transformation (see also **activity**).

radon, an element resulting from the α-decay of radium. It has one of the largest number (28) of **isotopes** on the chart of **nuclides**. If not otherwise specified, however, the term radon normally refers in this book to the isotope ^{222}Rn, a member of the natural decay chain of ^{238}U. Another isotope often encountered in radon work is ^{220}Rn, sometimes called **thoron** (Tn), a member of the natural decay chain of ^{232}Th.

radon progeny/daughter(s), the radioactive decay products of **radon** (especially the short-lived ones). See **daughter product(s)**.

recoil, the motion acquired by the residual nucleus ejecting a particle or photon; also, motion imparted by an energetic particle to another particle in a collision.

reference level, see **action level**.

relative biological effectiveness (RBE), compares the effectiveness of one radiation with that of another. It is the inverse ratio of the **absorbed doses** producing the same degree of a defined biological end-point.

remedial action, measures taken to rectify a hazardous situation, e.g. high levels of **dose** rates or **activity concentration of radon**. See **action level**, **intervention**, and **protective action**.

response of radon dosimeter, calibration coefficient, e.g. **track density** per unit radon **exposure** (tracks cm^{-2}/kBq m^{-3} h).

rift structure, is formed by the sinking of the ground between two roughly parallel faults.

sampler, radon, monitoring device for **radon**, utilizing **etched track detectors**, etc.

secondary porosity, is **porosity** caused by solutional activity after the rock has solidified.

sensitivity of radon dosimeters, is defined as the smallest detectable **radon concentration**.

shale, a laminated sediment, the constituent particles of which are clay grade. See also

alum shale, black shale.

shield volcano, characterized by large flat cones of lava with gentle slopes, built up of many flows. They are mostly **basaltic** in composition.

sievert (Sv), the SI unit of **equivalent dose**. 1 Sv = 1 J kg^{-1}.

sill, an intrusive tabular body of igneous rock relatively thin compared with its lateral extent, which has been emplaced parallel to the bedding or schistosity of the intruded rocks. See also **dike/dyke**.

skarn, rock types that occur together with iron and sulphide ores. Skarn is mainly composed of lime-bearing silicates. Uranium and rare earths are commonly present.

solid state nuclear track detectors (SSNTDs), insulating materials (e.g. **polymers**) in which heavy charged particles produce **damage trails**, etchable into **tracks** visible under an **optical microscope**.

solifluxion, the bodily flow of soil and rock debris down sometimes gentle slopes. Soil which is saturated with water will, under conditions of alternating freezing and thawing, flow downhill as a result of the changes in volume.

specific activity, the **activity** per unit mass of an element or compound containing one or more types of radioactive **nuclides**.

spectrometer, instrument used to measure energy distribution in a mixed field of radiation (e.g. **alpha particles** of various energies).

spectrometry/spectroscopy, the action performed by a **spectrometer** to resolve energies, and thus, possibly, to discriminate between the sources of the various particles.

speleotherapy, (putative) healing effect of a cave's atmosphere (e.g. by inhaling the high-concentration **radon** in it).

stopping power (S), energy loss rate (dE/dx), or linear energy transfer (**LET**), i.e., energy transferred by an incident **ion** per unit path length in the stopping medium.

tectonic, an adjective applied to rock structures and their resulting actions pertaining to earth movements.

Tertiary, the era of geological time during which strata from the Palaeocene through Eocene, Miocene, Oligocene and **Pliocene** were deposited. Approximate duration 65 to 2 million years before present.

thoron, see **radon**.

till, nonsorted, nonstratified sediment formed by crushing action of moving inland ice or glacier and carried and deposited by the ice. Till is the same soil type as morain(e).

track density, number of **tracks** per unit area (usually cm^{-2}) on the surface of a **track**

detector.

track detectors, see **solid state nuclear track detectors**.

track, etched, result of developing a latent trail of damage of a heavy charged particle in a **dielectric** solid, by using an appropriate chemical reagent.

track, latent, long-lived trail of a heavy charged particle in an insulating solid. Application of heat, before etching, can result in partial or total fading of the damage trail.

ultramafic rocks, igneous rocks in which the silica content is below 45%.

unattached fraction, that part of the solid **radon daughters** that do not attach themselves to **aerosol** particles. They form themselves into small or large clusters, and remain relatively reactive. These decay products eventually **plate-out** on solid surfaces, including the lungs of someone breathing the air containing them.

volcano, a stucture, e.g. a mountain, having opening(s) in **earth's crust** through which lava, cinders, steam, gases, etc., are or have been expelled contiuously or at intervals.

working level (WL), the sum of the total energy emitted as **alpha particles** during the complete decay of the short-lived **progeny** of ^{222}Rn, starting with ^{218}Po and down to, but excluding, ^{210}Pb, in a unit volume of air. This equals 1.3 x 10^5 MeV of alpha energy per litre of air. The unit is based on a ^{222}Rn concentration of 100 pCi per litre of air. In SI units the WL corresponds to 2.08 x 10^{-5} J m^{-3}.

working level month (WLM), a unit of **exposure** to **radon progeny**. 1 WLM = 170 WL h. In SI units, one WLM is equivalent to 3.54 mJ m^{-3} h.

Subject Index